Experimental Brain Research Series 16

Christine Heym (Ed.)

Histochemistry and Cell Biology of Autonomic Neurons and Paraganglia

With 195 Figures and 27 Tables

Springer-Verlag
Berlin Heidelberg New York
London Paris Tokyo

Professor Dr. Christine Heym

Anatomisches Institut der Universität, 1. Lehrstuhl
Im Neuenheimer Feld 307, 6900 Heidelberg, FRG

ISBN-13: 978-3-642-72751-1   e-ISBN-13: 978-3-642-72749-8
DOI: 10.1007/978-3-642-72749-8

2125/3140-543210

Preface

With the discovery of an ever increasing number of biogenetic polypeptides in the autonomic nervous system, it is evident that nervous transmission is controlled and modified not only by a series of classical transmitters but also by a variety of co-transmitters and modulators. Purification, characterization and synthesis of neuropeptides and peptide families allows the application of specific antibodies for radioimmunological, immunohistochemical and immunocytochemical localization. Transmitter interactions and phenotypic neuronal plasticity can be analyzed by pharmacological or surgical manipulations. Biochemical and autoradiographic determination of specific membrane receptors enables the recognition of peptide target structures. In addition, neurophysiological techniques in combination with histochemical methods result in functional interpretations. With regard to somatic manifestations of autonomic nervous disturbances these studies will offer fundamental insights for clinical medicine.

This volume contains the refereed proceedings of recent investigations on histochemistry and cell biology of autonomic neurons and paraganglia which were presented at an international colloquium on nervous transmission in Heidelberg in September 1986. The colloquium was originally planned to be held under the presidency of Olavi Eränkö but was continued by the editor after his death. It was sponsored by the German Research Foundation and took place on the occasion of the 600th anniversary of the Heidelberg University "Ruperto Carola" as part of the program in the international "Wissenschaftsforum".

Hopefully, the information provided in this book will stimulate new ideas and research strategies in the field of neuroscience.

We would like to thank Ms. H. Ehlers for her expertise in computerassisted word processing which was invaluable for the edition of this book. We also gratefully acknowledge the advice and active support of Dr. Thiekötter and his staff from the Springer-Verlag in publishing the present proceedings.

February 1987   Christine Heym

Contents

List of Contributors*

Adler, J.E. *269*[1]
Ahonen, M. *217*
Alho, H. *61*
Allen, T.G.J. *323*
Alm, P. *89*
Andersson-Forsman, C. *151*
Aoki, A. *43*
Baker, D.M. *310*
Best, K. *230*
Björklund, H. *39*
Black, I.B. *269*
Bloom, S.R. *11*
Böck, P. *125*
Bööj, S. *282*
Burnstock, G. *323*
Chen, I. *78, 163*
Chiba, T. *157, 174*
Christensen, J. *102*
Clowry, G.J. *297*
Colombo, M. *67*
Conlon, J.M. *195*
Costa, E. *61*
Costa, M. *23*
Coupland, R.E. *288*
Crabtree, J.B. *273*
Dafgard, E. *195*
Dahlström, A. *282*
Dail, W.G. *340*
Dalsgaard, C.-J. *39*
De Vente, J. *355*
Dixon, J.S. *243*
Dolivo, M. *334*

Dreifuss, J.J. *334*
Dun, N.J. *329*
Durbin, J. *288*
Elfvin, L.-G. *28, 151*
Eränkö, L. *3, 217, 305*
Falkmer, S. *195*
Folan, J.C. *266*
Forssmann, W.G. *43*
Fujita, T. *189*
Furness, J.B. *23*
Gabella, G. *315*
Gehrke, D. *120*
Gibbins, I.L. *23*
Goldstein, M. *282*
Gosling, J.A. *243*
Gratzl, M. *130*
Grothe, C. *213*
Guidotti, A. *61*
Haegerstrand, A. *39*
Häppölä, O. *3, 51, 145,*
  217, 305
Hanbauer, I. *61*
Hansen, J.T. *208*
Haskell, A. *260*
Hassal, C.J.S. *323*
Hervonen, A. *223, 236*
Heym, Ch. *34, 67, 134*
Hock, D. *43*
Hökfelt, T. *28, 39*
Hofmann, H.-D. *213*
Horn, M. *34*
Hoshino, M. *19*

---

* The address of the authors is given on the first page of each contribution
[1] Page, on which contribution commences

Introduction

Milestones in SIF Cell Research

L. Eränkö, H. Päivärinta, S. Soinila, and O. Häppölä

Department of Anatomy, University of Helsinki, Siltavuorenpenger 20A, 00170 Helsinki, Finland

The Discovery of SIF Cells

In the beginning of this century, Kohn (1903) observed chromaffin cells in the sympathetic ganglia and considered them, in addition to the principal nerve cells, to represent a second specific cell type of the sympathetic nervous system. During the following 50 years, several studies were published in which chromaffin cells were described in various ganglia and in nerve trunks of the sympathetic system in different animals. Also, physiological studies supported the idea that there was a second catecholamine-containing cell type in the sympathetic ganglia that affected ganglionic transmission. The transmission in the rabbit sympathetic ganglion was characterized by a slow positive postsynaptic wave, and it was suggested by Eccles and Libet (1961) that this wave was caused by liberation of adrenaline from chromaffin cells. The role of catecholamines in the modulation of ganglionic transmission through SIF cells is discussed elsewhere in detail (Eränkö 1978).

With the formaldehyde-induced fluorescence method (Eränkö 1967) for catecholamines, Eränkö and Härkönen (1963) detected principal nerve (PN) cells as well as a second cell type which was small and displayed an intense catecholamine fluorescence and was thus named Small Intensely Fluorescent (SIF) cell. A few years later this small sympathetic cell type was detected electron microscopically (Grillo 1966). Characteristically, these cells contained membrane-bound dense cored vesicles and were named Small Granule Containing (SGC) cells. Later, Grillo et al. (1974) demonstrated convincingly with a combined fluorescence histochemical method that these cells indeed represented SIF cells. During the following years SIF cells were found in several sympathetic ganglia of many species, including man. Other names have been proposed for the SIF cells, such as small catecholamine-containing cells, chromaffin cells, chromaffin-like cells, paraganglionic cells, small interneurons and paraneurons.

Several different subtypes of SIF cells were observed with the fluorescence technique. Some of them were in clusters around capillaries, while others were solitary and scattered between the PN cells and displayed long processes. The number of SIF cells and their subtypes differ considerably between different ganglia and between different species (Williams et al. 1976).

Development and Differentiation of SIF Cells

SIF cells, like the PN cells, are derivatives of the neural crest. The development of SIF cells has been extensively studied in the

Experimental Brain Research Series 16
© Springer-Verlag Berlin · Heidelberg 1987

sympathetic ganglion of the rat. During prenatal development, there is a wide range of cell types in relation to size and the intensity of the formaldehyde-induced catecholamine fluorescence. Prenatally, cells can be classified as either SIF or PN cells, and in the newborn rat ganglion intermediate forms between SIF and PN cells also can be observed. However, in the 7-day-old rat intermediate forms do not appear. Postnatally, the number of SIF cells significantly increases during the 4th postnatal week (Soinila 1984). Since mitoses have not been observed at this stage of development, the increased number of SIF cells probably results through differentiation from primitive cells or through increased catecholamine synthesis in pre-existing small cells.

A great deal of knowledge concerning factors affecting the development of the autonomic nervous system has been gathered from studies on SIF cells. Injections of glucocorticoids into newborn rats was found to increase the number of SIF cells dramatically (Eränkö and Eränkö 1972). Concomitantly, the amount of adrenaline increases in the ganglion and immunoreactivity to phenylethanolamine N-methyl-transferase, the enzyme synthesizing adrenaline, appears in glucocorticoid-induced SIF cells. Thus, glucocorticoids favour the differentiation of sympathetic cells towards the adrenergic direction. Also interesting was the observation of Aloe and Levi-Montalcini (1979) that nerve growth factor injected into prenatal rats and continued postnatally, causes the development of a small ganglion in place of the adrenal medulla. Glucocorticoids and the nerve growth factor thus seem to have opposite effect on the developing sympathetic cell: the former favours development to SIF cells and adrenal medullary-like cells, while the latter causes the primitive sympathetic cells to develop into PN cells.

Synaptology and other Contacts of SIF Cells

Abundant data are available on synaptic contacts to and from SIF cells whose final structural features are illustrated in Fig. 1. Afferent synapses containing empty synaptic vesicles were described on SIF cells in the superior cervical ganglion of the rat, indicating that these cells receive a cholinergic innervation from the preganglionic trunk (Grillo 1966). These synapses have been observed to degenerate after division of the preganglionic nerve (Matthews and Ostberg 1973). Afferent synapses containing empty synaptic vesicles also have been reported on SIF cells in all other species studied.

Efferent synapses where SIF cells or their processes are the presynaptic element, as indicated by clustering of large granular vesicles, and the dendrites or cell body of the nerve cell are the postsynaptic element have been reported (Williams 1967; Matthews and Nash 1970), thus suggesting that SIF cells can function as interneurons.

Neighbouring SIF cells are often in close contact and linked to each other with attachment plaques (Grillo 1966). The satellite cell sheath around the SIF cells often is lacking when the SIF cells are near fenestrated blood vessels (Matthews and Raisman 1969). The close relationship of SIF cells to blood vessels has suggested that they serve an endocrine function (Matthews and Raisman 1969), possibly acting on the nerve cells of the ganglion through a portal circulation (Siegrist et al. 1968). A proximity to blood vessels can also be associated with chemoreceptor function, and clusters of SIF cells that are in close relationship to each other or to sympathetic

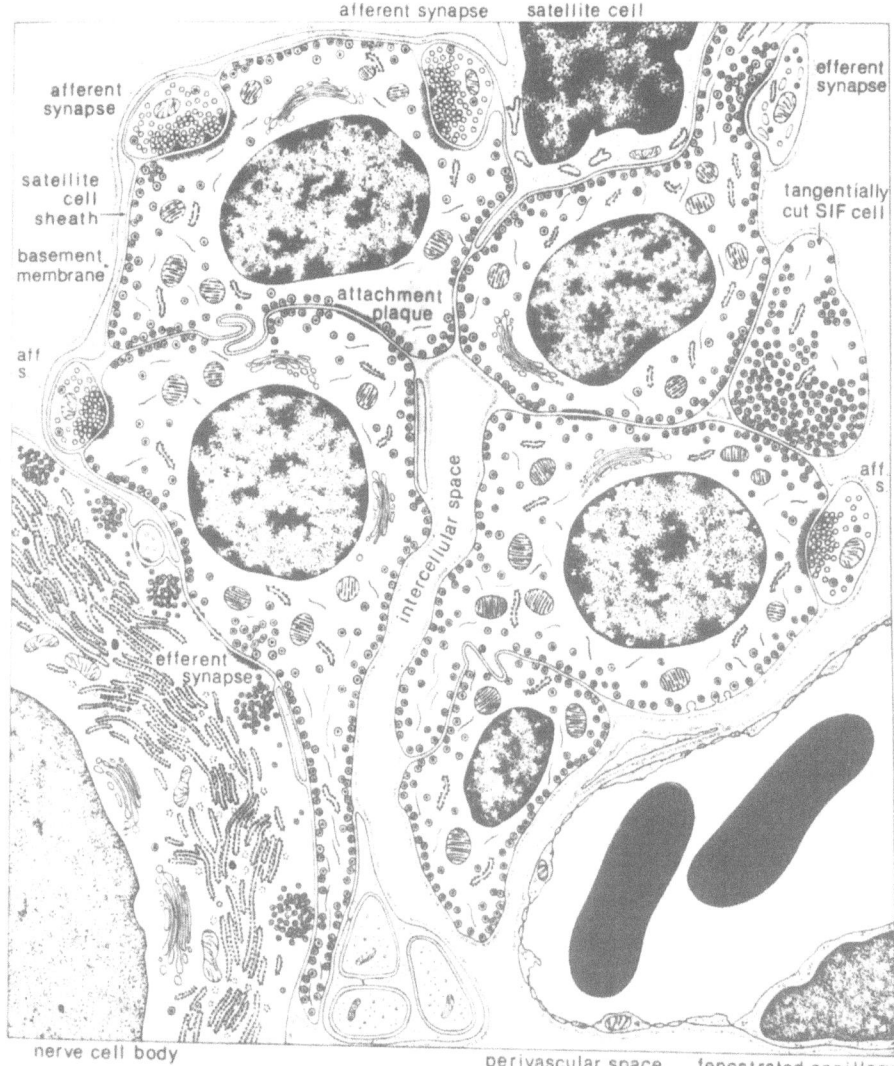

afferent synapse   satellite cell

efferent synapse

afferent synapse

satellite cell sheath

basement membrane

aff s

attachment plaque

tangentially cut SIF cell

intercellular space

aff. s

efferent synapse

nerve cell body

perivascular space   fenestrated capillary

Fig. 1. Schematic drawing illustrating the fine structure of a SIF cell group, a nerve cell body, a capillary and a satellite cell. Note the close apposition of neighbouring SIF cells as well as of a SIF cell and a nerve cell. One of the four afferent synapses (upper left) is opposite to a cluster of large granular vesicles, suggesting the presence of a reciprocal synapse. Two efferent synapses are drawn in the picture: somatosomatic synapse between a SIF cell and the nerve cell body and another from a thick SIF cell process to a nerve cell dendrite (upper right)

nerve cells and blood vessels may form small chemoreceptor organs resembling the carotid body (Eränkö and Eränkö 1971).

Neuroactive Substances in SIF Cells

Spectrophotofluorometric and immunohistochemical studies have shown that the SIF cells contain dopamine, noradrenaline or adrenaline (Eränkö et al. 1986). In addition to catecholamines, immunocytochemical studies have demonstrated 5-hydroxytryptamine- and histamine-immunoreactive SIF cells in the rat superior cervical ganglion (Häppölä et al. 1985, 1986). The SIF cells have also been found to be immunoreactive for the histamine-synthesizing enzyme, histidine decarboxylase. Further studies have demonstrated that SIF cells, which show 5-hydroxytryptamine and histidine decarboxylase immunoreactivity, are also immunoreactive for tyrosine hydroxylase (see Eränkö et al. 1986; Päivärinta et al. 1987) indicating coexistence of several amines in the same SIF cells. In addition to amines, several neuropeptides have been described immunohistochemically in SIF cells, e.g., enkephalins, bombesin/gastrin releasing peptide and dynorphin (see Eränkö et al. 1986).

The SIF cells seem to be a heterogeneous population of neuronal cells as regards their neuroactive substances. Since electrophysiological and biochemical studies have shown that 5-hydroxytryptamine, histamine and different peptides affect nervous transmission in the sympathetic ganglion, it is possible that in addition to catecholamines, the SIF cells may modulate ganglionic transmission through different neuroactive substances in a manner more complex than was previously understood.

References

Aloe L, Levi-Montalcini R (1979) Nerve growth factor in vivo-induced transformation of immature chromaffin cells in sympathetic neurons: effect of antiserum to the nerve growth factor. Proc Natl Acad Sci USA 76: 1246-1250

Eccles RM, Libet B (1961) Origin and blockade of the synaptic responses of curarized sympathetic ganglia. J Physiol 157: 484-503

Eränkö L, Eränkö O (1972) Effect of hydrocortisone on histochemically demonstrable catecholamines in the sympathetic ganglia and extra-adrenal chromaffin tissue of the rat. Acta Physiol Scand 84: 125-133

Eränkö L, Päivärinta H, Soinila S, Häppölä O (1986) Transmitters and modulators in the superior cervical ganglion of the rat. Med Biol 64: 75-83

Eränkö O (1967) The practical histochemical demonstration of catecholamines by formaldehyde-induced fluorescence. J Royal Microsc Soc 87: 259-276

Eränkö O (1978) Small intensely fluorescent (SIF) cells and nervous transmission in sympathetic ganglia. Ann Rev Pharmacol Toxicol 18: 417-430

Eränkö O, Eränkö L (1971) Small, intensely fluorescent granule-containing cells in the sympathetic ganglion of the rat. Progr Brain Res 34: 39-51

Eränkö O, Härkönen M (1963) Histochemical demonstration of fluoroge-
nic amines in the cytoplasm of sympathetic ganglion cells of the
rat. Acta Physiol Scand 58: 285-286

Grillo M (1966) Electron microscopy of sympathetic tissues. Pharmac
Rev 18: 387-399

Grillo MA, Jacobs L, Comroe JH (1974) A combined fluorescence histochemical and electron microscopic method for studying special monoamine-containing cells (SIF cells). J Comp Neurol 153: 1-14

Häppölä O, Soinila S, Päivärinta H, Panula P, Eränkö O (1985) Histamine-immunoreactive cells in the superior cervical ganglion and in
the coeliac-superior mesenteric ganglion complex of the rat. Histochemistry 82: 1-3

Häppölä O, Päivärinta H, Soinila S, Steinbusch H (1986) Pre- and
postnatal development of 5-hydroxytryptamine-immunoreactive cells in
the superior cervical ganglion of the rat. J Auton Nerv Syst 15: 21-
31

Kohn A (1903) Die Paraganglien. Arch Mikrosk Anat Entwicklungsmech
62: 263-265

Matthews MR, Nash JRG (1970) An afferent synapse from a small granule-containing cell to a principal neurone in the superior cervical
ganglion. J Physiol (Lond) 210: 11P-14P

Matthews MR, Ostberg A (1973) Effects of preganglionic nerve section
upon the afferent innervation of the small granule-containing cells
in the rat superior cervical ganglion. Acta Physiol Pol 24: 215-224

Matthews MR, Raisman G (1969) The ultrastructure and somatic efferent synapses of small granule-containing cells in the superior
cervical ganglion. J Anat 105: 255-282

Päivärinta H, Häppölä O, Joh TH, Panula P, Steinbusch H, Watanabe T
(1987) Coexistence of histamine, histidine decarboxylase, 5-hydroxy-
tryptamine and tyrosine hydroxylase in sympathetic cells of the rat.
Histochem J (submitted)

Siegrist G, Dolivo M, Dunant Y, Foroglou-Kerameus C, De Ribaupierre
F, Rouiller C (1968) Ultrastructure and function of the chromaffin
cells in the superior cervical ganglion of the rat. J Ultrastruct
Res 25: 381-407

Soinila S (1984) Pre- and postnatal development of the small intensely fluorescent cells in the rat superior cervical ganglion. Int J
Devl Neurosci 2: 65-76

Williams TH (1967) Electron microscopic evidence for an autonomic
interneuron. Nature (Lond) 214: 309-310

Williams TH, Chiba T, Black AC Jr, Bhalla RC, Jew J (1976) Species
variation in SIF cells of superior cervical ganglia: are there two
functional types? In: Eränkö O (ed) SIF Cells. Structure and
Function of the Small Intensely Fluorescent Sympathetic Cells. DHEW
Publication No (NIH) 76-942. U.S. Government Printing Office, Washington D.C., pp 143-162

Neurotransmitters
and Neuromodulators

Regulatory Peptides in the Autonomic and Sensory Nervous Systems

J. M. Polak[1] and S. R. Bloom[2]

[1] Department of Histochemistry, Royal Postgraduate Medical School, Hammersmith Hospital,
 Du Cane Road, London W12 0HS, Great Britain
[2] Department of Medicine, Royal Postgraduate Medical School, Hammersmith Hospital, Du Cane Road,
 London W12 0HS, Great Britain

Introduction

Numerous bioactive peptides have been found to be present in the so-called "diffuse neuroendocrine system". The generic term of regulatory peptides has been proposed for these peptides which are found in nerves and/or endocrine cells and act either as neurotransmitters or endocrine/paracrine hormones. In this review it is intended to describe the distribution of active peptides in the autonomic and sensory nervous systems.

Technology (Polak and Van Noorden 1986; Polak and Varndell 1984; Clarke and Hall 1986)

Our knowledge of neuropeptide production and distribution, localisation of binding sites, visualisation of messenger RNA and their possible involvement in pathology has been advanced greatly by the application of a number of modern microscopical imaging techniques.

Immunocytochemistry has been instrumental in determining the precise localisation of characteristic peptide proteins in the nervous (Fig. 1) and endocrine components of the "diffuse neuroendocrine system". Furthermore, application of the technique at the ultrastructural level has allowed the demonstration of peptide storage sites (secretory granules) (Fig. 2) and determination of intra- or intrergranular co-existence of more than one peptide and/or amine as well as the topographic segregation of peptide molecular forms. A number of "general neuroendocrine markers" have been recently proposed. These include the use of antibodies to neuron specific enolase, to chromogranins, to protein gene product 9.5 and to enzymes involved in catecholamine synthesis (e.g. tyrosine hydroxylase) (Fig. 3) or acetylcholine synthesis. Numerous techniques of hybridohistochemistry have recently been employed, including the use of complimentary DNA or RNA sequences, as well as the use of synthetic oligonucleotides (Fig. 4). Autoradiography has been used increasingly for the localisation of peptide binding sites (Fig. 5). The combined use of retrograde tracing methods with immunocytochemistry has allowed the delineation of a novel neurochemical anatomy linking peripheral organs with the central nervous system.

Distribution of Regulatory Peptides Throughout the Body

This section will be covered by a number of tables summarizing the main data so far available in terms of the distribution of regulatory peptides in the enteric nervous system (Table 1), autonomic/sen-

Experimental Brain Research Series 16
© Springer-Verlag Berlin · Heidelberg 1987

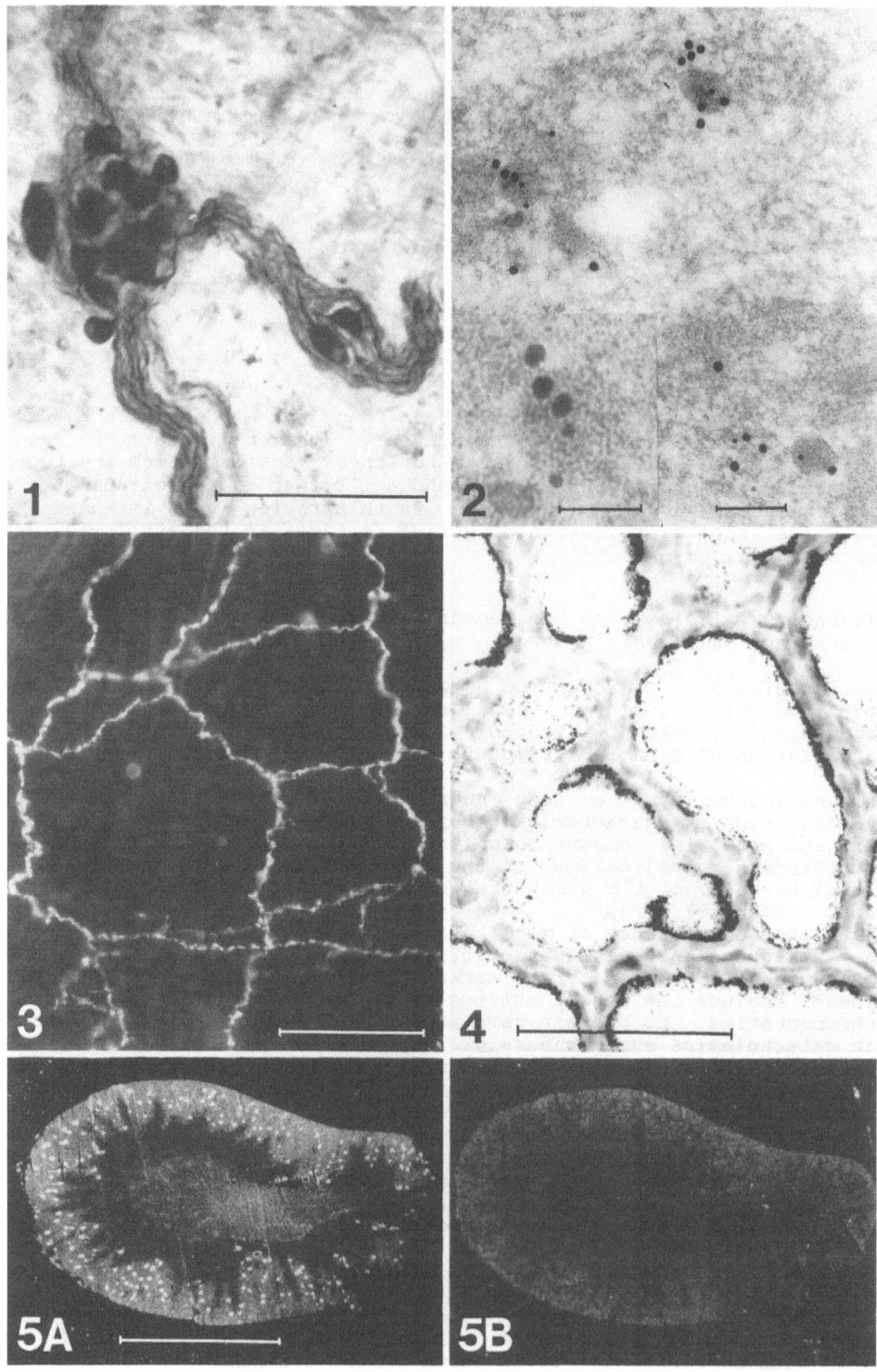

sory nerves of the cardiovascular (Table 2), genitourinary (Table 3) and respiratory systems (Table 4), as well as the skin (Table 5). Data on peptides localised to the adrenal gland, carotid body and sympathetic chain is also presented (Tables 6, 7, 8).

Involvement of Regulatory Peptides in Disease (Polak and Bloom 1985a)

Increasing evidence indicates that the regulatory peptide system (diffuse neuroendocrine system) is abnormal in a number of diseases of animals and humans. Tumours of the regulatory peptide containing system have been reported frequently (Polak and Bloom 1985b), including tumours of the gastrointestinal tract, pancreas, respiratory system (carcinoids and small cell carcinomas), and catecholamine-containing tissues (adrenal glands, carotid bodies, sympathetic chain), in addition to the well recognised tumours of the pituitary gland and thyroid (medullary carcinoma). Abnormalities of peptide containing nerves in non-tumour pathology have also been reported increasingly. These include the depletion of nerves in the genital tract of impotent males, unstable bladder, intractable constipation (Hirschsprung's disease, Chagas' disease, grass sickness of horse) asthma, cystic fibrosis, as well as diseases in which proliferation of peptide containing nerves have been shown including the diabetic eye and gut and Crohn's disease.

◄ **Fig. 1.** Ganglion cells and nerve fibres in Meissner's plexus revealed by immunostaining of neurofilament triplet proteins in a whole mount preparation of human colonic submucosa. Benzoquinone solution-fixed tissue stained by the peroxidase-antiperoxidase technique. Scale bar = 100 um

Fig. 2. Ultrastructural co-localisation of substance P and calcitonin gene-related peptide immunoreactivities to the same secretory vesicles in the adventitia of the superior mesenteric artery adventitia of the guinea pig. Double immunogold staining procedure. Substance P, 10 nm gold particles, calcitonin gene-related peptide 5 nm particles counterstained with uranyl acetate and lead citrate. Scale bars: main figure = 100 um, inset = 50 um

Fig. 3. Tyrosine hydroxylase (TH) immunostaining in benzoquinone solution-fixed whole mount preparation of guinea-pig abdominal vena cava. Network of fine varicose TH-immunoreactive nerve fibres encircling the vein at the adventitial-medial junction. Scale bar = 100 um

Fig. 4. Casein A _in situ_ hybridisation — mouse mammary gland. SP6-cDNA probe. 10 um paraformaldehyde fixed tissue counterstained with haematoxylin. Scale bar = 100 um

Fig. 5. A) Distribution of atrial natriuretic peptide receptors in a section of rat kidney incubated with 200_i+(1-28) atrial natriuretic peptide. B) Adjacent section to A) incubated with unlabelled h-ANP included in the incubation buffer. Scale bar = 5 mm

Table 1. Neuropeptides in the gastrointestinal tract

Peptide	Distribution	Species differences	Coexistence	Origin	Main actions	Refs
VIP/PHM	Most common type. Less in stomach than intestine. Present in all layers.		NPY Galanin Ach	Mainly intrinsic	Vasodilation Muscle relaxation	1,4
Galanin	Most in ileum. Present in all layers.	Mostly in pig	VIP/PHM	Mainly intrinsic	Secretion. Muscle contraction	2,9
Sub P..	Present in all layers. Also occurs in EC cells of some species.		CGRP in sensory nerves	Dual extrinsic (sensory)& intrinsic	Sensory. Muscle contraction. Vasodilatation.	8
NPY....	Present in all layers. Prominent around blood vessels.		VIP catecholamines	Dual extrinsic (sympathetic) & intrinsic	Vasoconstriction	11
CGRP...	Present in all layers. Prominent around blood vessels.	In rat, innervation of stomach is entirely sensory, that of colon sensory & intrinsic.	Sub P in sensory neurons	Dual extrinsic (sensory) & intrinsic	Sensory vasodilation	10
Met-enk	Mainly in muscle & myenteric plexus. Only minor population.			Intrinsic	Opiate effects	5,6
Somatostatin	Present in all layers. Only minor population.		Catecholamines	Intrinsic	Inhibition of release of several peptides	3,7

Key to abbreviations in Tables

VIP: vasoactive intestinal polypeptide, Sub P: substance P, NPY: neuropeptide Y, CGRP: calcitonin gene-related peptide, Met-enk: methionine-enkephalin, NT: neurotensin, ENK: enkephalin, SOM: somatostatin, ANP: atrial natriuretic peptide, PHI: peptide with histidine isoleucine, SP/NKA: substance P/neurokinins, PHI(M): peptide with histidine isoleucine (methionine), GAL: galanin, Ach: acetylcholine

References – Table 1

1) Bishop AE et al. (1984) Peptides 5: 255–259
2) Bishop AE et al. (1985) Gut 27: 849–857
3) Costa M et al. (1980) Neuroscience 5: 841–852
4) Costa M, Furness JB (1983) Neuroscience 8: 665–676
5) Elde R et al. (1976) Neuroscience 1: 349–351
6) Furness JB et al. (1983) Neuroscience 8: 653–664
7) Hökfelt T et al. (1975) Experientia 31: 852–854
8) Llewellyn-Smith IJ et al. (1984) Gastroenterology 86: 421–435
9) Melander T et al. (1985) Cell Tissue Res 239: 253–270
10) Mulderry PK et al. (1985) Regul Pept 12: 133–143
11) Polak JM, Bloom SR (1984) Peptides 5: 79–89

Table 2. Peptides in the mammalian cardiovascular system

Peptide	Localisation	Co-existence	Actions	Refs
NPY..	Perivascular & cardiac nerve fibres. Intracardiac neurones.	With TH in noradrenergic nerves.	Vasoconstrictor. ?Inotropic & chronotropic.	7,8,9 11
VIP..	" " "	------	Vasodilator, positive inotropic & chronotropic actions.	16,17 5
CGRP.	Perivascular & cardiac nerve fibres.	)CGRP & sub P co-exist to)a varying extent in)primary afferent neurones	Vasodilator, positive inotropic & chronogropic actions.	12,18
Sub P	" " "	))	Vasodilator	3,6, 13,19
NT..	" "	)	Inotropic & chronotropic actions	14
ENK..	" " "	------	" " "	4,15
SOM..	" " "	------	" " "	
ANP..	Atrial myoctyes.	------	Renal salt & water excretion. Vasocontriction. Renal actions of ADH. Aldosterone synthesis & release. Renin release.	1,2

References – Table 2

1) Cantin M et al. (1983) Histochemistry 80: 113–127
2) Chapeau C et al. (1985) J Histochem Cytochem 33: 541–550
3) Dalsgaard C-J et al. (1986) Cell Tissue Res 243: 477–485
4) Day SM et al. (1985) Br Heart J 63: 153–157
5) Della NG et al. (1983) Neuroscience 9: 605–619
6) Furness JB et al. (1982) Neuroscience 7: 447–459
7) Gu J et al. (1984) J Histochem Cytochem 32: 367–482
8) Gu J et al. (1983) Lancet 11: 1008–1010
9) Hassall CJS, Burnstock G (1984) Neurosci Lett 52: 111–115
10) Lundberg JM et al. (1983) Neurosci Lett 42: 167–172
11) Mulderry PK et al. (1985) Neuroscience 14: 941–954
12) Papka RE et al. (1981) Neurosci Lett 27: 47–53
13) Reinecke M (1985) Prog Histochem Cytochem 16: 1–175
14) Weihe E et al. (1983) Life Sci 33: 711–714
15) Weihe E, Reinecke M (1981) Neurosci Lett 26: 283–288
16) Weihe E et al. (1984) Cell Tissue Res 236: 527–540
17) Wharton J et al. (1986) J Auton Nerv Syst (in press)
18) Wharton J et al. (1981) Neuroscience 6: 2193–2204

Table 3 – Peptides in the Genitourinary System

Peptide	Main distribution	Origin of nerves	Main actions	Refs
VIP..	Nerve fibres in lamina propria, muscle layers, & close to blood vessels.	Pelvic plexus	Vasodilation. Smooth muscle relaxation. Secretion.	8,9
PHI..	Nerve fibre distribution closely mimicks that for VIP.	Pelvic plexus	Smooth muscle relaxation. Secretion.	4,3
NPY..	Nerve fibres in muscle layers & around blood vessels.	Pelvic plexus. Inferior mesenteric ganglion. Aortico-renal plexus. Sympathetic chain.	Vasoconstriction. Smooth muscle contraction.	1,7
Sub P	Nerve fibres in subepithelial locations & around blood vessels & muscle.	Dorsal root ganglia: T_{11}-L_3- hypogastric nerve. L_6-S_1- pelvic nerve.	Sensory neurotransmission Vasodilation. Smooth muscle contraction.	3,10

References – Table 3

1) Adrian JE et al. (1984) Life Sci 35: 2643–2648
2) Ghatei MA et al. (1985) Peptides 6: 809–815
3) Gu J et al. (1983) J Urol 130: 386–391
4) Huang WM et al. (1984) Histochem J 16: 1297–1310
5) Inyama CO et al. (1986) Neurosci Lett 69: 13–18

6) Lamano Carvalho TL et al. (1986) J Anat (in press)
7) Lundberg JM et al. (1982) Acta Physiol Scand 116: 477-480
8) Ottesen B (1983) Am J Obstet Gynecol 147: 208-224
9) Polak JM, Bloom SR (1984) Peptides 5: 225-230
10) Traurig H et al. (1984) Naunyn-Schmiedeberg's Arch Pharmacol
    326: 343-346

Table 4. Peptides in the respiratory tract

Peptide	Distribution	Actions	Species	Co-existence	Refs
CGRP..	Intra- & sub-epithelial nerves. Smooth muscle. Blood vessels. Glands. Mucosal endocrine cells (rat).	Sensory. Vasodilator. Bronchoconstrictor.	Most in rat & guinea pig. Less in man & monkey.	? SP	1,2,6
SP/ NKA...	Intra- & sub-epithelial. Smooth muscle. Blood vessels.	Sensory. Vasodilator. Bronchoconstrictor.	Most species including man.		4,6,8, 10
VIP/ PHI(M)	Smooth muscle. Blood vessels. Glands. Intrinsic ganglion.	Autonomic. Vasodilator. Bronchodilator. Secretomotor.	"	Cholinergic	9
NPY...	Blood vessels.	Autonomic. Vasoconstrictor	"	Catecholamines	5,7
GAL...	Smooth muscle. Blood vessels. Glands. Intrinsic ganglion.	Autonomic. Enhances muscle tone.	Pig and dog especially.	VIP	3

References - Table 4

1) Barnes PJ et al. (1986) Br J Pharmacol (in press)
2) Cadieux A et al. (1986) Neuroscience (in press)
3) Cheung A et al. (1985) Thorax 40: 889-896
4) Lundberg JM et al. (1984) Cell Tissue Res 235: 251-261
5) Lundberg JM et al. (1983) Neurosci Lett 42: 167-172
6) Lundberg JM et al. (1985) Eur J Pharmacol 108: 315-319
7) Sheppard MN et al. (1984) Thorax 39: 326-330
8) Sundler F et al. (1977) In: Von Euler US, Pernow E (eds) Sub-
    stance P. Raven, New York, pp 271-273
9) Uddman R et al. (1978) Acta Otalyryngol 86: 443-448
10) Wharton J et al. (1979) Invest Cell Path 2: 3-10

Table 5. Peptides in the skin

CGRP..	Intra-& sub-epithelial nerves. Smooth muscle. Touch receptors. Hair follicles. Blood vessels.	Sensory. Vasodilator - potently causes long-lasting flare.	Most in rat	? SP	1,2
SP/ NKA...	As for CGRP but also around sweat glands.	Sensory. Vasodilator - wheal and flare.	Most species		1,3,5
VIP/ PHI(M)	Free nerve endings - dermis, epidermis. Blood vessels. Sweat glands.	Autonomic & sensory. Vasodilatory - wheal & flare. Secretomotor.	Most species	Cholinergic	4,5
NPY...	Blood vessels. Sweat glands.	Autonomic. Vasoconstrictor.	Most species	Catecholamines	
SOM...	Sub-epidermal nerves. Touch receptors.	Sensory & autonomic.	Rat, human		

References - Table 5

1) Björklund H et al. (1986) Cell Tissue Res 243: 51-57
2) Brain SD et al. (1985) Nature 313: 54-56
3) Hökfelt T et al. (1975) Brain Res 100: 235-252
4) Lundberg JM et al. (1979) Neuroscience 4: 1539-1559
5) O'Shaugnessy DJ et al. (1983) Life Sci 32: 2827-2836

Table 6. Peptides in the mammalian adrenal gland

Peptide	Localisation	Co-existence	Refs
ENK...	Adrenal medullary cells & nerve fibres.	In a subpopulation of NPY (& catecholamine) containing adrenal medullary cells.	3,9,6
NPY...	" " " "	In a subpopulation of catecholamine containing adrenal medullary cells.	13 4,11
Sub P.	Nerve fibres.	----	3
VIP...	Nerve fibres.	----	3
SOM...	Adrenal medullary cells.	In a subpopulation of catecholamine containing cells.	6
Neuro-tensin	Adrenal medullary cells.	Noradrenaline containing cells.	10

Table 7. Peptides in the mammalian carotid body

ENK...	Type I cells.	Catecholamine containing type I cells.	7,13
VIP...	Nerve fibres and ganglion cells.	----	7,12
Sub P.	Nerve fibres & type I cells.	----	14,7,12
NPY...	Nerve fibres.	Noradrenaline.	2

References - Tables 6, 7

1) Cuello AC, McQueen DS (1980) Neurosci Lett 17: 215-219
2) Dalsgaard C-J et al. (1983) Neuroscience 9: 191-211
3) Helen P et al. (1984) Neuroscience 12: 907-916
4) Kondo H et al. (1986) Brain Res 372: 353-356
5) Linnoila RI et al. (1980) Neuroscience 5: 2247-2259
6) Lundberg JM et al. (1983) Neurosci Lett 42: 167-172
7) Lundberg JM et al. (1982) Acta Physiol Scand 116: 477-480
8) Lundberg JM et al. (1979) PNAS USA 76: 4079-4083
9) Lundberg JM et al. (1979) Acta Physiol Scand 157: 279-281
10) Schultzberg M (1983) Neuroscience 8: 363-374
11) Schultzberg M et al. (1979) Neuroscience 4: 249-270
12) Schultzberg M et al. (1978) Neuroscience 3: 1169-1186
13) Terenghi G et al. (1983) Endocrinology 112: 226-233
14) Varndell IM et al. (1984) Endocrinology 114: 1460-1462

Table 8. Peptides in mammalian sympathetic ganglia

NPY	Sympathetic neurones SIF cells	Noradrenaline	7,10
ENK	Sympathetic neurones SIF cells	Noradrenaline	1,11,23
VIP	Sympathetic neurones	Acetylcholine	4,11
SOM	Sympathetic neurones	Noradrenalines	5,12
Dynorphin	Sympathetic neurones SIF cells		8,23
NT	SIF cells and nerve fibres		4,12
Sub P	SIF cells and nerve fibres		4,6
Bombesin	Nerve fibres		1,16,17
Cholecystokinin	Nerve fibres		9

References - Table 8

1) Cuello AC, McQueen DS (1980) Neurosci Lett 17: 215-219
2) Dalsgaard C-J et al. (1983) Neuroscience 9: 191-211
3) Helen P et al. (1984) Neuroscience 12: 907-916
4) Heym Ch et al. (1984) Cell Tissue Res 235: 411-418
5) Hökfelt T et al. (1977) Proc Natl Acad Sci USA 74: 3587-3591
6) Hökfelt T et al. (1977) Brain Res 132: 29-41
7) Kondo H et al. (1986) Brain Res 372: 353-356

8) Kummer W et al. (1986) Anat Embryol 174: 401–405
9) Larsson L-I et al.. (1979) Brain Res 165: 201–218
10) Linnoila RI et al. (1980) Neuroscience 5: 2247–2259
11) Lundberg JM et al. (1979) PNAS USA 76: 4079–4083
12) Lundberg JM et al. (1982) Proc Natl Acad Sci USA 79: 1303–1307
13) Lundberg JM et al. (1979) Acta Physiol Scand 157: 279–281
14) Lundberg JM et al. (1979) Neuroscience 4: 1539–1559
15) Lundberg JM et al. (1982) Acta Physiol Scand 114: 153–155
16) Lundberg JM et al. (1983) Neurosci Lett 42: 167–172
17) Lundberg JM et al. (1982) Acta Physiol Scand 116: 477–480
18) Schultzberg M (1983) Neuroscience 8: 363–374
19) Schultzberg M et al. (1979) Neuroscience 4: 249–270
20) Schultzberg M et al. (1978) Neuroscience 3: 1169–1186
21) Terenghi G et al. (1983) Endocrinology 112: 226–233
22) Varndell IM et al. (1984) Endocrinology 114: 1460–1462
23) Varndell IM et al. (1982) J Histochem Cytochem 30: 682–690
24) Vincent SR et al. (1984) Neuroscience 11: 973–987
25) Wharton J et al. (1980) Nature 282: 269–271

Conclusion

A range of techniques are available at hand, including immunocytochemistry, electron microscopy, hybridohistochemistry and in vitro
autoradiography, which allow the precise determination of the distribution and cellular localisation of regulatory peptides and their
binding sites. Novel techniques of retrograde tracing and hybridohistochemistry also permit assessment of novel peptide containing
neurochemical pathways as well as assessment (mRNA) of the functional state of neural/endocrine elements synthesising and releasing
active peptides. The information obtained by analysis of disease
states allows further identification of the possible mode of actions
of regulatory peptides (Polak and Bloom 1983) and further understanding of the nature of the disease process.

References

Clarke CR, Hall MD (1986) Emerging technique. Hormone receptor
autoradiography: recent developments. TIBS 11: 195–200

Polak JM, Bloom SR (1983) Regulatory peptides: key factors in the
control of bodily functions. Brit Med J 286: 1461–1466

Polak JM, Bloom SR (1985a) Pathophysiology of the diffuse endocrine
system. In: St C Symmers W (ed) Systemic pathology. Churchill Livingstone, Edinburgh

Polak JM, Bloom SR (1985b) Endocrine tumours. Churchill Livingstone,
Edinburgh

Polak JM, Van Noorden S (eds) (1986) Immunocytochemistry: modern
methods and applications, 2nd edn. J Wright, Bristol

Biochemical and Immunological Aspects of Gastrointestinal Hormones

N. Yanaihara[1, 2], Ch. Yanaihara[1], T. Mochizuki[1], M. Hoshino[2], and K. Iguchi[1]

[1] Laboratory of Bioorganic Chemistry, Shizuoka College of Pharmacy, 2-2-1 Oshika, Shizuoka, Japan
[2] Laboratory of Cellular Metabolism, National Institute for Physiological Sciences, Okazaki, Japan

Introduction

Rapid advances in the field of peptide chemistry and gene technology have resulted in a burst of determinations about the molecular structures not only of regulatory peptides in mammalian and avian gastrointestinal tract but also of their precursors. Based on the known information about precursor and mature peptide structures, a number of regulatory peptides and other peptides with related amino acid sequences have been synthesized. Certain purpose-designed synthetic peptides have enabled the investigation of such questions as the molecular basis of receptor binding and the active roles of peptides (Yanaihara N 1980). Synthetic replicates of regulatory peptides and their analogues, fragments, and active peptide products derived from their precursors provide us with important immunogens for producing region-specific antisera. Specific antibodies against a regulatory peptide or its precursor-related peptides serve well for demonstrating the post-translational biosynthetic processing in the cells and for identifying the steps in the metabolic pathway of the regulatory peptide in tissues and tissue fluids (Yanaihara C et al. 1984).

This report provides examples of our biochemical and immunological studies, on various gastrointestinal regulatory peptides and related synthetic peptides, including vasoactive intestinal polypeptide (VIP)/peptide histidine isoleucine (PHI)/secretin/helodermin, galanin/substance P and glicentin/glucagon family peptides.

VIP/PHI Secretin Family

PHI is a heptadecapeptide amide first isolated from porcine intestine by Tatemoto and Mutt (1980). Based on our study on PHI synthesis, we were able to produce antisera specific to this peptide (Yanaihara N et al. 1984). With use of a PHI carboxy-terminal specific antiserum, we demonstrated immunocytochemically the coexistence of PHI- and VIP-like antigens (IR-PHI and IR-VIP) in mammalian intestinal nerves (Yanaihara N 1983). This finding was compatible with the observation that a VIP precursor molecule (Itoh et al. 1983) contained a PHI-like sequence. The proposed molecular structure of the precursor seems to allow for the probability that PHI-27 and VIP-28 are derived on an equimolar basis from the precursor. With use of the region-specific antisera, we examined radioimmunologically crude extracts of neuroblastoma cells and mammalian tissues by gel filtration and high performance liquid chromatography (HPLC). This enabled us to demonstrate various product peptides of PHI-IR in the extracts (Hoshino et al. 1984b; Yanaihara N et al. 1984; Ogino et al. 1986).

This provided evidence that, although mammalian VIP-producing cells generate PHI-27 as well as VIP-28, this did not occur on an equimolar basis. VIP-28 is a major product peptide of VIP-related peptides produced. In the VIP precursor, the VIP sequence is fringed at both ends with a pair of basic amino acids, while the PHI sequence is preceded by only one Arg residue. This suggests the possibility that multiple PHI components or modifications may occur.

Galanin/Substance P Family

Tissue distribution: Galanin is a peptide isolated from porcine intestine (Tatemoto et al. 1983) by a chemical assay technique detecting the C-terminal amide structure. Immunohistochemical and radioimmunological studies have revealed immunoreactive (IR) galanin not only in the gut but also in the central nervous system of several species (Ch'ng et al. 1985; Melander et al. 1985). IR-galanin also exists in the spinal cord of the rat and pig. However, galanin has structure similarity, especially in its C-terminal region, to other neuropeptides present in the spinal cord, such as substance P and neurokinins. These post-translational molecular variations must not be overlooked in immunochemical and immunohisto-chemical studies on galanin. In order to investigate this peptide in the spinal cord, we have produced galanin monoclonal antibodies which can discriminate galanin immunochemically from other structurally-related peptides (Yanaihara C et al. 1986). The distri-bution pattern of galanin positive fibers in laminae I and II of the dorsal horn is consistent with the reported data, but we found them more restricted in distribution to these areas than previously reported. Our radioimmunological measurements of pig spinal cord extracts revealed IR-galanin is almost exclusively restricted to the dorsal horn, where it occurs throughout the length of the spinal cord.

Biological activity: Having obtained the above immunohistochemical and immunochemical data, we extended the study of galanin to its electrophysiological role on the spinal cord of newborn rat. It was found that galanin produced a definite inhibitory effect on the monosynaptic reflex produced by a single shock stimulation of a lumbar dorsal root in the newborn rat (Yanagisawa et al. 1986). In addition, galanin significantly reduced the glucose-induced insulin release from the isolated perfused rat pancreas (Takeda et al. 1986). These suppressive effects of galanin-like peptide, actually produced by using highly purified synthetic peptides, were events regarded as unique and remarkable.

Glucagon/Glicentin Family

Our current studies on the glucagon/glicentin system (Yanaihara C et al. 1984, 1985a, 1985c) have established the existence of glicentin carboxy-terminal hexapeptide in mammalian pancreas. A glicentin C-terminal hexapeptide specific radioimmunoassay, using synthetic hexapeptide as standard, revealed in rat pancreas a peptide identi-fied with synthetic rat proglucagon (64-69). The hexapeptide was released concomitantly with glucagon by arginine stimulation from the isolated perfused rat pancreas (Yanaihara C et al. 1985b). The results indicate that the pancreas co-stores and possibly co-releases the hexapeptide, with glucagon as one of the peptide pro-cessing routes taken by rat proglucagon.

The findings described in this chapter demonstrate clearly that synthetic peptides are extremely important for exploring biological and immunological aspects of gastrointestinal hormones, to help answer important remaining questions about their physiology.

Acknowledgement. This work was supported in part by Grants-in-Aid for Scientific Research from the Ministry of Education, Science and Culture of Japan.

References

Ch'ng JLC, Christofides ND, Anand P, Gibson SJ, Allen YS, Su HC, Tatemoto K, Morrison JFB, Polak JM, Bloom SR (1985) Distribution of galanin immunoreactivity in the central nervous system and the responses of galanin-containing neuronal pathways to injury. Neuroscience 16: 343-354

Hoshino M, Yanaihara C, Ogino K, Iguchi K, Sato H, Suzuki T, Yanaihara N (1984) Production of VIP- and PHM (human PHI)-related peptides in human neuroblastoma cells. Peptides 5: 155-160

Itoh N, Obata K, Yanaihara N, Okamoto H (1983) Human preprovasoactive intestinal polypeptide contains a novel PHI-27-like peptide, PHM-27. Nature 304: 547-549

Melander T, Hökfelt T, Rökaeus A, Fahrenkrug J, Tatemoto K, Mutt V (1985) Distribution of galanin-like immunoreactivity in the gastrointestinal tract of several mammalian species. Cell Tissue Res 239: 253-270

Ogino K, Hoshino M, Nokihara K, Iguchi K, Yanaihara C, Yanaihara N (1986) Processing products of VIP/PHM precursor in human neuroblastoma cultured cells. In: Kiso Y (ed) Peptide chemistry 1985. Protein Research Foundation, Osaka, pp 385-390

Takeda Y, Yanaihara C, Hashimoto Y, Yamamoto Y, Takeda K, Tatemoto K, Mutt V, Yanaihara N (1986) Galanin effect on insulin and C-peptide release. Can J Physiol Pharmacol. Abstracts of the Sixth International Symposium on Gastrointestinal Hormones, p 95

Tatemoto K, Mutt V (1980) Isolation of two novel candidate hormones using a chemical method for finding naturally occurring polypeptide. Nature 285: 417-418

Tatemoto K, Rökaeus A, Jörnvall H, McDonald TJ, Mutt V (1983) Galanin - a novel biologically active peptide from porcine intestine. FEBS Lett 164: 124-128

Yanagisawa M, Yagi H, Otsuka M, Yanaihara C, Yanaihara N (1986) Inhibitory effects of galanin on the isolated spinal cord of the newborn rat. Neurosci Lett (in press)

Yanaihara C, Matsumoto T, Hong Y-M, Yanaihara N (1985a) Isolation and chemical characterization of glicentin C-terminal hexapeptide in porcine pancreas. FEBS Lett 189: 50-56

Yanaihara C, Matsumoto T, Kadowaki M, Iguchi K, Yanaihara N (1985b) Rat pancreas contains the proglucagon (64-69) fragment and arginine stimulates its release. FEBS Lett 187: 307-310

Yanaihara C, Matsumoto T, Nishida T, Uchida S, Kobayashi A, Moody AJ, Orci L, Yanaihara N (1984) Chemical approach to develop glicentin C-terminal specific radioimmunoassay. Biomed Res 5 (Suppl): 19-32

Yanaihara C, Matsumoto T, Tanaka M, Yamamoto Y, Yanaihara N (1985c) Radioimmunological demonstration of the existence of glicentin C-terminal hexapeptide in porcine pancreas. Biomed Res 6 (Suppl): 33-38

Yanaihara C, Yagi H, Okamura H, Ibata Y, Yanagisawa M, Otsuka M, Yanaihara N (1986) Galanin in spinal cord. Can J Physiol Pharmacol, Abstracts of the Sixth International Symposium on Gastrointestinal Hormones, p 161

Yanaihara N (1980) Immunochemical application of synthetic peptides to studies on the prohormone-hormone system. Biomed Res 1: 105-116

Yanaihara N, Nokihara K, Yanaihara C, Iwanaga T, Fujita T (1983) Immunocytochemical demonstration of PHI and its co-existence with VIP in intestinal nerves of the rat and pig. Arch Histol Jpn 46: 575-581

Yanaihara N, Yanaihara C, Nokihara K, Iguchi K, Fukata S, Tanaka M, Yamamoto Y, Mochizuki T (1984) Immunochemical study on PHI/PHM with use of synthetic peptides. Peptides 5: 247-254

Chemical Coding of Autonomic Neurons

I. L. Gibbins[1], J. L. Morris[2], J. B. Furness[1], and M. Costa[3]

[1] Department of Anatomy and Histology, School of Medicine, Flinders University,
Bedford Park, S.A. 5042, Australia
[2] Department of Physiology, School of Medicine, Flinders University, Bedford Park, S.A. 5042, Australia
[3] Centre for Neuroscience, Flinders University, Bedford Park, S.A. 5042, Australia

Introduction

Recent studies of autonomic and sensory neurons have shown that neurons commonly contain several substances with the potential to participate in neurotransmission (Lundberg and Hökfelt 1983). Our investigations of such neurons in a variety of vertebrate species have led us to propose that the neurons are "chemically coded" by the particular combinations of peptide and non-peptide transmitters contained within them (Costa et al. 1986; Furness et al. 1986). The chemical codes mark specific populations of neurons in a precise way which is related to the functions and the peripheral targets of those neurons. The presence of a large number of transmitter substances in a single neuron implies that neurotransmission is likely to be a plurichemical process, dependent upon the release and actions of several different transmitters (Furness et al. 1986). In this article we will briefly describe some examples of the chemical coding of autonomic neurons, which have been demonstrated by double-labelling immunofluorescence.

Materials and Methods

The procedures used for double-labelling immunofluorescence have been described in detail elsewhere (Costa et al. 1986).

Fixed tissue was exposed to a mixture of two primary antibodies which were raised in different species (such as a rat and a rabbit), followed by indirect fluorescence labelling with species-specific secondary antibodies raised in a third species (such as sheep, in this example). One of the species-specific secondary antibodies is labelled with fluorescein isothiocyanate (FITC), whilst the other one is labelled with tetramethyl rhodamine isothiocyanate (TRITC). Selective visualization of one or the other of the fluorophores in the same tissue sample is achieved with selective filters and dichromic mirrors in a Leitz epifluorescence microscope.

Examples of Chemical Coding of Autonomic Neurons

The pathway-specific chemical coding of autonomic neurons may be manifested in many ways, some of which can be illustrated by the following examples obtained from recent work in our laboratory.

1) Neurons Originating in the Same Ganglion May Have Different Chemical Codes Depending on their Peripheral Projections

Experimental Brain Research Series 16
© Springer-Verlag Berlin · Heidelberg 1987

The superior cervical ganglion of the guinea-pig contains noradrenergic neurons with immunoreactivity to both neuropeptide Y (NPY) and
dynorphin (Fig. 1a); NPY but not dynorphin; dynorphin but not NPY;
or neither of these peptides. Neurons containing both NPY and dynorphin project to the iris (Fig. 1b); neurons with NPY but not dynorphin project to blood vessels; whilst noradrenergic neurons with
dynorphin but not NPY probably project to piloerector muscles.
Similar target-related chemical coding has been described recently
for sympathetic neurons in guinea-pig coeliac ganglion (Costa and
Furness 1984) and for guinea pig sensory neurons containing different combinations of substance P, calcitonin gene-related peptide
(CGRP), cholecystokinin (CCK) and dynorphin (Gibbins et al. 1987a).

**2) Neurons of the Same Ganglion Projecting to the Same Target Tissue
May Have Different Chemical Codes**

In the submucous ganglia of the guinea-pig ileum, some secretomotor
neurons contain dynorphin, galanin and vasoactive intestinal peptide
(VIP), whilst other ones contain CGRP, CCK, galanin, NPY and somatostatin together (Furness et al. 1985; Costa et al. 1986). These two
populations of secretomotor neurons receive different synaptic
inputs (Bornstein et al. 1986) and there is good evidence that they
mediate physiological responses controlled by different neuronal
pathways (Furness and Costa 1986).

**3) Differently Coded Axons Make Connections with Specific Popula-
tions of Post-Synaptic Neurons**

In human sympathetic ganglia, pre-ganglionic fibers containing immunoreactivity for pro-enkephalin-derived peptides surround non-noradrenergic post-ganglionic neurons. Most of these neurons are also
supplied by a second population of fibers in which VIP (Fig. 1c) and
gastrin-releasing peptide (GRP) co-exist. Conversely, pre-ganglionic
axons containing NPY are associated only with noradrenergic post-

Fig. 1a, a'. Guinea-pig superior cervical gangion double-labelled ▶
for immunoreactivity to NPY (a) and dynorphin (DYN, a'). One cell
body (1) is immunoreactive for NPY but not dynorphin, whilst another
binucleate cell (2) contains both NPY- and dynorphin-like immunoreactivity

Fig. 1b, b'. Wholemount of guinea-pig iris double-labelled for
immunoreactivity to NPY (b) and dynorphin (b'). Arrows indicate
examples of varicose axons with both NYP and dynorphin immunoreactivities

Fig. 1c, c'. Human lumbar sympathetic ganglion double-labelled for
VIP (c) and tyrosine hydroxylase (TH; c') immunoreactivities. VIP
boutons surround one cell (1) which does not contain TH, whilst no
VIP boutons are associated with a nearby TH-immunoreactive neuron
(2). L, autofluorescent lipofuscin granules

Fig. 1d, d'. Human lumbar sympathetic ganglion double-labelled for
NPY (d) and TH (d'). NPY boutons are associated with one neuron (1)
which is TH-immunoreactive, but not with another neuron (2) which
lacks TH. L, autofluorescent lipofuscin granules

Scale bars in all illustrations represent 20 um

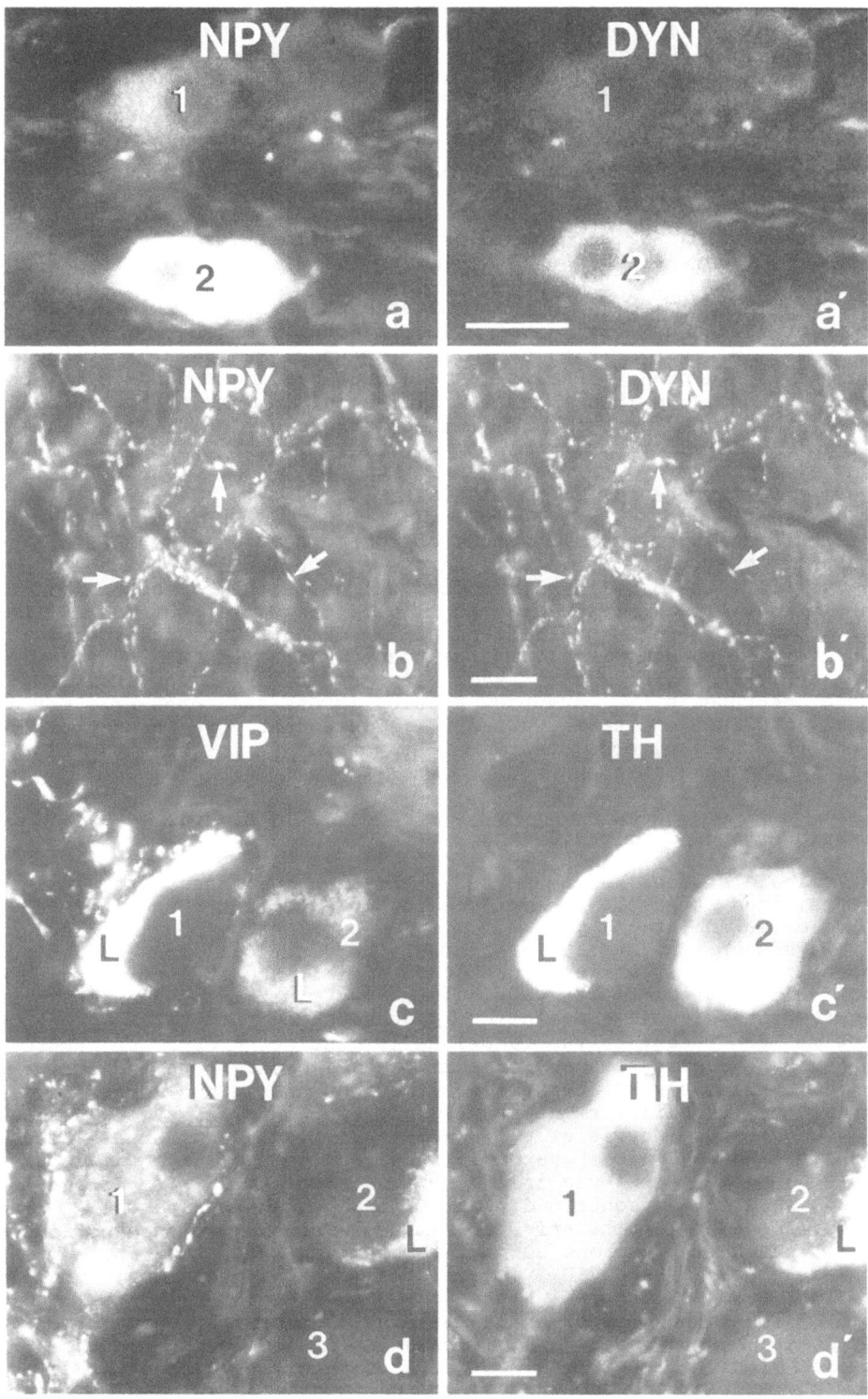

ganglionic   neurons (Fig.   1d).   We also have found precise connections   between immunohistochemically identified populations of   pre-
and post-ganglionic neurons   in the vagus nerve and sympathetic
ganglia of toads (Morris et al. 1986a). Chemically coded associations between autonomic neurons and their synaptic inputs also occur
in prevertebral, enteric and pelvic ganglia of guinea-pigs (Bornstein et al. 1986; Costa et al. 1986; Macrae et al. 1986; and
unpublished results of the authors).

4)   Neurons with Similar Physiological Functions in Different Parts
of the Body May Have Different Chemical Codes

Perivascular non-noradrenergic neurons containing VIP have been
implicated in the mediation of vasodilation responses in many vascular beds. However, in guinea-pigs, non-noradrenergic perivascular
neurons containing VIP also contain both NPY and dynorphin in the
pelvic vasculature (Morris et al. 1985), but in most of the cephalic
vasculature, they contain VIP without these other peptides.

Species Differences in Chemical Coding

Whilst the principles of chemical coding of autonomic neurons outlined above seem to be applicable across all vertebrate species examined so far, the details of the chemical codes themselves may differ
in unpredictable ways (Furness et al. 1986; Gibbins et al. 1987b).
For example, although NPY is found in sympathetic cardiovascular
neurons in species as diverse as toads (Morris et al. 1986a) and
guinea-pigs (Morris et al. 1986b), it is largely absent from comparable neurons in two marsupial species (brush tail opossum: Morris
et al. 1986c; and tammar wallaby). Consequently, we cannot assume
that the chemical code observed in a pathway in one species always
will be the same in another.

Consequences of Chemical Coding

Anatomical studies, such as those briefly described above, have
demonstrated that the chemical coding of autonomic neurons provides
a valuable analytical tool allowing us to identify specific neural
pathways with considerable precision. The functional consequences of
these patterns of co-existence of potential neurotransmitters are
difficult to predict. In particular, two circumstances arise which
require considerable care in interpretation. First, the same neuropeptide can occur in two functionally distinct groups of neurons
supplying the same tissue. For example, NPY occurs both in noradrenergic (presumably vasoconstrictor) neurons and in non-noradrenergic
(presumably vasodilator) neurons supplying the pelvic blood vessels
of guinea-pigs (Morris et al. 1985). Similarly, dynorphin occurs
both in noradrenergic sympathetic axons and in unmyelinated sensory
axons in the guinea-pig iris (Gibbins et al. 1987a). Second, a
single neuron may contain different peptides which apparently have
opposing actions on the effector tissue. For example, non-noradrenergic autonomic neurons supplying the guinea-pig uterine artery
contain both NPY and VIP. In isolated preparations of this vessel,
NPY is a vasoconstrictor, whilst VIP is a vasodilator (Morris et al.
1985). Thus, the effects of stimulating an identified population of
autonomic neurons to a target tissue are likely to be mediated by
complex plurichemical processes, even though, at present, we do not
know how most of the substances forming the chemical codes take part
in neurotransmission.

Acknowledgements. The work desribed here was supported by grants from the National Health and Medical Research Council of Australia, the National Heart Foundation of Australia and the Flinders Medical Centre Research Foundation. We thank Rae Tyler for typing the manuscript.

References

Bornstein JC, Costa M, Furness JB (1986) Synaptic inputs to immuno-histochemically identified neurones in the submucous plexus of the guinea-pig small intestine. J Physiol (Lond) (in press)

Costa M, Furness JB (1984) Somatostatin is present in a subpopulation of noradrenergic nerve fibres supplying the intestine. Neuro-science 13: 911-920

Costa M, Furness JB, Gibbins IL (1986) Chemical coding of enteric neurons. Prog Brain Res 68: 217-240

Furness JB, Costa M (1986) The enteric nervous system. Churchill Livingstone, Edinburgh

Furness JB, Costa M, Gibbins IL, Llewellyn-Smith IJ, Oliver JR (1985) Neurochemically similar myenteric and submucous neurons directly traced to the mucosa of the small intestine. Cell Tissue Res 241: 155-163

Furness JB, Costa M, Morris JL, Gibbins IL (1986) Novel neurotransmitters and the chemical coding of neurons. Adv Physiol (in press)

Gibbins IL, Furness JB, Costa M (1987a) Pathway-specific patterns of coexistence of substance P, calcitonin gene-related peptide, cholecystokinin and dynorphin in dorsal root ganglion neurons of the guinea-pig. Cell Tiss Res (in press)

Gibbins IL, Morris JL, Furness JB, Costa M (1987b) Innervation of systemic blood vessels. In: Burnstock G, Griffith S (eds) Non-adrenergic innervation of blood vessels. CRC Press, Boca Raton, Florida (in press)

Lundberg JM, Hökfelt T (1983) Coexistence of peptides and classical neurotransmitters. Trends in Neuroscience 6: 325-333

Macrae I, Furness JB, Costa M (1986) Distribution of subgroups of noradrenaline neurons in the coeliac ganglion of the guinea-pig. Cell Tissue Res 244: 173-180

Morris JL, Gibbins IL, Furness JB, Costa M, Murphy R (1985) Co-localization of NPY, VIP and dynorphin in non-noradrenergic axons of the guinea-pig uterine artery. Neurosci Lett 62: 31-37

Morris JL, Gibbins IL, Campbell G, Murphy R, Furness JB, Costa M (1986a) Innervation of the large arteries and heart of the toad (Bufo marinus) by adrenergic and peptide-containing neurons. Cell Tissue Res 243: 171-184

Morris JL, Murphy R, Furness JB, Costa M (1986b) Partial depletion by neuropeptide Y from noradrenergic perivascular and cardiac axons by 6-hydroxydopamine and reserpine. Regul Peptides 13: 147-162

Morris JL, Gibbins IL, Murphy R (1986c) Neuropeptide Y-like immuno-reactivity is absent from most perivascular noradrenergic axons in a marsupial, the brush-tailed possum. Neurosci Lett 71: 264-270

Peptides and Transmitters in Para- and Prevertebral Ganglia and Their Projections

B. Lindh[1], T. Hökfelt[2], and L.-G. Elfvin[1]

[1] Department of Anatomy, Karolinska Institutet, P.O. Box 60400, 10401 Stockholm, Sweden
[2] Department of Histology, Karolinska Institutet, P.O. Box 60400, 10401 Stockholm, Sweden

Introduction

Acetylcholine (ACh) and noradrenaline (NA) are the classical transmitters in the autonomic nervous system, but more recent research
has established the presence also of several peptides in peripheral
systems. For example, in the celiac superior mesenteric ganglion
(CSMG) of the guinea-pig, noradrenergic neurons containing immunoreactivity to somatostatin (SOM) (Hökfelt et al. 1977a), vasoactive
intestinal polypeptide (VIP) (Hökfelt et al. 1977b), as well as a
pancreatic polypeptide-like peptide, presumably neuropeptide Y (NPY)
(Lundberg et al. 1982, 1983), have been identified, whereby SOM- and
NPY-like immunoreactivity (LI) coexist with NA in separate neuron
populations (Lundberg et al. 1982).

Peptide immunoreactivity has also been demonstrated in nerve fibers,
including a heterogeneously distributed VIP-immunoreactive (IR)
plexus and a more evenly distributed substance P (SP)-IR fiber
network in the CSMG (Hökfelt et al. 1977b, 1977c). Furthermore,
networks of enkephalin (ENK)- (Schultzberg et al. 1978, 1979), cholecystokinin (CCK)- (Larsson and Rehfeld 1979), bombesin (BOM)-
(Schultzberg 1983), dynorphin (DYN)-positive (Vincent et al. 1983)
fibers have been observed in the CSMG and calcitonin gene-related
peptide (CGRP)-IR fibers in the inferior mesenteric ganglion (Franco-Cereceda et al. 1986).

In this study we have used immunohistochemical techniques to further
analyze the topography of the peptidergic neuronal subpopulations
and the distribution of afferent peptidergic nerve fibers in the
CSMG of the guinea pig. Finally, some sympathetic ganglia have been
studied using monoclonal antibodies to the ACh-synthesizing enzyme
choline acetyltransferase (ChAT).

Materials and Methods

Male guinea pigs (200–300 g) were deeply anesthetized, and after
whole body perfusion with Tyrode's solution the appropriate ganglia
were rapidly dissected and fixed in 10% formalin containing picric
acid (Zamboni and de Martino 1967). Cryostat sections were processed
for indirect immunohistochemistry according to Coons and collaborators (see Coons 1958). Rabbit antisera to the following antigens
were used: tyrosine hydroxylase (TH), dopamin-beta-hydroxylase
(DBH), SOM, NPY, VIP and polypeptide HI (PHI), SP, ENK, CCK, BOM,
DYN and CGRP. A monoclonal antibody to ChAT was also used. After
rinsing the sections were incubated with fluorescein isothiocyanate
(FITC)-conjugated swine anti-rabbit antibodies or donkey anti-rabbit

antibodies. For the ChAT antiserum FITC-conjugated sheep anti-mouse antiserum was used. As a control for the peptide immunoreactivity sections were incubated with antisera preabsorbed with an excess of the respective peptide (10 nmol/ml diluted antisera). Normal mouse serum served as control for ChAT antiserum and normal rabbit serum for TH and DBH antisera. As a complement to the double staining techniques, the elution-restaining technique of Tramu et al. (1978) was used to show co-localization of two or more antigens. All sections were examined in a Zeiss fluorescence microscope equipped with appropriate filter combinations.

Results and Discussion

SOM- and NPY-LI were found in separate populations of noradrenergic neurons, although occasional cells contained both; moreover, they occupied different domains of the ganglion. About 65% of the principal ganglion cells contained NPY-LI, whereas 25% were SOM-IR. The NPY-positive cells were mainly found bilaterally in the posterior superior parts of the ganglion representing the celiac poles. The SOM-positive cells predominated in the anterior-inferior part, i.e. the superior mesenteric pole. A third population of ganglion cells reacted with antisera to VIP and PHI (Fig. 1a). The VIP/PHI-containing cell bodies comprised less than 1% of the ganglion cells and were preferentially located in the celiac poles. Many of the VIP/PHI-positive neurons contained NPY-LI and SOM-LI (cf. Fig. 1a, b) and occasionally TH-LI. A small population of DYN-containing cell bodies were located in the celiac poles, and it has been reported that VIP-LI and DYN-LI are co-localized in ganglion cells of the guinea pig CSMG (Macrae et al. 1986). In a small population of the NA neurons no peptide immunoreactivity has been found so far.

The projections of the different noradrenergic neuron populations in the CSMG of the guinea pig to the gastrointestinal wall have been investigated after denervation (Furness et al. 1983; Costa and Furness 1984) and with retrograde axonal tracing (Lindh et al. 1986a). Costa, Furness and collaborators (Furness et al. 1983; Costa and Furness 1984) showed that these subgroups supply different tissue components in the wall of the small intestine. Thus, the NA/SOM neurons mainly innervate the submucosal ganglia and the mucosa and would primarily be involved in the regulation of mucosal functions. The NA/NPY neurons mainly innervate blood vessels and would thus be involved in the regulation of intestinal blood flow. The NA neurons lacking SOM- and NPY-LI project to the myenteric ganglia and would primarily be involved in the regulation of intestinal motility.

SOM-LI could also be observed in varicose nerve fibers in areas where the SOM-containing cell bodies dominated (Fig. 2a). These varicose nerve fibers often formed basket-like structures surrounding the SOM-positive perikarya. A DBH-positive fiber network of a similar morphology is observed in the SOM-dominated part of the ganglion (Fig. 2b). Double staining experiments revealed that these two fiber networks in some cases overlapped, and presumably were identical (cf. Fig. 2a, b).

A very dense network of VIP/PHI-positive fibers was observed in well delimited areas of the CSMG. Networks of similar morphology and distribution were seen in sections incubated with antisera to CCK, BOM, DYN and CGRP. All these fiber networks generally were confined to areas where the SOM-containing principal ganglion cells predominated. They disappeared after transection of the mesenteric nerves,

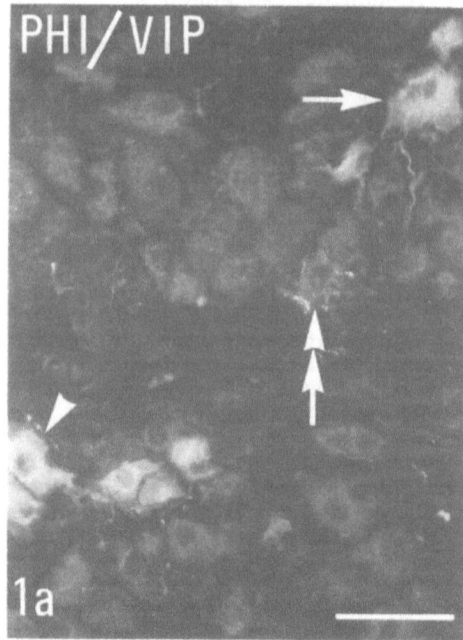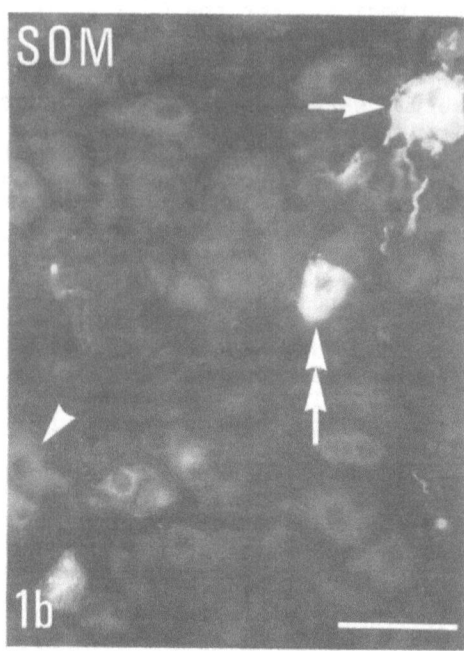

Fig. 1. Immunofluorescence micrographs of the celiac superior mesenteric ganglion after incubation with antisera to PHI (a) and SOM
(b). A double staining technique was used to show co-localization of
the two antigens. Arrows point to a cell containing immunoreactivity
to PHI/VIP and SOM. Arroheads point to a PHI/VIP-positive neuron
which lacks SOM-LI. Double arrows point to a SOM-containing neuron
devoid of PHI/VIP-LI. Bars indicate 50 um

indicating that they arise in the gastrointestinal tract and presumably participate in the reflex arc between the intestine and the
CSMG. These results suggest chemical coding of the circuitry, possibly by messenger molecules related to functional differentiation
of neuron populations. In this context, the studies of Costa and
Furness (Furness et al. 1983; Costa and Furness 1984) discussed
above are very interesting. Thus, the VIP/PHI-, CCK-, BOM-, DYN-,
and CGRP-positive fibers projecting to the ganglion would primarily
be of importance in controlling mucosal functions via NA/SOM neurons. The fibers would not be involved in the regulation of the
intestinal blood flow, which is exerted by NA/NPY neurons.

A second type of afferents were sensory fibers containing SP- and
CGRP-LI which formed moderately dense varicose networks. Finally, a
sparse network of ENK-positive fibers, as well as a few weakly
labelled ChAT nerve fibers were also observed in the CSMG, probably
originating in the spinal cord. These nerve fiber networks were
found in all parts of the ganglion and did not disappear after
transection of the mesenteric nerves.

The distribution of ChAT- and ENK immunoreactivities has been analyzed in two paravertebral ganglia, the superior cervical ganglion
(SCG) and the stellate ganglion (SG) (Lindh et al. 1986b). In the
SCG a dense network of fine varicose ChAT-positive nerve fibers was
seen in all parts of the ganglion, with more intensely fluorescent

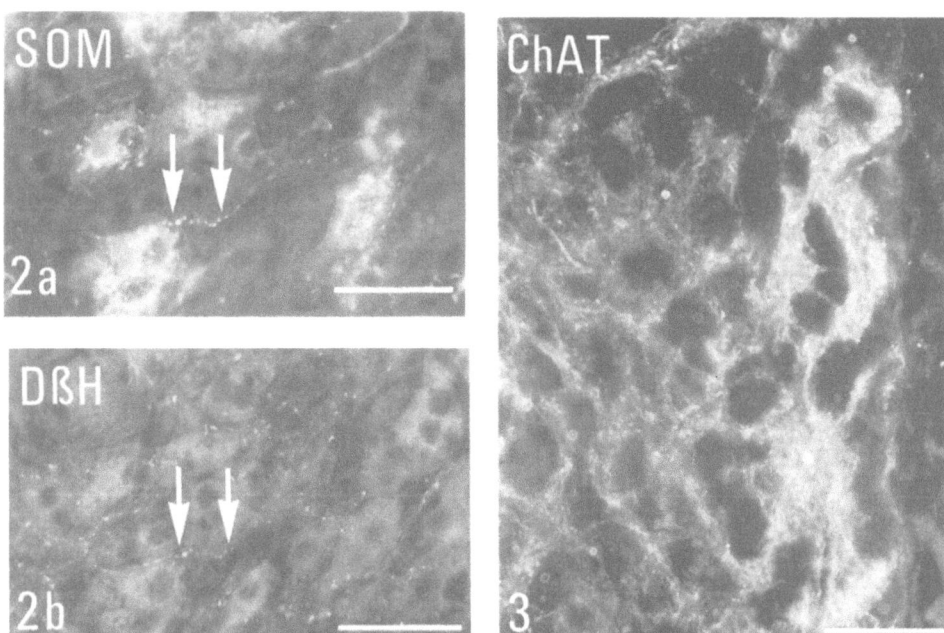

Fig. 2. Immunofluorescence micrographs of a section of the celiac superior mesenteric ganglion after incubation with antisera to SOM (a) and DBH (b). Double staining technique was used. Arrows point to a varicose nerve fiber containing both SOM- and DBH-LI. Bars indicate 50 um

Fig. 3. Immunofluorescence micrographs of the superior cervical ganglion after incubation with antiserum to ChAT. The ChAT-positive fibers are fine and have a varicose appearance, and they form intensely fluorescent networks with a patchy appearance and with higher number of fibers around certain cells. Bar indicates 50 um

networks around certain groups of ganglion cells resulting in a patchy staining pattern (Fig. 3). Transection of the cervical sympathetic trunk caudal to the SCG caused disappearance of ChAT-positive fibers in this ganglion. In the SG a dense network of varicose ChAT-positive fibers was found, here also with a somewhat regional distribution. A sparse network of ENK-positive fibers was seen in the SG. This network had a complementary distribution to that of the ChAT-positive one. The most dense ChAT-positive nerve fiber networks were found in regions poor in ENK-immunoreactivity, and vice versa. The majority of the SG cells were NPY-IR, and they seemed to be more prevalent in the ENK-rich regions.

The findings may be considered in relation to earlier speculations of a possible co-localization of ACh and an ENK-like peptide in preganglionic neurons (Schultzberg et al. 1978, 1979) and the actual demonstration of ChAT- and ENK-LI in the same neurons in the sympathetic lateral column of the rat (Kondo et al. 1985). A conclusive answer as to whether the preganglionic sympathetic neurons innervating SCG and SG in the guinea-pig contain both substances or whether ACh and the ENK-like peptide at least sometimes are in separate neurons, has to await further studies at the spinal cord level.

Acknowledgements. The present study was supported by grants from the Karolinska Institutet, the Swedish Medical Research Council (12X-5189, 04X-2887), the Swedish Society for Medical Sciences, Ruth and Richard Julins Stiftelse and Alice and Knut Wallenbergs Stiftelse. For generous supply of antisera we thank Drs. Robert Elde, Dept. of Anatomy, University of Minnesota, Minneapolis, MI, USA (SOM), Piers Emson, Agricultural and Food Research Council, Institute of Animal Physiology, Babraham, Cambridge, UK (SP), Jan Fahrenkrug, Dept. of Clinical Chemistry, Bispebjerg Hospital, Copenhagen, Denmark (VIP and PHI), Peter Frey, Wander Research Institute (Sandoz), Bern, Switzerland (CCK), Menek Goldstein, Dept. of Psychiatry, New York, NY, USA (TH and DBH), Paul M. Salvaterra, Beckman Research Institute of the City of Hope, Duarte, CA, USA (ChAT), and Lars Terenius, Uppsala University, Uppsala, Sweden (DYN, ENK and NPY).

References

Coons AH (1958) Fluorescent antibody methods. In: Danielli JF (ed) General cytochemical methods. Academic Press, New York, pp 399–422

Costa M, Furness JB (1984) Somatostatin is present in a subpopulation of noradrenergic nerve fibers supplying the intestine. Neuroscience 13: 911–919

Franco-Cereceda A, Henke H, Lundberg JM, Petermann JB, Hökfelt T, Fischer JA (1986) Calcitonin gene-related peptide (CGRP) in capsaicin-sensitive substance P-containing sensory neurons in animals and man: distribution and release by capsaicin. Peptides (in press)

Furness JB, Costa M, Emson PC, Hakanson R, Moghimzadeh E, Sundler F, Taylor IL, Chance RE (1983) Distribution, pathways and reaction to drug treatment of nerves with neuropeptide Y- and pancreatic polypeptide-like immunoreactivity in the guinea pig digestive tract. Cell Tissue Res 234: 71–92

Hökfelt T, Elfvin LG, Elde R, Schultzberg M, Goldstein M, Luft R (1977a) Occurrence of somatostatin-like immunoreactivity in some peripheral sympathetic noradrenergic neurons. Proc Natl Acad Sci USA 74: 3587–3591

Hökfelt T, Elfvin LG, Schultzberg M, Fuxe K, Said SI, Mutt V, Goldstein M (1977b) Immunohistochemical evidence of vasoactive intestinal polypeptide-containing neurons and nerve fibers in sympathetic ganglia. Neuroscience 2: 885–896

Hökfelt T, Elfvin LG, Schultzberg M, Goldstein M, Nilsson G (1977c) On the occurrence of substance P-containing fibers in sympathetic ganglia: immunohistochemical evidence. Brain Res 132: 29–41

Kondo H, Kuramoto H, Wainer BH, Yanaihara N (1985) Evidence for the coexistence of acetylcholine and enkephalin in the sympathetic preganglionic neurons of rats. Brain Res 335: 309–314

Larsson LI, Rehfeld JF (1979) Localization and molecular heterogeneity of cholecystokinin in the central and peripheral nervous system. Brain Res 165: 201–218

Lindh B, Hökfelt T, Elfvin LG, Terenius L, Fahrenkrug J, Elde R, Goldstein M (1986a) Topography of NPY-, somatostatin-, VIP-immunoreactive, neuronal subpopulations in the guinea pig celiac-superior mesenteric ganglion and their projection to the pylorus. J Neurosci 6: 2371-2383

Lindh B, Staines W, Hökfelt T, Terenius L, Salvaterra PM (1986b) Immunohistochemical demonstration of choline acetyltransferase-immunoreactive preganglionic nerve fibers in guinea pig autonomic ganglia. Proc Natl Acad Sci USA 83: 5316-5320

Lundberg JM, Hökfelt T, Änggard A, Terenius L, Elde R, Markey K, Goldstein M, Kimmel J (1982) Organizational principles in the peripheral sympathetic nervous system: subdivision by coexisting peptides (somatostatin-, avian pancreatic polypeptide-, and vasoactive intestinal polypeptide-like immunoreactive materials). Proc Natl Acad Sci USA 79: 1303-1307

Lundberg JM, Terenius L, Hökfelt T, Goldstein M (1983) High levels of neuropeptide Y in peripheral noradrenergic neurons in various mammals including man. Neurosci Lett 42: 167-172

Macrae IM, Furness JB, Costa M (1986) Distribution of subgroups of noradrenaline neurons in the coeliac ganglion of the guinea-pig. Cell Tissue Res 244: 173-180

Schultzberg M (1983) Bombesin-like immunoreactivity in sympathetic ganglia. Neuroscience 8: 363-374

Schultzberg M, Hökfelt T, Lundberg JM, Terenius L, Elfvin LG, Elde R (1978) Enkephalin-like immunoreactivity in nerve terminals in sympathetic ganglia and adrenal medulla and in adrenal medullary gland cells. Acta Physiol Scand 103: 475-477

Schultzberg M, Hökfelt T, Terenius L, Elfvin LG, Lundberg JM, Brandt J, Elde RP, Goldstein M (1979) Enkephalin immunoreactive nerve fibers and cell bodies in sympathetic ganglia of the guinea-pig and rat. Neuroscience 4: 249-270

Tramu G, Pillez A, Leonardelli J (1978) An efficient method of antibody elution for the successive or simultaneous localization of two antigens by immunocytochemistry. J Histochem Cytochem 26: 322-324

Vincent SR, Dalsgaard CJ, Schultzberg M, Hökfelt T, Christensson I, Terenius L (1984) Dynorphin-immunoreactive neurons in the autonomic nervous system. Neuroscience 11: 973-987

Zamboni L, de Martino C (1967) Buffered picric acid formaldehyde: a new rapid fixative for electron microscopy. J Cell Biol 35: 148A

Peptide Immunohistochemistry of Guinea Pig Lumbar Sympathetic Ganglia

R. H. Webber[1], M. Horn[2], and Ch. Heym[2]

[1] Department of Anatomical Sciences, Faculty of Health Sciences, State University of New York at Buffalo, 317 Farber Hall, Buffalo, NY 14214, USA
[2] Anatomisches Institut, Universität Heidelberg, Im Neuenheimer Feld 307, 6900 Heidelberg 1, FRG

Introduction

The discovery of neuropeptides in the autonomic nervous system has resulted in a resurgence of investigation on sympathetic pathways and how they relate to these possible neurotransmitters or neuromodulators. In particular, guinea pig prevertebral ganglia have been shown to be richly supplied by an array of peptides (cf. Schultzberg et al. 1983; Heym and Lang 1986). There are comparatively few data available on the respective lumbar paravertebral ganglia. We have applied immunohistochemical methods to investigate the pathways of the second, third and fourth left lumbar sympathetic trunk ganglia in guinea pigs after selective surgical interruption for the purpose of establishing the location of perikarya of neuropeptide–containing fibers in the ganglia.

Materials and Methods

Sixteen female guinea pigs (appr. 250 g body weight) were used in this study. The animals were anesthetized and a) all visceral branches from the left lumbar ganglia 2, 3 and 4, b) all communicating rami to the respective spinal nerves were transected. Additionally, in both procedures, the sympathetic trunk was cut rostral to the second and caudal to the fourth lumbar ganglion in order to isolate the investigated ganglia from descending and ascending peptidergic nerve fibers. Four control guinea pigs were kept under the same laboratory conditions as experimental animals. Five days after surgery, the animals in anesthesia were perfused with Zamboni's fluid (Zamboni and de Martino 1967). Three lumbar ganglia of both experimental and control sides were dissected and embedded in paraffin. 7 um tissue sections were immunostained according to the biotin–streptavidin–peroxidase method (Bonnard et al. 1984) applying primary antisera directed to substance P (SP; INC, lot 27231, dilution 1:3000), calcitonin gene–related peptide (CGRP; Peninsula, lot 000114, dilution 1:4000), met–enkephalin (met–ENK; INC, lot 8416015, dilution 1:5000) and vasoactive intestinal polypeptide (VIP; INC, lot 8403015, dilution 1:5000). Specificity tests were carried out (Forssmann et al. 1981), including preabsorption of each antiserum with the respective antigene (20–100 ug peptide/ml antiserum) and replacement of specific antisera by pre-immune serum. All controls indicated the specificity of the obtained immunoreactions. Quantification of immunoreactive (IR) fiber density was obtained by computer–enhanced electronic scanning (IBAS II–Kontron).

Results and Discussion

No differences in the density of peptide-IR nerve fibers are observed between right lumbar sympathetic chain ganglia of experimental animals and lumbar ganglia of controls. SP-IR fibers form varicose terminals around a number of principal ganglionic neurons (PGN; Fig. 1). CGRP-IR fibers are distributed in the same areas as SP-IR fibers. They are more numerous than SP-IR fibers (Fig. 2). When visceral branches from the sympathetic trunk ganglia including splanchnic nerves to the inferior mesenteric ganglion are surgically transected there is little or no change in the density of SP- and CGRP-IR fibers, but many of the fibers appear to be swollen. Selectively cutting the communicating rami interrupts all visible SP- and CGRP-IR fibers in the ganglia (Figs. 3 and 4). Previous evidence has been presented (Hökfelt et al. 1977; Matthews and Cuello 1982) that SP-IR fibers in sympathetic ganglia are processes of primary afferent neurons and that SP in these neurons is often co-localized with CGRP (Gibbins et al. 1985). Recently, CGRP-IR has also been found in postganglionic neurons of cat cervical and thoracic paravertebral ganglia, and few of these neurons were also SP-IR (Kummer and Heym, to be published). From diminuition of SP-IR and CGRP-IR nerve fibers after decentralization it can be inferred that in guinea pig lumbar paravertebral ganglia these fibers do not derive from sympathetic PGN but rather from sensory neurons.

ENK-IR fibers in control ganglia are relatively few, forming varicose terminals around some of the PGN (Fig. 5). There is no visible change in ENK-IR fibers when visceral branches from the trunk ganglia are surgically transected. After cutting the communicating rami, however, very few ENK-IR fibers remain (Fig. 6). Previous studies have shown that ENK-IR fibers in guinea pig prevertebral ganglia are of preganglionic origin with cell bodies in the spinal cord (Dalsgaard et al. 1983). Those few fibers, remaining visible after decentralization, may represent processes of small, paraganglionic cells which in the superior cervical ganglion have been demonstrated to be of intrinsic character (Kummer et al. 1986).

In accordance with previous findings (cf. Heym and Triepel 1985) VIP-IR is present in lumbar sympathetic trunk ganglia, both in nerve fibers and in some PGN. Varicose nerve fibers are few in number. After transecting the visceral branches there is no change in number or staining intensity of VIP-IR perikarya, but the number of varicose fibers is somewhat increased and some of the VIP-IR fibers appear to be swollen (Fig. 7). When the communicating rami are surgically cut VIP-IR cell bodies are darkly stained and there is a distinct increase in density of the VIP-IR fiber networks in the ganglion. Nearly all PGN are surrounded by VIP-IR varicose terminals (Fig. 8). This visual estimation was confirmed by measurement of the % fiber density in the VIP antiserum-reacted sections of control and experimental ganglia by computer-enhanced electronic scanning (Table 1). Numerically increased and enlarged VIP-IR fibers after transection of the visceral branches can derive from three sources:
(1) Axons of PGN in lumbar and paravertebral ganglia not only join the respective spinal nerves but also contribute to the visceral (e.g. renal) fiber supply (McLachlan 1985).
(2) Preganglionic nerve fibers originating in the substantia intermedia of the spinal cord traverse paravertebral ganglia on their way to prevertebral ganglia (cf. Heym and Triepel 1985).
(3) VIP-IR sensory neurons, being located in dorsal root ganglia, may provide sympathetic ganglia with collaterals (Schultzberg et al. 1980).

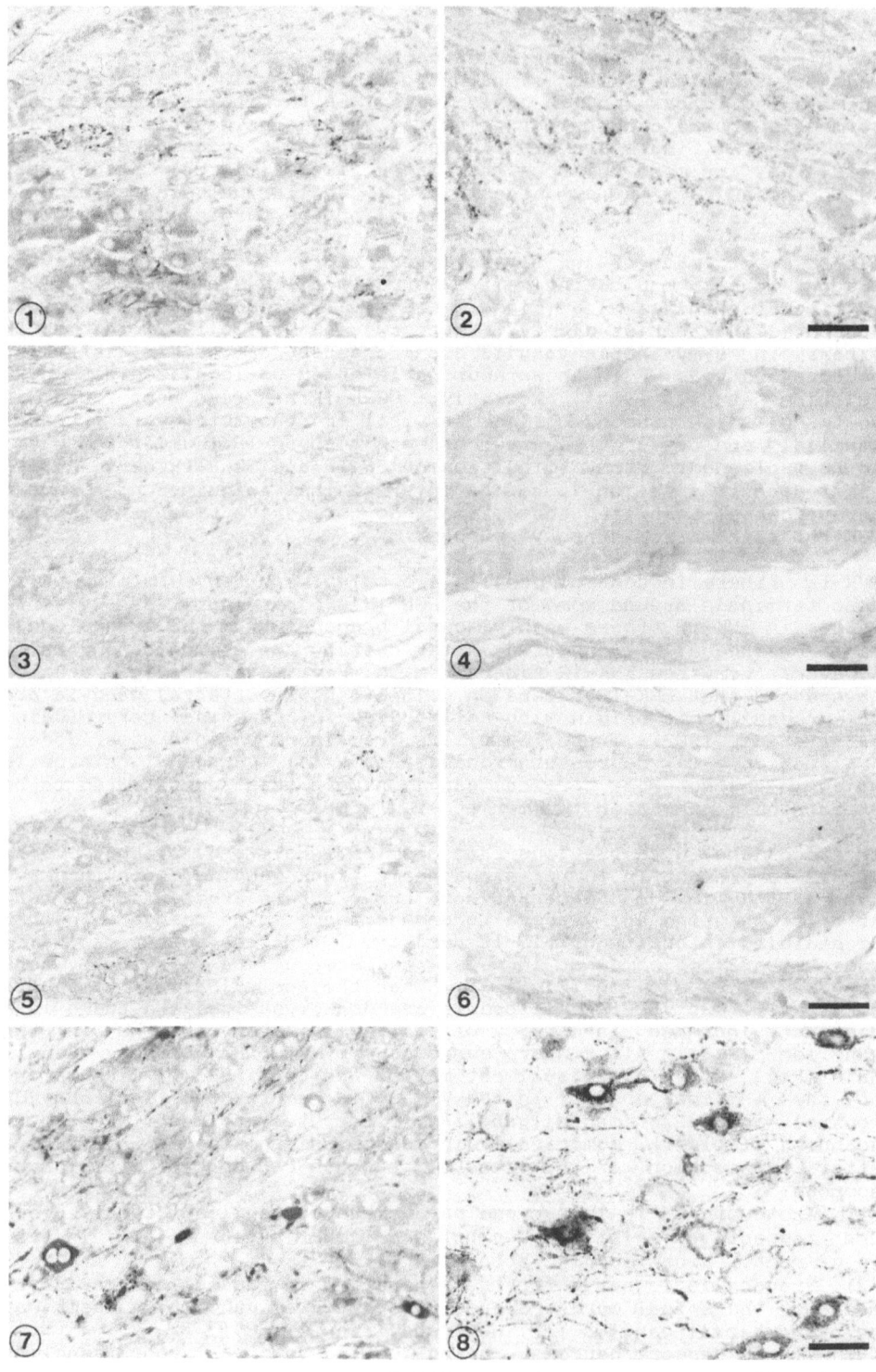

Table 1. Area occupied,by VIP-IR fibers (expressed as area %) in guinea pig second lumbar ganglia, as obtained by computer-enhanced electronic scanning

controls	transs. visc. br.	transs. c. rami
X±S.D.=0.03 ±0.01 n=4	X±S.D.=0.61 ±0.85 n=4	X±S.D.=1.04 ±1.55 n=4

Preganglionic denervation is known to cause a rise in the content of VIP-IR material in postganglionic neurons (Gu et al. 1983). By cutting both communicating rami, however, additonally all postganglionic efferences to the lumbar plexus are interrupted. Therefore, it is conceivable that the numerous VIP-IR varicosities surrounding non-reactive perikarya after transection of the communicating rami represent both axonal and dendritic processes of double effected intraganglionic PGN.

Acknowledgements. This study was supported by the DFG, grant He 919/6-1. We thank Prof. W. Höpker for providing the IBAS-II-Kontron for quantitative evaluations.

References

Bonnard C, Papermaster DS, Kraehenbuhl JP (1984) The streptavidin-biotin bridge technique: application in light and electron microscope immunocytochemistry. In: Polak JM, Varndell IM (eds) Immunolabelling for electron microscopy. Elsevier, Amsterdam, pp 95-112

Dalsgaard CJ, Hökfelt T, Schultzberg M, Lundberg JM, Terenius L, Dockray GJ, Goldstein M (1983) Origin of peptide-containing fibers in the inferior mesenteric ganglia of the guinea pig: immunohistochemical studies with antisera to substance P, enkephalin, vasoactive intestinal polypeptide, cholecystokinin and bombesin. Neuroscience 9: 191-211

Forssmann WG, Pickel V, Reinecke M, Hock D, Metz J (1981) Immunohistochemistry and immunocytochemistry of nervous tissue. In: Heym Ch, Forssmann WG (eds) Techniques in neuroanatomical research. Springer, Berlin, pp 171-206

Gibbins TL, Furness JB, Costa M, MacIntyre I, Hillyard CJ, Girgis S (1985) Co-localization of calcitonin gene-related peptide-like immunoreactivity with substance P in cutaneous, vascular and visceral sensory neurons of guinea pigs. Neurosci Lett 57: 125-130

◄ **Fig. 1-8.** Paraffin sections of third and fourth lumbar sympathetic trunk ganglia, respectively. Bars indicate 50 um. **(1)** SP-IR control. **(2)** CGRP-IR control. **(3)** SP-IR after transection of the communicating rami. **(4)** CGRP-IR after transection of the communicating rami. **(5)** Met-ENK-IR control. **(6)** Met-ENK-IR after transection of the communicating rami. **(7)** VIP-IR after transection of all visceral rami. **(8)** VIP-IR after transection of the communicating rami

Gu J, Huang WM, Blank M, Morrison JFB, Bloom SR, Polak JM (1983) Measurement of VIP and origin of its innervation in the rat urinary bladder. Reg Pept 6: 305

Heym Ch, Lang R (1986) Transmitters in sympathetic postganglionic neurons. In: Panula P, Päivärinta H, Soinila S (eds) Neurohistochemistry: modern methods and applications. Alan R. Liss, New York, pp 493-525

Heym Ch, Triepel J (1984) Immunohistochemical localization of vasoactive intestinal polypeptide in the mammalian central and peripheral nervous systems. In: Parvez H, Parvez S, Gupta D (eds) Neuroendocrinology of hormone transmitter interaction. VNU International Science Press, Utrecht, pp 99-125

Hökfelt T, Elfvin LG, Schultzberg M, Goldstein M, Nilsson G (1977) On the occurrence of substance P containing fibers in sympathetic ganglia: immunohistochemical evidence. Brain Res 132: 29-41

Kummer W, Heym Ch, Colombo M, Lang R (1986) Immunohistochemical evidence for extrinsic and intrinsic opioid systems in the guinea pig superior cervical ganglion. Anat Embryol 174: 401-405

Matthews MR, Cuello AC (1982) Substance P-immunoreactive peripheral branches of sensory neurons innervate guinea pig sympathetic neurons. Proc Natl Acad Sci USA 79: 1668-1672

McLachlan EM (1985) The components of the hypogastric nerve in male and female guinea pigs. J Auton Nerv Syst 13: 327-342

Schultzberg M, Hökfelt T, Lundberg JM, Dalsgaard CJ, Elfvin LG (1983) Transmitter histochemistry of autonomic ganglia. In: Elfvin LG (ed) Autonomic ganglia. Wiley, Chichester, pp 205-233

Schultzberg M, Hökfelt T, Lundberg JM, Fuxe K, Mutt V, Said S (1980) Distribution of VIP neurons in the peripheral and central nervous system. Endocrinol Jap 1: 23-30

Zamboni L, de Martino C (1967) Buffered picric-acid formaldehyde: a new rapid fixation for electron microscopy. J Cell Biol 35: 148

Sensory Transmitter Candidates and Their Role in the Periphery

C.-J. Dalsgaard[1], W. Stains[2], H. Björklund[2], A. Haegerstrand[1], A. Hultgardh-Nilsson[3], J. Kjartansson[4], J. Nilsson[2], and T. Hökfelt[2]

[1] Department of Anatomy, Karolinska Institutet, P.O.Box 60400, 10401 Stockholm, Sweden
[2] Department of Histology, Karolinska Institutet, P.O.Box 60400, 10401 Stockholm, Sweden
[3] Department of Medical Cell Genetics, Karolinska Institutet, P.O.Box 60400, 10401 Stockholm, Sweden
[4] Department of Plastic Surgery, Karolinska Hospital, P.O.Box 60400, 10401 Stockholm, Sweden

Introduction

A number of biologically active peptides have been identified in primary sensory neurons in various species, including substance P (SP), neurokinin A (NKA), somatostatin (SOM), vasoactive intestinal polypeptide (VIP), cholecystokinin (CCK) and calcitonin gene-related peptide (CGRP) (Hökfelt et al. 1975a,b; Lundberg et al. 1978; Dalsgaard et al. 1982, 1985; Mason et al. 1984). It seems from biochemical studies that the major proportion of peptides synthesized in the primary afferent neurons are transported to the periphery and not to the spinal cord. Sensory neuropeptides have also been shown to exert certain peripheral effects, such as vasodilation (SP, NKA and CGRP) and plasma extravasation (SP and NKA) (Lembeck and Holzer 1979; Hua et al. 1984; Brain et al. 1985). In a recent study a dramatic sprouting of sensory nerve fibers has been observed in the skin during wound healing, suggesting a role for sensory neurotransmitters also in processes related to wound healing (Hermanson et al. 1986).

Although many studies have indicated possible roles for sensory neuropeptides in the periphery, very little is known about the distribution of peptidergic sensory nerve fibers in the human skin. In the present study this issue was addressed and furthermore, the growth stimulatory effects of sensory neuropeptides on cultured arterial smooth muscle cells and skin fibroblasts was investigated.

Materials and Methods

In the present study, skin biopsies from healthy volunteers (the authors) were taken with a 2 mm punch, fixed in 0.4% picric acid 10% formalin solution and processed for indirect immunohistochemistry. Briefly, cryostat sections (14 um) were incubated with antisera to SP, NKA, SOM, VIP and CGRP or a combination of antisera (SP-CGRP; SOM-CGRP; SP-SOM) at 4°C in a humid atmosphere for 24 h, rinsed in PBS, incubated with FITC- or rhodamine-conjugated species specific anti-IgG at 37°C for 30 min, rinsed, mounted in glycerol:PBS (3:1) and examined in a Zeiss fluorescence microscope. By comparison of consecutive sections or by combining primary antisera raised in different species and secondary antisera with different fluorochromes, it was possible to investigate whether or not any coexistence of peptide-like immunoreactivities (-LI) occurred in sensory fibers in the skin. In control sections, antisera preabsorbed with an excess (10 nmoles/ml) of their respective peptide and in other sections secondary antisera against IgG of other species used in the double staining experiments were tested for crossreactivity.

Experimental Brain Research Series 16
© Springer-Verlag Berlin · Heidelberg 1987

Smooth muscle cells were isolated enzymatically from the aortic media of adult Sprague-Dawley rats and fibroblasts were isolated by outgrowth of human skin explants (Nilsson et al. 1985). Secondary cultures of muscle cells and cultures of 10th to 12th passage were growth-arrested by transfer to serum-free medium (MCDB 104/0.1% bovine serum albumin). They were then incubated with 2.5 uCi/ml of ^{3}H-thymidine and 10^{-8} and 10^{-6}M of SP (Bachem) NKA, SOM, VIP (all Peninsula) and CGRP (generously supplied by Dr. I. McIntyre) for 24 h, fixed in buffered 3% glutaraldehyde and processed for autoradio-graphy. The rate of entrance of cells into S-phase of the cell cycle was then determined by counting the number of labelled nuclei in 500 randomly selected cells in each specimen. As a control, Spantide (3 x 10^{-6}M – 3 x 10^{-5}M; Bachem), a tachykinin receptor antagonist, was added to some of the cultures incubated with SP and NKA.

Results and Discussion

In the human skin SP-LI was found in free nerve endings in the dermis and epidermis, in a few Meissner's corpuscles and in fibers connected to blood vessels, hair follicles and sweat glands as described previously (Dalsgaard et al. 1983). NKA-LI was found with the same distribution pattern, although the fibers were not as dense or as intense as the SP-positive fibers. By comparison of consecuti-ve sections it was found that the NKA fibers also displayed SP-LI. SOM-LI fibers were seen in the dermis running parallel to the der-mis-epidermis boundary. A few SOM-positive fibers were also observed close to the sweat glands and hair follicles. Preliminary data achieved by double staining techniques indicate that the SP- and SOM-LI fibers are separate. CGRP-LI fibers were seen in the epider-mis and dermis as free nerve endings, in some Meissner's corpuscles and in fibers related to blood vessels, hair follicles and sweat gland ducts. The majority of CGRP-LI fibers were also SP-positive and some were SOM-positive as revealed by double labelling techni-ques. In fact, all CGRP-positive fibers observed were found to display either SP- or SOM-LI. The possibility that a fiber may display all three peptides has not yet been fully investigated. VIP-LI was seen mainly in fibers around sweat glands and hair follicles but a few VIP-positive free nerve endings also were encountered (Björklund et al. 1986). Whether or not the VIP-LI fibers display any other immunoreactivity has not yet been investigated.

Serum-free medium only initiated DNA synthesis in 19% of the smooth muscle cells and in 17% of the fibroblasts. Addition of 10^{-8}M NKA resulted in 44% labelled nuclei and 10^{-6}M NKA resulted in 62% label-led nuclei in smooth muscle cells and 34% (10^{-6} M) and 38% (10^{-8} M) labelled nuclei in the fibroblasts. SP was less potent than NKA and induced significant increase in DNA synthesis in smooth muscle cells (33%) and fibroblasts (27%) only at 10^{-6} M. Spantide completely blocked the effect of NKA and SP on initiation of DNA synthesis. SOM, VIP or CGRP were not able to induce stimulation of DNA synthe-sis in arterial smooth muscle cells or in human skin fibroblasts.

The present study shows a specific distribution pattern of sensory peptidergic nerve fibers in the human skin. Furthermore, some fibers contain at least three different peptides at the same time. The distribution of different peptides gives a hint as to the possible peripheral function. The peptides in perivascular fibers have been shown to cause vasodilation and plasma extravasation (see above) and, as reported previously (Nilsson et al. 1985), stimulate in vitro DNA synthesis in arterial smooth muscle cells. The fact that sensory nerve fibers innervating vessels, hair follicles and gland ducts show a dramatic sprouting during healing of superficial skin wounds (Hermanson et al. 1986) makes it tempting to suggest that sensory transmitters may be involved in wound healing processes. Studies on the effect of sensory neuropeptides on the initiation of DNA synthesis in other cell types of the skin involved in wound healing are in progress.

In conclusion, SP-, NKA-, SOM-, CGRP- and VIP-LI have been identified in sensory nerve fibers in the human skin. SP-, NKA- and CGRP-LI as well as SOM- and CGRP-LI have been shown to coexist in these fibers. NKA and SP, but not SOM, CGRP or VIP, induced stimulation of DNA synthesis in cultured arterial smooth muscle cells and human skin fibroblasts in a dose-dependent manner. The growth stimulation by NKA and SP was blocked by addition of Spantide to the medium, suggesting a mediation via tachykinin receptors.

The peptidergic sensory nerve fibers are known to be involved in vasodilation and inflammatory response and may also have a role in wound healing.

Acknowledgements. This study was supported by grants from Karolinska Institutet, the Swedish MRC (2887; 6537; 7126) and Marcus and Amalia Wallenbergs Found.

References

Björklund H, Dalsgaard C-J, Jonsson C-E, Hermanson A (1986) Sensory and autonomic innervation of non-hairy and hairy human skin. An immunohistochemical study. Cell Tissue Res 243: 51-57

Brain SD, Williams TJ, Tippins JR, Morris HR, MacIntyre I (1985) Calcitonin gene-related peptide is a potent vasodilator. Nature 313: 54-56

Dalsgaard C-J, Vincent SR, Hökfelt T, Lundberg JM, Dahlström A, Dockray GJ, Cuello AC (1982) Coexistence of cholecystokinin- and substance P-like peptides in neurons of the dorsal root ganglia of the rat. Neurosci Lett 33: 159-163

Dalsgaard C-J, Jonsson C-E, Hökfelt T, Cuello C (1983) Localization of substance P-immunoreactive nerve fibres in the human digital skin. Experientia 39: 1018-1020

Dalsgaard C-J, Haegerstrand A, Theodorsson-Norheim E, Brodin E, Hökfelt T (1985) Neurokinin A-like immunoreactivity in rat primary sensory neurons; coexistence with substance P. Histochemistry 83: 37-39

Hermanson A, Dalsgaard C-J, Björklund H, Lindblom U (1986) Sensory reinnervation and sensibility after superficial skin wounds. Neurosci Lett (in press)

Hua X-Y, Lundberg JM, Theodorsson-Norheim E, Brodin E (1984) Comparison of cardiovascular and bronchoconstrictor effects of substance P, substance K and other tachykinins. Naunyn-Schmiedeberg's Arch Pharmacol 328: 196-201

Hökfelt T, Elde R, Johansson O, Luft R, Arimura A (1975a) Immunohistochemical evidence of the presence of somatostatin, a powerful

inhibitory peptide, in some primary sensory neurons. Neurosci Lett
1: 231–235

Hökfelt T, Kellerth J–O, Nilsson G, Pernow B (1975b) Substance P.
Localization in the central sensory neurons. Science 190: 889–890

Lembeck F, Holzer P (1979) Substance P as neurogenic mediator of
antidromic vasodilation and neurogenic plasma extravasation. Naunyn-
Schmiedeberg's Arch Pharmacol 310: 175–183

Lundberg JM, Hökfelt T, Nilsson G, Terenius L, Renfeld J, Elde R,
Said S (1978) Peptide neurons in the vagus, splanchnic and sciatic
nerves. Acta Physiol Scand 104: 499–501

Mason RT, Peterfreund RA, Sawchenko PE, Corrigan AZ, River JE, Vale
WW (1984) Release of the predicted calcitonin gene-related peptide
from cultured rat trigeminal ganglion cells. Nature 308: 653–655

Nilsson J, von Euler AM, Dalsgaard C–J (1985) Stimulation of connective tissue cell growth by substance P and substance K. Nature 315:
61–63

Cardiodilatin as a Neuropeptide
(Cardiac Polypeptide Hormones Are also Neuropeptides)

W. G. Forssmann[1], R. E. Lang[2], A. Aoki[3], M. Reinecke[1], G. Rippegather[1], and D. Hock[1]

[1] Anatomisches Institut, Universität Heidelberg, Im Neuenheimer Feld 307, 6900 Heidelberg 1, FRG
[2] Abteilung für Pharmakologie, Universität Heidelberg, Im Neuenheimer Feld 366, 6900 Heidelberg 1, FRG
[3] Centro de Microscopia Electronica, Universidad de Cordoba, Cordoba, Argentina

Introduction

During the 1970s, we applied repeatedly histochemical techniques to elucidate the significance of the secretory granules in the heart atria, studies which mainly resulted in the detection that peptidergic cardiac innervation is particularly strong in the endocrine areas (Goebel et al. 1986). These special atrial cells were first recognized by Kisch (1956) and later by Rouiller and his research group (Bompiani 1959) who described cells of the guinea-pig and rat atria to contain electron-dense granules, now well-known under the name of myoendocrine cells, representing special myocardiocytes which had already been analyzed thoroughly in earlier electron-microscopy studies (Jamieson and Palade 1964; McNutt and Fawcett 1969; Bencosme and Berger 1971). Discussing these results with Professor Viktor Mutt he offered us collaboration at the Karolinska Institute of Stockholm and thus to extend our work on this project to the isolation of cardiac peptide hormones.

In the meantime, Hatt had published the first functional impact of the myoendocrine cells, showing reactions of the secretory granules to changes in the body fluid and electrolyte balance (Marie et al. 1976). Later, Debold et al. (1981) described a diuretic effect obtained by the injection of extracts from atrial tissue into rats, and subsequently a relaxant bioactivity on vascular smooth muscle induced by atrial extracts from various mammals was found (Deth et al. 1982; Currie et al. 1983; Forssmann et al. 1983, 1984). Thus, the isolation of the bioactive substance became possible by using a biotest to identify the presumed peptide during the different steps of isolation.

Isolation of Cardiodilatin (CDD)

In 1983, we isolated some 100 micrograms of a peptide from 20 kg of porcine right atria. Application of this peptide to different vaso-relaxation tests showed that renal artery and aortic smooth muscle strips were particularly sensitive while the relaxation of mesenteric and femoral artery muscle strips after precontraction with catecholamines was very weak (Forssmann et al. 1983). The polypeptide we obtained by the use of this biotest was sequenced and shown to consist of 126 amino acids (Fig. 1). On the basis of its tissular origin and its biological activity, we called it cardiodilatin (CDD).

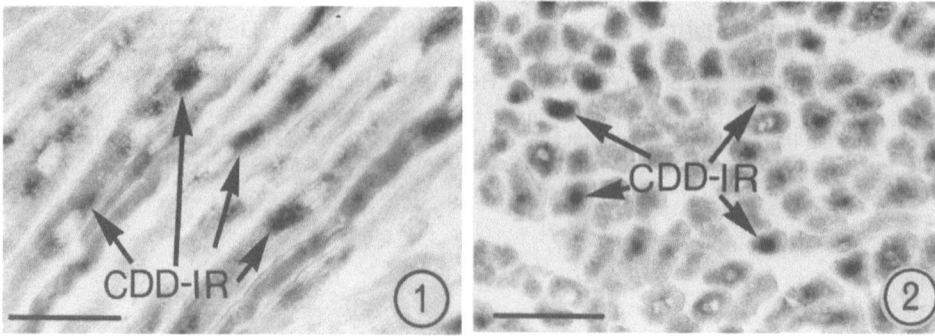

Fig. 1. CDD—immunoreactivity in porcine atrial myoendocrine cells revealed by peroxidase—antiperoxidase (PAP) immunocytochemistry. The immunoreactive CDD is seen in longitudinal sections related to the Golgi zone at the perinuclear areas. Bar indicates 50 um

Fig. 2. Similar preparation as in Fig. 1, however, the CDD—containing myoendocrine cells are seen in a transverse section. Note the CDD—immunoreactivity in the center of the myoendocrine cells. Bar indicates 50 um

Immunohistochemical Localization of CDD in the Heart

A few months later we had raised some specific antibodies against synthetic fragments of the N—terminus and the C—terminus of the molecule. First, we confirmed the localization of CDD in the porcine atrial myoendocrine cells by immunofluorescence as well as by the peroxidase—antiperoxidase method (Metz et al. 1964; Forssmann et al. 1984). The immunoreactive (IR) substance was found in the peri-nuclear regions where the secretory granules are located (Figs. 1 and 2). The same results were obtained by Cantin et al. (1984) for rat atria. Further light microscopical immunohistochemical and bio-logical studies revealed the existence of this polypeptide in the hearts of all vertebrate species which exhibit atrial myoendocrine cells by ultrastructural identification (Reinecke et al. 1985, 1987). Finally, the application of the immunogold method (Maldonado et al. 1986) established the criteria, demanded by a morphologist, to confirm CDD and the myoendocrine cells to be a polypeptide hor-mone system. In the meantime, numerous groups, e.g. Debold and co-workers (Flynn et al. 1983), isolated a homologous peptide contai-ning the 28 C—terminal amino acids and called it cardionatrin. Kangawa and Matsuo (1984) then characterized the homologous polypep-tide of the human heart and called it alpha—atrial natriuretic polypeptide. This is by now the most common name for the same pep-tide, known also under many other names: CDD—28, cardionatrin, atriopeptin, auriculin, etc. (Cantin and Genest 1985; De Bold 1985; Forssmann 1986). The 28 residues—containing polypeptide is the cir-culating form which we isolated recently as the posttranslationally processed hormone from human blood (Forssmann et al. 1986) cleaved from the prohormone CDD—126 during the release process form atrial myoendocrine cells.

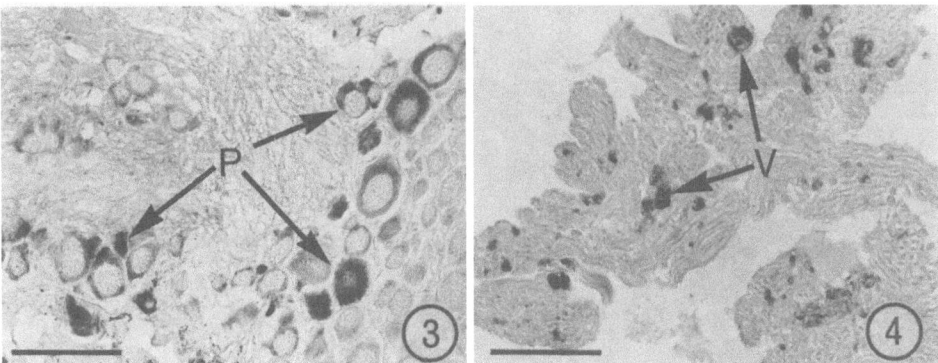

Fig. 3. CDD-IR neurons in the subesophageal ganglion of the snail. Note the stained perikarya (P). Bar indicates 50 um

Fig. 4. CDD-IR profiles in the snail heart. The immunohistochemical stain is seen in large neurosecretory varicosities (V) identified by electron microscopy. Bar indicates 50 um

CDD Localization in Nervous Tissues of Molluscs

In 1984, in our research group the question was raised as to whether CDD is also a neuronal peptide, since this had been substantiated as a rule for most of the hormonal peptides which were known then. We obtained a first indication which confirmed this postulate when we found that the CDD-like immunoreactivity was also present in the snail heart (Nehls et al. 1985), where ultrastructurally no myoendocrine cells but large neurosecretory endings were known to occur (Fig. 3). Our assumption that the peptide was located in nerves was confirmed by tracing the immunoreactivity to the subesophageal ganglion where the corresponding CDD-IR perikarya are located (Fig. 4). Furthermore, CDD-like, vasorelaxant bioactivity obtained from crude extracts of the subesophageal ganglia convinced us that a CDD-containing neurosecretory system along a neurocardiac axis existed in this species (Nehls et al. 1985). The use of HPLC revealed that the CDD-IR substance of the snail neurocardiac axis existed mainly in a large molecular form in the perikarya and in a small molecular form in the neurosecretory endings of the heart atrium. This indicates processing during the axonal transport. Surprisingly, in the snail this peptide seems primarily to be involved in the regulation of the heart functions as seen in studies of the isolated snail heart (unpublished results).

CDD in Nervous Tissue of Vertebrates

Hence, the neuronal localization of cardiac polypeptides in the phylogenetic tree was the goal of our further studies, and, as expected, we found CDD-IR in the brain of primitive vertebrates where it is particularly frequent, e.g. in the brain of the myxine glutinosa (Reinecke et al. 1987) and plecostoma cordobensis (Aoki and Forssmann, unpublished), CDD-IR peptides were found in many perikarya and varicosities of the brain stem and the spinal cord (Figs. 5 and 6). In further studies we found CDD in the mammalian brain (Forssmann and Mutt 1985), which in the same year was also

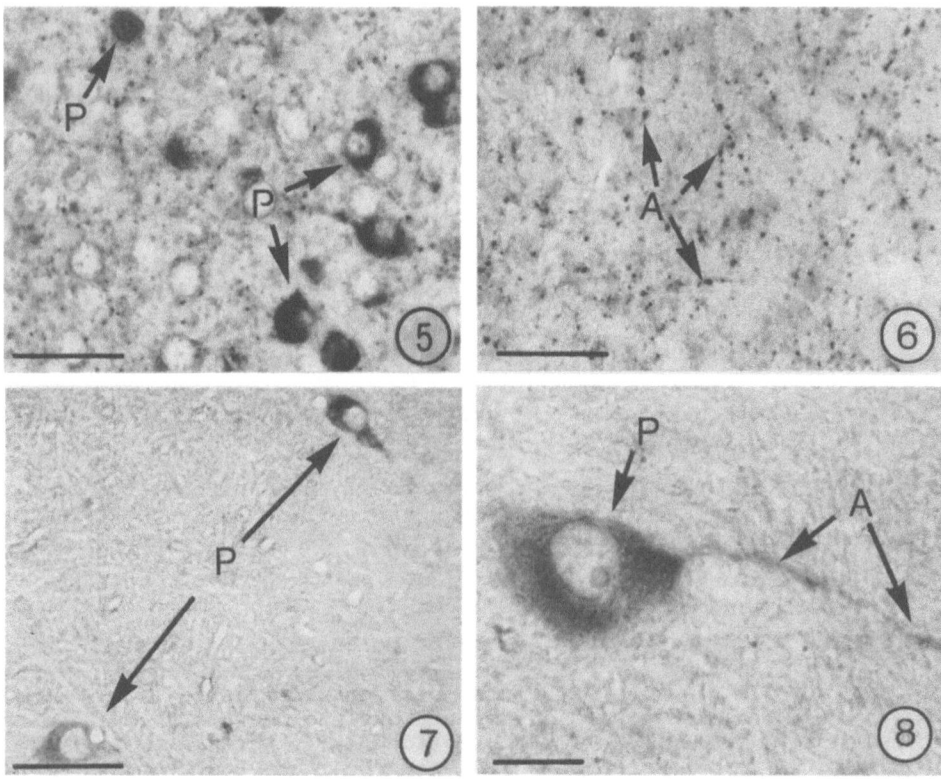

Fig. 5. CDD-IR neurons in the brain of myxine glutinosa. Note the numerous CDD-IR perikarya localized in the brain stem (P). Bar indicates 50 um

Fig. 6. CDD-immunoreactivity in varicosities of the lower brain stem of myxine glutinosa. Note the high density of CDD-IR axons (A). Bar indicates 50 um

Fig. 7. CDD-IR neurons of tupaia belangeri. Note the IR perikarya localized in the lateral periventricular area of the hypothalamus (P). Bar indicates 50 um

Fig. 8. Higher magnification of a CDD-IR perikaryon (P) of the dorsal hypothalamic area exhibiting an axonal process (A) which is also stained with the antiserum against CDD. Bar indicates 20 um

reported by other research groups for the rat brain (Kangawa and Matsuo 1984; Kawata 1985; Saper et al. 1985; Skofitsch et al. 1985; Zamis et al. 1985). In the primitive primate, tupaia belangeri, we detected CDD-IR perikarya mainly in three areas of the brain (Figs. 7 and 8): (1) in the periventricular nucleus of the area dorsalis hypothalami, (2) in the central amygdaloid nucleus, and (3) in the supraoptic nucleus. Varicosities are mainly observed in the septal region, in the periaqueductory grey, in the medulla oblongata and in the spinal cord. Most of these tracts, which have yet to be traced correctly, are apparently involved in the cardiovascular regulation or the body fluid and electrolyte balance control. The central

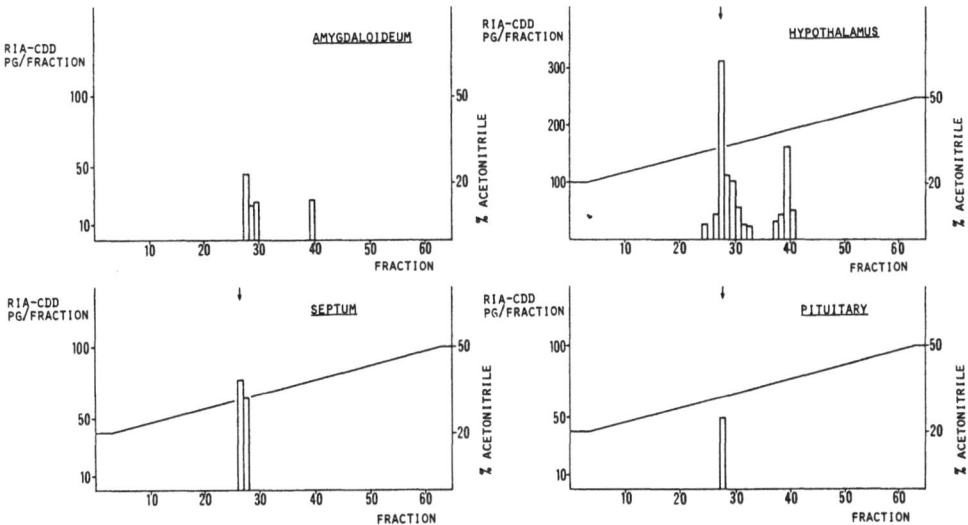

Fig. 9. Combined RIA and HPLC analysis of radioimmunoassayable CDD (RIA-CDD). Note that brain areas where CDD-IR neurons have been localized (amygdaloideum and hypothalamus) contain large molecular CDD (eluting mainly in fraction 40) whereas the small molecular CDD is found in areas where CDD-IR varicosities were detected (septum or pituitary with infundibulum)

connections of the cardiovascular neuronal regulation coincide with most of the mentioned areas containing CDD-IR neurons.

Further investigations recently carried out in collaboration with other research groups gave further results indicating the great importance of cardiac polypeptides as putative neuronal transmitters: CDD-immunoreactivity was found to be stored in neurons together with other well-known neuropeptides or in neurons containing specific receptors for steroid hormones. (1) In rat brain CDD-immunoreactivity was localized in a population of perikarya of the nucleus supraopticus which contains oxytocin immunoreactivity (Jirikowski et al. 1986). (2) Opiate peptides such as dynorphin were detected in the CDD-IR perikarya of the neurons of the nucleus periventricularis and the nucleus supraopticus in tupaia (Kummer et al., unpublished). (3) Rat hypothalamic neurons exhibiting CDD-immunoreactivity contain also specific steroid hormone receptors (Bach et al., unpublished).

Biochemical Identification of CDD in the Brain

To further substantiate our findings we performed combined HPLC-RIA (high performance liquid chromatography-radioimmunoassay) studies of the rat brain by extracting the peptides from the areas which were seen to contain CDD immunoreactivity in immunohistochemistry. The results were as follows: CDD in small and large molecular forms is located in those areas where also CDD-IR perikarya are found (Fig. 9). By contrast, brain areas exhibiting only CDD-IR varicosities apparently store the smaller molecular form of this polypeptide. This small CDD-like substance exhibits the same chromatographic and immunochemical properties as the 28 residues-containing form of CDD, i.e. CDD 99-126 or alpha-ANP. We therefore conclude that the precursor form of CDD is produced in the perikarya and then cleaved into the small molecular form during axonal transport or that it is stored in the varicosities or nerve terminals.

Conclusions

According to our results CDD exists (1) as a neuronal peptide in the brain, (2) as a neuroendocrine substance in the hypothalamic-hypophyseal tract, and (3) as an endocrine signal transmitter in the heart myoendocrine cells. In the phylogenetic tree the morphological and functional importance of this peptide family may change and divert from a neurosecretory substance to a triplex system including the mammalian neuronal-neurosecretory-endocrine axis. The most analogous impact of this polypeptide within the phylogenetic tree is its contribution to the cardiovascular and water-electrolyte regulation.

From our studies presented here, the following conclusions may be summarized:
(1) CDD is not only an endocrine polypeptide hormone but also a neuropeptide and a neurosecretory substance.
(2) Neuronal CDD occurs in a large molecule in the perikarya and in a small molecular form in the varicosities and nerve terminals.
(3) The small molecular form seems to be identical to the circulating hormone.
(4) CDD is predominantly found in brain areas related to cardiovascular regulation.

Acknowledgements. We acknowledge the skillful technical help of B. Brühl, B. Herbold and I. Stenull and the careful preparation of the manuscript by R. Botz and G. Sürig. Supported by Orpegen GmbH, Heidelberg.

References

Bencosme SA, Berger JM (1971) Specific granules in mammalian and non-mammalian vertebrate cardiocytes. In: Bajusz E, Jasmin G (eds) Meth Achievm exp Path 5: 173-213. Karger, Basel

Bompiani GD, Rouiller C, Hatt PY (1959) Le tissue de conduction du coeur chez le rat. Etude au microscope électronique. Arch Mal Coeur 52: 1257-1274

Cantin M, Genest J (1985) The heart and the atrial natriuretic factor. Endocr Rev 6: 107-127

Cantin M, Gutkowska J, Thibault G, Milne RE, Ledoux S, MinLi S, Chapeau C (1984) Immunocytochemical localization of atrial natriuretic factor in the heart and salivary glands. Histochemistry 80: 113-127

Currie MG, Geller DM, Cole BR, Boylan JG, YuSheng W, Holmberg SW, Needleman P (1983) Bioactive cardiac substances: potent vasorelaxant activity in mammalian atria. Science 221: 71-73

De Bold A (1985) Atrial natriuretic factor: a hormone produced by the heart. Science 230: 767-770

De Bold AJ, Borenstein HB, Veress AT, Sonnenberg H (1981) A rapid and potent natriuretic response to intravenous injection of atrial extracts in rats. Life Sci 28: 89–94

Deth RC, Wong K, Fukozawa S, Rocco R, Smart JL, Lynch CJ, Awad R (1982) Inhibition of rat aorta contractile response by natiuresis-inducing extract of rat atrium. Fed Proc 41: 983

Flynn TG, De Bold ML, De Bold AJ (1983) The amino acid sequence of an atrial peptide with potent diuretic and natriuretic properties. Biochem Biophys Res Comm 117: 859–865

Forssmann WG (1987) Cardiac hormones: I. Review on the morphology, biochemistry and molecular biology of the endocrine heart. Eur J Clin Invest (in press)

Forssmann WG, Birr C, Carlquist M, Christmann M, Finke R, Henschen A, Hock D, Kirchheim H, Kreye V, Lottspeich F, Metz J, Mutt V, Reinecke M (1984) The auricular myocardiocytes of the heart constitute an endocrine organ. Characterization of a porcine cardiac peptide hormone, cardiodilatin–126. Cell Tissue Res 238: 425–430

Forssmann K, Hock D, Herbst F, Schulz–Knappe P, Talartschik J, Scheler F, Forssmann WG (1986) Isolation and structural analysis of the circulating human cardiodilatin (alpha ANP). Klin Wochenschr 64: 1276–1280

Forssmann WG, Hock D, Lottspeich F, Henschen A, Kreye V, Christmann M, Reinecke M, Metz J, Carlquist M, Mutt V (1983) The right auricle of the heart is an endocrine organ. Cardiodilatin as a peptide hormone candidate. Anat Embryol 168: 307–313

Forssmann WG, Mutt V (1985) Cardiodilatin–immunoreactive neurons in the hypothalamus of tupaia. Anat Embryol 172: 1–5

Goebel J, Metz J, Forssmann WG (1986) Korrelation zwischen myoendokrinen Zellen und peptiderger Innervation im Herz. Verh Anat Ges 80: 551–553

Jamieson JD, Palade GE (1964) Specific granules in atrial muscle cells. J Cell Biol 23: 151–172

Jirikowski GF, Back H, Forssmann WG, Stumpf W (1986) Coexistence of atrial natriuretic factor (ANF) and oxytocin in neurons of the rat hypothalamus. Neuropeptides 8: 243–249

Kangawa K, Matsuo H (1984) Purification and complete amino acid sequence of alpha–human atrial natriuretic polypeptide (alpha–hANP). Biochem Biophys Res Comm 118: 131–139

Kawata M, Nakao K, Morii N, Kiso Y, Yamashita H, Imura H, Sano Y (1985) Atrial natriuretic polypeptide: topographical distribution in the rat brain by radioimmunoassay and immunohistochemistry. Neuroscience 16: 521–546

Kisch B (1956) Electron microscopy of the atrium of the heart. I. Guinea pig. Exp Med Surg 14: 99–112

Maldonado CA, Sagga W, Forssmann WG (1986) Cardiodilatin–immunoreactivity in specific atrial granules of human heart revealed by the immunogold stain. Anat Embryol 173: 295–298

Marie JP, Guillemot H, Hatt PY (1976) Le degré de granulation des cardiocytes auricularies. Etude planimétrique au cours de différents apports d'eau et de sodium chez le rat. Path Biol 24: 549-554

Metz J, Mutt V, Forssmann WG (1964) Immunohistochemical localization of cardiodilatin in myoendocrine cells of the cardiac atria. Anat Embryol 170: 123-127

McNutt NS, Fawcett DW (1969) The ultrasture of the cat myocardium. II. Atrial muscle. J Cell Biol 42: 46-67

Nehls M, Reinecke M, Lang RE, Forssmann WG (1985) Biochemical and immunological evidence for a cardiodilatin-like substance in the snail neurocardiac axis. Proc Natl Acad Sci 82: 7762-7766

Reinecke M, Betzler D, Forssmann WG (1987) Immunocytochemistry of cardiac polypeptide hormones (cardiodilatin/atrial natriuretic poly-peptide) in brain and hearts of Myxine glutinosa (cyclostomata). Histochemistry 86: (in press)

Reinecke M, Nehls M, Forssmann WG (1985) Phylogenetic aspects of cardiac hormones as revealed by immunocytochemistry, electronmicro-scopy, and bioassay. Peptides 6, Suppl 3: 321-331

Saper CB, Standaert DG, Currie MG, Schwartz D, Geller DM, Needleman P (1985) Atriopeptin-immunoreactive neurons in the brain: presence in cardiovascular regulatory areas. Science 227: 1047-1049

Skofitsch G, Jacobowitz DM, Eskay RL, Zamir N (1985) Distribution of atrial natriuretic factor-like immunoreactive neurons in the rat brain. Neuroscience 16: 917-948

Zamir N, Skofitsch G, Eskay RL, Jacobowitz DM (1985) Distribution of immunoreactive atrial natriuretic peptides in the central nervous system of the rat. Brain Res 365: 105-111

Immunohistochemistry of Histamine and Histidine Decarboxylase in SIF Cells and Intestinal Nerves

P. Panula[1], O. Häppölä[1], T. Watanabe[3], H. Wada[2], and H. Päivärinta[1]

[1] Department of Anatomy, University of Helsinki, Siltavuorenpenger 20A, 00170 Helsinki, Finland
[2] Department of Pharmacology II, Osaka University School of Medicine, 3–57, Nakanoshima 4, Kita-ku, Osaka 530, Japan
[3] Department of Pharmacology, Tohoku University School of Medicine, Sendai 980, Japan

Introduction

Histamine and its synthesizing enzyme are present in neurons of the posterior hypothalamic region in the rat (Watanabe et al. 1984; Panula et al. 1984), and physiological evidence suggests that histamine may act as neurotransmitter in Aplysia (McCaman and Weinrich 1985). Immunohistochemical studies on the peripheral nervous system suggest that histamine may be present in nerve fibers of mammalian intestine, some adrenal chromaffin cells and small intensely fluorescent cells of the sympathetic ganglia (Häppölä et al. 1985a,b; Panula et al. 1985). To characterize the peripheral histamine-containing cells, more sensitive immunocytochemical methods have now been developed for detection of histamine. To study the histamine-synthesizing capability of the histamine-containing cells in the sympathetic ganglion, antibodies against L-histidine decarboxylase were used.

Materials and Methods

Antisera against histamine were produced as described earlier (Panula et al. 1984, 1985) by injecting histamine conjugated to succinylated hemocyanin with 1-ethyl-3(3-dimethylaminopropyl)-carbodiimide intradermally in rabbits. The specificity of the antisera was established by standard radioimmunoassay, solidphase immunoabsorption tests (Panula et al. 1984, 1985) and with model experiments with cellulose ester filters. In these tests, the antisera appeared to detect histamine coupled to succinylated hemocyanin or bovine serum albumin with almost equal potency, while histamine coupled to these carrier proteins with paraformaldehyde gave a considerably weaker reaction. The model experiments led to the use of carbodiimide in the perfusion mixture and resulted in superior immunostaining as compared to freshly prepared paraformaldehyde. Immunohistochemical blocking controls with histamine, L-histidine, noradrenaline, adrenaline, dopamine and histidine-containing peptides showed that the reaction was specific for histamine. Antiserum against histidine decarboxylase (HDC) was prepared as described in detail (Watanabe et al. 1984). The enzyme was purified from rat liver (Taguchi et al. 1984) and produced in rabbits (Watanabe et al. 1984). The specificity was established by immunoinhibition and immunodiffusion tests and absorption controls.

Antiserum against tyrosine hydroxylase was made in rabbits by injecting purified bovine adrenal enzyme into rabbits as described (Joh

Experimental Brain Research Series 16
© Springer-Verlag Berlin · Heidelberg 1987

et al. 1973). The specificity was established by double immunodiffusion, immunoelectrophoresis and immunohistochemical blocking controls.

Rats were perfused through the left ventricle with saline followed
by 4% paraformaldehyde in 0.1 M sodium phosphate buffer. Some animals were perfused with saline followed by freshly prepared 4% 1-
ethyl-3(3-dimethylaminopropyl)carbodiimide (50 ml) followed by 200
ml of 4% paraformaldehyde. Pieces of stomach, ileum and superior
cervical ganglia were removed and immersed in the same fixative for
additional 2 h, then washed with 0.1 M sodium phosphate buffer for
at least 24 h. Ten um thick cryostat sections were incubated with
normal swine serum (20%) to inhibit the background staining, then
with the specific antisera produced in rabbits for 48 h at 4°C.
After washing with phosphate buffered saline containing 0.25% Triton
X-100 they were incubated with fluorescein isothiocyanate (FITC)
labeled swine anti-rabbit immunoglobulins (DAKO, Copenhagen, Denmark) diluted 1:40 for 60 min at room temperature. For double staining experiments, sections were treated according to the elution
method of Tramu et al. (1978) with $KMnO_4$ and sulphuric acid to
remove the antibodies before incubation with the second antiserum.
Before the second incubation, the sections were incubated with the
FITC-conjugated antiserum to make sure that all immunoglobulins from
the first incubation had been removed. The samples were examined
under a Leitz Ortholux II fluorescence microscope equipped with
appropriate filter combinations.

Results and Discussion

Only one extensively characterized histamine antiserum (Panula et
al. 1984, 1985, 1986; Häppölä 1985a,b) gave strong and specific
immunofluorescence both with and without carbodiimide in the fixative. Other antisera gave much stronger reaction with carbodiimide and
paraformaldehyde fixation than with paraformaldehyde alone in rat
gastric enterochromaffin-like cells (Figs. 1A and B). The use of
carbodiimide allowed dilution of the histamine antiserum 1:5000-
1:20000 as compared to 1:500 with conventional paraformaldehyde
fixation. The results suggest that carbodiimide may be successfully
used in tissue fixation, and it may be the fixative of choice when
it has been used in preparing the conjugate for immunization.

The small intensely fluorescent (SIF) cells, first observed in the
sympathetic ganglion of the rat due to their high catecholamine
content (Eränkö and Härkönen 1963), were immunoreactive for histamine (Fig. 1E). After fixation with carbodiimide, some of these cells
appeared to have processes (Fig. 1C), which were only occasionally
seen after paraformaldehyde fixation (Häppölä et al. 1985b). Histamine-containing cells appeared singly or in small clusters. They
were identified as SIF cells on the basis of high nucleus/cytoplasm
ratio and small size (10-20 um in diameter). The histamine-synthesizing enzyme, HDC, was present in all histamine-immunoreactive cells.
This suggests that some SIF cells are capable of synthesizing histamine. Furthermore, all histamine-immunoreactive cells contained
immunoreactivity for the rate-limiting enzyme of catecholamine synthesis, tyrosine hydroxylase (TH) (Fig. 1C and D), which shows that
the catecholamine-storing SIF cells may synthesize and store histamine. The number of TH- and HDC-immunoreactive cells was larger than
that of histamine-immunoreactive cells, which may be due to different sensitivities of the antisera or lack of histamine in some SIF
cells.

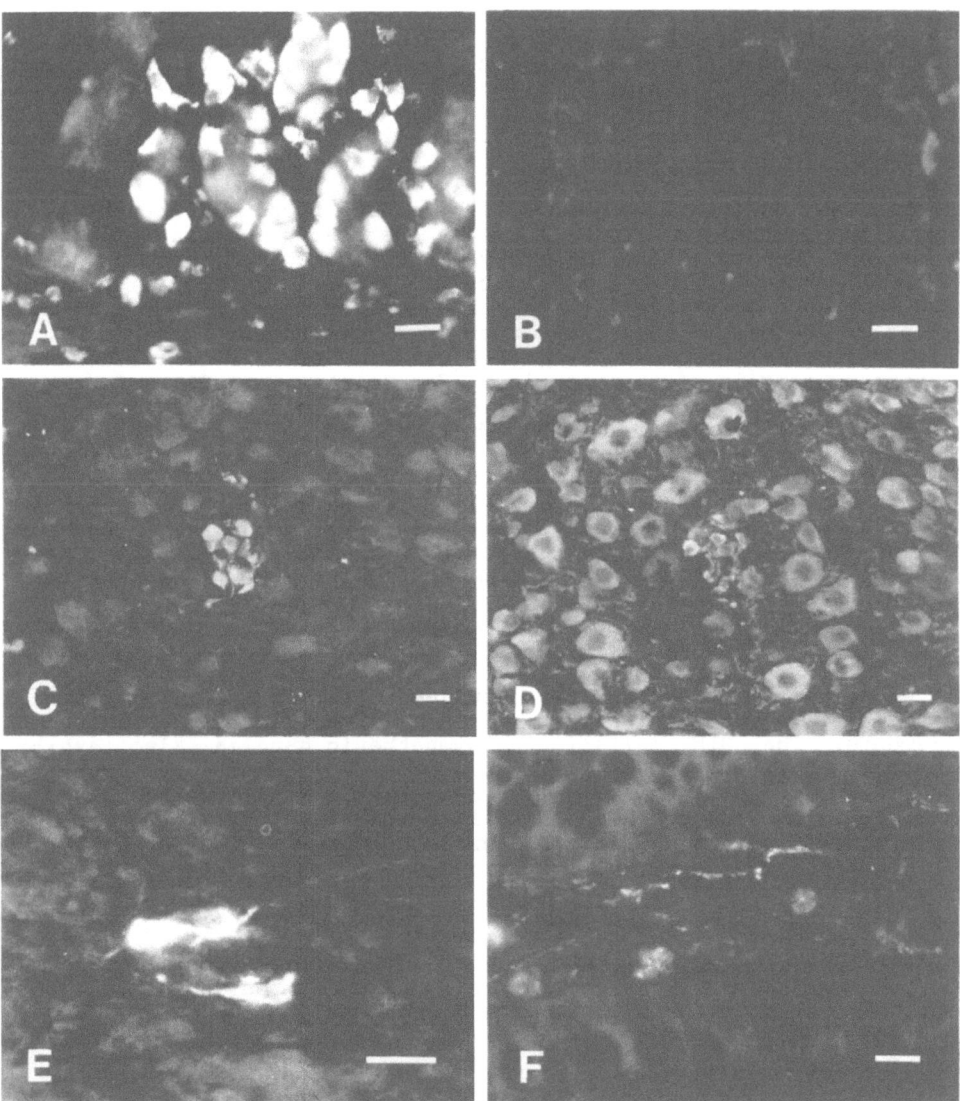

Fig. 1. A) Bright histamine—immunofluorescence in enterochromaffin-like cells of the rat stomach after fixation with carbodiimide followed by paraformaldehyde. B) Similar section from a rat fixed with paraformaldehyde and incubated with the same dilution of histamine—antiserum shows only traces of fluorescence. C) Histidine decarboxylase—like immunoreactivity in a group of SIF cells in the rat superior cervical ganglion. D) The same SIF cell group exhibits immunoreactivity for tyrosine hydroxylase. The surrounding principal ganglion cells are also immunoreactive for this enzyme. E) Three SIF cells in the rat superior cervical ganglion are immunoreactive for histamine after carbodiimide—paraformaldehyde fixation. Immunoreactive processes are also seen (arrowheads). F) Histamine—immunoreactive nerve fibers in the rat ileum under the epithelium after paraformaldehyde fixation. This antiserum detects histamine without carbodiimide fixation.
Scale bar = 20 um

Electrophysiological studies on the possible histaminergic synapses between the histamine-containing SIF cells and principal ganglion (PN) cells are not available. Hence, the significance of histamine in these cells is unclear. However, the histaminergic neurons in _Aplysia_ CNS mediate excitation on some followers, inhibition on others and mixed functions on some cells (McCaman and Weinreich 1985).

Histamine-immunoreactive nerve fibers were found in the rat ileum under the epithelium (Fig. 3). They are also present in developing human gut (Panula 1986). The fibers in the rat gut did not originate from the spinal sensory ganglia, because these were devoid of hista-mine-immunoreactive cells. Only immunofluorescent mast cells were found in the spinal ganglia. After local colchicine treatment, a few histamine-immunoreactive nerve cell bodies were detected in the submucous ganglia of the rat ileum (Panula et al. 1985). Some epi-thelial endocrine cells of the ileum were also immunoreactive for histamine (Panula et al. 1985). Histamine accumulates on both sides of ligated vagus nerve in the rat, and (^{3}H)histamine is taken up by guinea pig gut wall, probably by neuronal fibers (Hakanson et al. 1983). However, the histamine synthesizing enzyme HDC has not yet been demonstrated in these nerves. Therefore, these nerves can take up histamine, they store histamine in normal conditions without loading with L-histidine or histamine, and they can release histami-ne upon electrical stimulation in a Ca^{2+}-dependent manner (Hakanson et al. 1983). However, the production of histamine by these nerves awaits documentation.

In conclusion, histochemical studies suggest that non-mast cell histamine in the gut wall is at least in part located in neuronal elements. The SIF cells in the sympathetic ganglia also exhibit immunoreactivity for HDC, which suggests that histamine may act as a neurotransmitter in the mammalian sympathetic ganglion and gut wall. Electron microscopic studies will be required to reveal the subcel-lular distribution and possible granular storage sites of histamine in these neurons.

Acknowledgements. The authors gratefully acknowledge excellent tech-nical help of Ms. Pirjo Kotilainen, Ms. Sinikka Törmälä and Ms. Marja-Leena Piironen and secretarial work of Ms. Maija-Leena Johans-son. Supported by grants from the Finnish Medical Research Council and the University of Helsinki.

References

Eränkö O, Härkönen M (1963) Histochemical demonstration of fluoroge-nic amines in the cytoplasm of sympathetic ganglion cells of the rat. Acta Physiol Scand 58: 285-288

Hakanson R, Wahlestedt C, Westlin L, Vallgren S, Sundler F (1983) Neuronal histamine in the gut wall releasable by gastrin and chole-cystokinin. Neurosci Lett 42: 305-310

Häppölä O, Soinila S, Päivärinta H, Joh TH, Panula P (1985a) Hista-mine-immunoreactive endocrine cells in the adrenal medulla of the rat. Brain Res 339: 393-396

Häppölä O, Soinila S, Päivärinta H, Panula P, Eränkö O (1985b) Histamine-immunoreactive cells in the superior cervical ganglion and in the coeliac-superior mesenteric ganglion complex of the rat. Histochemistry 82: 1-3

Joh TH, Gegham C, Reis DJ (1973) Immunochemical demonstration of increased tyrosine hydroxylase protein in sympathetic ganglia and adrenal medulla elicited by reserpine. Proc Natl Acad Sci USA 70: 2767-2773

McCaman RE, Weinreich D (1985) Histaminergic synaptic transmission in the cerebral ganglia of _Aplysia_. J Neurophysiol 53: 1016-1037

Panula P (1986) Histamine in nervous tissue. In: Panula P, Päivärinta H, Soinila S (eds) Neurohistochemistry: modern methods and applications. Alan R. Liss, New York, p 425

Panula P, Kaartinen M, Mäcklin M, Costa E (1985) Histamine-containing peripheral neuronal and endocrine systems. J Histochem Cytochem 33: 933-941

Panula P, Yang H-YT, Costa E (1984) Histamine-containing neurons in the rat hypothalamus. Proc Natl Acad Sci USA 81: 2572-2576

Taguchi Y, Watanabe T, Kubota H, Hayashi H, Wada H (1984) Purification of histidine decarboxylase from liver of fetal rats and its immunochemical and immunohistochemical characterization. J Biol Chem 259: 5214-5221

Tramu G, Pillez A, Leonardelli J (1978) An efficient method of antibody elution for the successive or simultaneous localization of two antigens by immunocytochemistry. J Histochem Cytochem 26: 322-324

Watanabe T, Taguchi Y, Shiosaka S, Tanaka J, Kubota H, Terano Y, Tohyama M, Wada H (1984) Distribution of histaminergic neuron system in the central nervous system of rats: a fluorescent immunohistochemical analysis with histidine decarboxylase as a marker. Brain Res 295: 13-25

The Enteric Nervous System, with Particular Reference to the Enkephalin and VIP-PHI Neurons

S. Kobayashi[1], M. Suzuki[1], T. Nishisaka[2], and N. Yanaihara[3]

[1] Department of Anatomy, Yamanashi Medical School, Tamaho, Yamanashi, Japan
[2] Health Science Center, Tokyo University of Agriculture and Technology, Koganei, Tokyo, Japan
[3] Laboratory of Bioorganic Chemistry, Shizuoka College of Pharmacy, 2-2-1 Oshika, Shizuoka, Japan

Immunocytochemical and silver-impregnation studies were performed in the guinea-pig small intestine. The framework of the enteric nerve plexuses was found to consist of enteroglial cells which were immunopositive for S-100b protein (Kobayashi et al. 1986). The fine structure of individual enteroglial cells was determined by Golgi's silver-impregnation method. They could be classified into three types: 1) a ganglion type which occurred in the myenteric and submucous ganglia surrounded the somata of enteric neurons; 2) there was a strand type in the nerve strands, which embraced bundles of neuronal processes; and 3) a groundplexus type was seen in the autonomic groundplexus. The "interstitial cells (neurons)" illustrated by Ramon y Cajal (1893) were found to include a peculiar artifact inherent to the Golgi's method.

There were many somata of Met-enkephalin-Arg^6-Gly^7-Leu^8-immunopositive neurons (enkephalin neurons) in the myenteric ganglia (Kobayashi et al. 1984, 1985). Processes of the enkephalin neurons twined themselves within the grooves on the enteroglial cell sheath in the longitudinal and circular muscle layers, in the submucous layer and in the muscularis mucosae.

Many somata of vasoactive intestinal polypeptide-peptide histidine isoleucine (VIP-PHI)-immunopositive neurons were seen in the submucous ganglia (Suzuki et al. 1986). Most of them were multipolar in morphology. VIP-PHI-immunopositive neuronal processes together with enteroglial cells were found in the villous and periglandular autonomic ground plexuses of the mucous coat, where no enkephalin-immunopositive elements were detected.

Argon laser irradiation provided a simple and reliable methodology for the study of the projections of the enteric neurons (Kobayashi and Nishisaka 1985). Accumulations of material immunoreactive for Met-enkephalin-Arg^6-Gly^7-Leu^8 and VIP-PHI were found in the terminal expansions of the severed neuronal processes. There were myenteric enkephalin neurons with numerous orally-directed thin processes and an anally-directed thick process (Kobayashi and Nishisaka 1985). Most of the processes of the VIP-PHI neurons were directed anally in the myenteric plexus, whereas those of the submucous plexus showed no directional polarity (Suzuki et al. 1986).

Basic Structure of the Enteric Nerve Plexuses

Figure 1 is a schematic representation of the enteroglial cell components of the enteric nerve plexuses. These include a delicate network of enteroglial cells extending along the longitudinal muscle layer, and networks in Auerbach's plexus, in the circular muscle layer including the deep muscular plexus, in the submucous layer including Meissner's plexus, and in the lamina muscularis mucosae and lamina propria mucosae. This network of enteroglial cells provided a skeleton or framework for supporting and guiding the somata and processes of all the enteric neurons (Kobayashi et al. 1986).

Figure 2 is a simplified model of the enteric nerve plexuses. The somata of the enteric neurons usually occurred in the myenteric and submucous ganglia. For the neuronal processes it was impossible to distinguish morphologically between axons and dendrites. The autonomic groundplexus consisted of a three-dimensional network, of which the unit comprised an enteroglial cell and neuronal processes which were twisted into a cord. Thus, in the autonomic groundplexus, as proposed by Hillarp (1959), there was a convergence of processes which derived from different neurons with various chemical and physiological characteristics.

Enteroglial Cells and Interstitial Cells of Cajal

Golgi's method was useful for determining the fine structure of the enteric nerve plexus. As illustrated in Figure 3, each of the three types of enteroglial cells exhibited different features. The ganglion type was characterized by an astrocyte-like appearance. The strand type was identical with Schwann cells. The groundplexus type was multipolar in shape and projected cytoplasmic processes measuring up to 100 um in length.

In the Golgi preparations, an artifact with an enteroglial cell body and neuronal processes was frequently formed. Figure 4 illustrates examples of this artifact. Careful examinations of the illustrations of Ramon y Cajal (1893) demonstrated that the "interstitial cells of Cajal" were not true cells but the silver-impregnated chimera of an enteroglial cell and fragments of neuronal processes (Kobayashi et al. 1986). Thus, Ramon y Cajal's neuron theory that the "interstitial cells" which he observed in the interstitial nerve plexuses were neurons is refuted by this evidence. Hillarp (1959) who proposed the autonomic groundplexus theory overlooked the thin neuronal processes associated with the "interstitial cells of Cajal".

Enteric Enkephalin Neurons

Kobayashi and Nishisaka (1985) applied laser-photocoagulation techniques for the immunocytochemical investigations of the enkephalin neurons in the enteric nerve plexuses. Features of neuronal regrowth containing enkephalin-like immunoreactivity were as follows: 1) swelling and budding at the severed proximal ends; 2) elongation and branching of newly-formed neuronal processes; and 3) formation of a tangled overgrowth. It was demonstrated in the guinea-pig jejunum that the immunoreaction of enkephalin neurons become stronger after the laser microsurgery and that the anally-directed thick process of the enkephalin neurons transported a larger amount of the immunoreactivity than the orally-directed thin branches did. The direction of the enkephalin neuron processes may be related to the neuronal control of the peristaltic and antiperistaltic movement of the gut (Kobayashi et al. 1985).

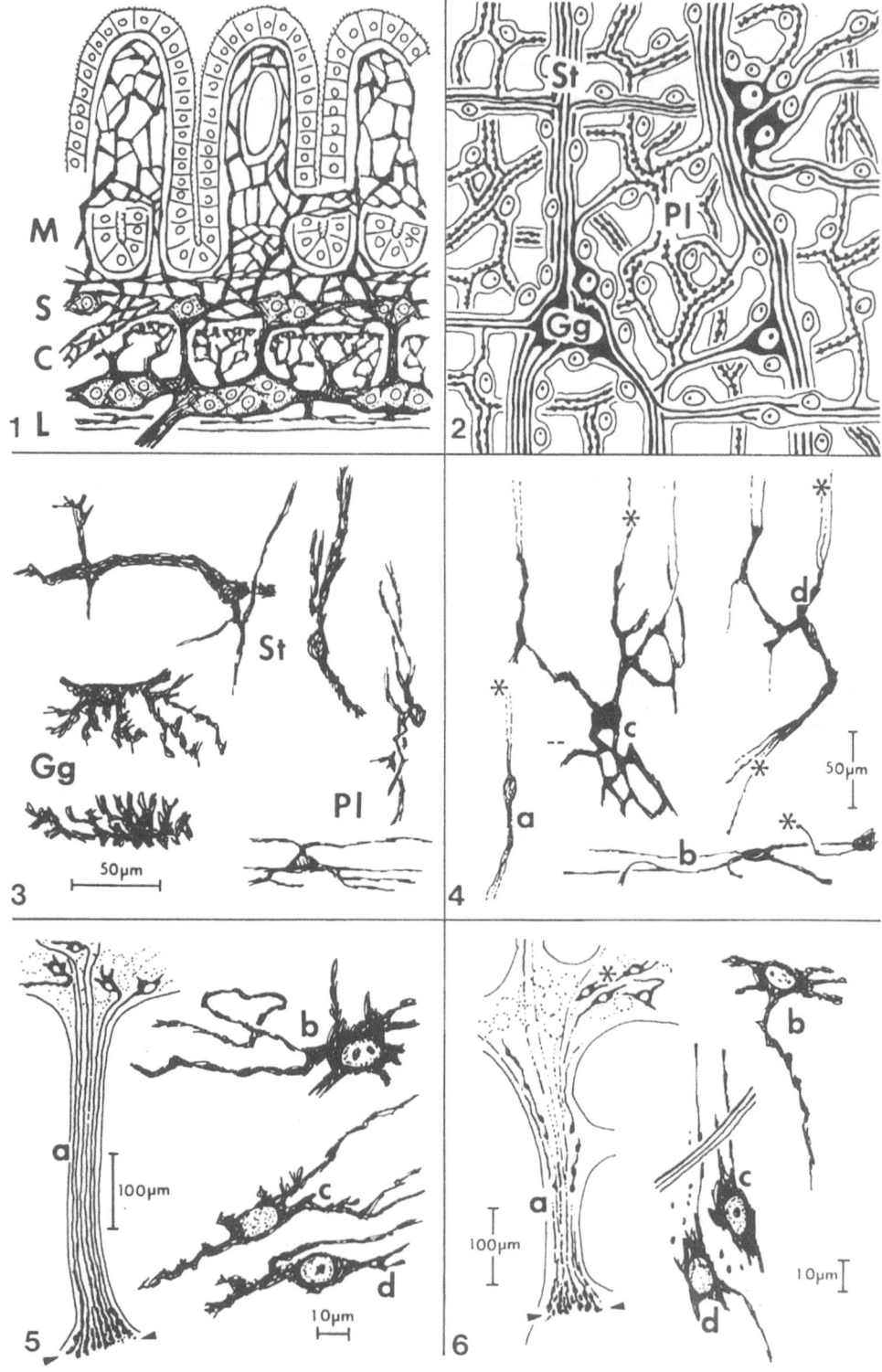

Enteric VIP-PHI Neurons

Nucleotide sequence analysis of the cloned DNA complementary to the mRNA coding for VIP revealed that the VIP precursor protein contained a PHI-like peptide sequence, designated peptide histidine methionine (Itoh et al. 1983). The term VIP-PHI neurons indicated enteric neurons with VIP- and/or PHI-like immunoreactivities.

Laser irradiation combined with immunocytochemistry has been used for the morphological investigation of the VIP-PHI neurons in the guinea-pig small intestine (Suzuki et al. 1986). The results obtained were in agreement with those of Costa and Furness (1983) concerning the origins, pathways and termination of the VIP-PHI neurons. Figure 6 illustrates examples of the VIP-PHI neurons in the myenteric and submucous plexuses after the laser-microsurgery. Most, if not all, of the VIP-PHI neurons of both the myenteric and submucous ganglia possessed a single thick process which ran in the primary fasciculus of the myenteric plexus in the anal direction. Somata of the VIP-PHI neurons in the submucous ganglia projected several long, tapering processes in addition to this thick process. These processes showed no directional polarity and seemed to enter the autonomic groundplexus of the lamina propria mucosae.

Immunocytochemistry Combined with Laser-Microsurgery as a New Strategy for Enteric Neuron Research

Glial elements of the enteric nervous system could be found by immunocytochemistry using anti-S-100b protein (Kobayashi et al.

◄ **Fig. 1.** Schematic representation of the framework of the enteric nerve plexus. Enteroglial cells form the skeleton of all the enteric nerve plexuses. M: mucous coat. S: submucous coat. C: circular muscle layer. L: longitudinal muscle layer and serous coat

Fig. 2. Schematic representation of the autonomic groundplexus. Gg: ganglion containing enteric neuronal somata and enteroglial cells. St: nerve strand containing neuronal processes surrounded by an enteroglial cell sheath. Pl: autonomic groundplexus

Fig. 3. Types of enteroglial cells. Golgi preparations. Gg: ganglion type. St: strand type. Pl: groundplexus type

Fig. 4. Silver-impregnated chimera of an enteroglial cell and fragments of neuronal processes. Golgi preparation. They were found in the myenteric plexus (a), deep muscular plexus (b), and submucous plexus (c, d). Asterisks indicate fragments of neuronal processes

Fig. 5. Myenteric enkephalin neurons on the oral side of the laser irradiated area. Immunocytochemistry. They were laser-cut at the arrowheads (a). After laser-microsurgery, the fine structure of the enkephalin neurons becomes distinct (b, c, d)

Fig. 6. VIP-PHI neurons. In the myenteric plexus there are VIP-PHI immunopositive swellings only on the oral side of the laser irradiation. Hence, it is likely that the VIP-PHI neurons send an anally-directed thick process (a). A few neuronal somata in the myenteric ganglia contain VIP-PHI-like immunoreactivity (asterisk in a). They are multipolar in shape (b). Submucous VIP-PHI neurons project several long, tapering processes which show no directional polarity (d)

1986). Enteric neurons contain various neuropeptides (Furness and Costa 1980): enkephalin neurons, in particular, predominate in the muscular coats (Kobayashi et al. 1984), whereas VIP-PHI neurons represent the greatest number of the submucous plus mucous neurons (Costa and Furness 1983).

Laser-photocoagulation provided a simple access to nerve cutting experiments in the enteric nerve plexuses (Kobayashi and Nishisaka 1985). Thus, immunocytochemistry using antisera for S-100b protein and for specific neuroactive peptides combined with laser-microsurgery may be one of the most effective strategies for the morphological investigation of the enteric nervous system.

References

Costa M, Furness JB (1983) The origins, pathways and terminations of neurons with VIP-like immunoreactivity in the guinea-pig small intestine. Neuroscience 8: 665-676

Furness JB, Costa M (1980) Types of nerves in the enteric nervous system. Neuroscience 5: 1-20

Hillarp N-A (1959) The contruction and functional organization of the autonomic innervation apparatus. Acta Physiol Scand 46, suppl 157: 1-38

Itoh N, Obata K, Yanaihara N, Okamoto H (1983) Human preprovasoactive intestinal polypeptide contains a novel PHI-27-like peptide, PHM-27. Nature 304: 547-549

Kobayashi S, Nishisaka T (1985) Myenteric enkephalin neurons around the laser-photocoagulation necrosis: an immunocytochemical investigation in the guinea pig jejunum and proximal colon. Arch histol jap 48: 239-254

Kobayashi S, Suzuki M, Endo T, Tsuji S, Daniel EE (1986) Framework of the enteric nerve plexuses: an immunocytochemical study in the guinea pig jejunum using an antiserum to S-100 protein. Arch histol jap 49: 159-188

Kobayashi S, Suzuki M, Uchida T, Yanaihara N (1984) Enkephalin neurons in the guinea pig duodenum: a light and electron microscopic immunocytochemical study using an antiserum to methionine-enkephalin-Arg^6-Gly^7-Leu^8. Biomed Res 5: 489-506

Kobayashi S, Suzuki M, Yanaihara N (1985) Enkephalin neurons in the guinea pig proximal colon: an immunocytochemical study using an antiserum to methionine-enkephalin-Arg^6-Gly^7-Leu^8. Arch histol jap 48: 27-44

Ramon y Cajal S (1893) Sur les ganglions et plexus nerveux d'intestin. Cr Soc Biol Paris 45: 217-223

Suzuki M, Kobayashi S, Yanaihara N (1986) VIP-PHI neuron: immunocytochemical study in the guinea pig small intestine. In: Miyoshi A (ed) Gut hormones (VI). Igaku-Tosho-Shuppan, Tokyo, pp 322-328

Gamma-Aminobutyric Acid (GABA) Signaling in the Adrenal Medulla

H. Alho[1], I. Hanbauer[2], A. Guidotti[1], and E. Costa[1]

[1] Fidia-Georgetown Institute for the Neurosciences, Georgetown University, 3900 Reservoir Road, N.W., Washington D.C. 20007, USA
[2] National Heart, Lung and Blood Institute, NIH, Bethesda, MD 20205, USA

Introduction

Gamma-aminobutyric acid (GABA) is known to be an inhibitory neuro-transmitter in the CNS. The presence of GABA in several peripheral organs has been reported (see Erdo and Bowery 1986). In bovine adrenal gland, chromaffin cells have been shown to contain, release, store and take-up ^{3}H-GABA (Kataoka et al. 1984; Alho et al. 1986). Glutamic acid decarboxylase (GAD) has been localized in cultured bovine adrenal chromaffin cells by indirect immunofluorescence (Kataoka et al. 1984) and peroxidase-antiperoxidase (PAP) techniques (Alho et al. 1986). Borman and Clapham (1985) recently demonstrated that the GABA recognition sites located in chromaffin cells are coupled to Cl⁻-channels and to the benzodiazepine recognition sites. Moreover, it has been shown that GABAergic mechanisms are involved in catecholamine secretion from the adrenal medulla (Kitayama et al. 1984; Kataoka et al. 1986). In this study, the localization of GAD and GABA in the adrenal gland and the modulation of catecholamine release from the adrenal gland by GABA is presented, and the possible functions of GABA are discussed.

Materials and Methods

Immunohistochemistry. Canine, bovine and rat adrenals were fixed with 4% formaldehyde, 0.1% glutaraldehyde, 0.1 M lysine HCl and 0.01 M sodium periodate in 0.1 M phosphate buffered saline (PBS; pH 7.4) for 12–16 h at 4°C. After fixation, the adrenals were washed in PBS and 20–30 um vibratome sections were cut. Free floating sections were stained using the peroxidase-antiperoxidase (PAP) method according to Sternberger (1979). The antisera used were polyclonal sheep anti-GAD (Oertel et al. 1981) and polyclonal rabbit anti-GABA (Seguela et al. 1984). GAD and GABA antisera were diluted 1:200 and 1:1500, respectively. The reaction product was visualized by 3,3'-diaminobenzidine reaction (Sternberger 1979). Preabsorption of the GABA antiserum with 1 mM free GABA diminished the GABA-staining and 5 mM abolished it. No specific staining was observed after incubating the sections with pre-immune serum or after omission of either second antiserum or PAP complex.

Catecholamine release experiments. In situ perfusion of the adrenal gland was carried out in female foxhounds as described by Hilton et al. (1958). In brief, the left carotid artery and the left renal artery were irrigated through an aortic pouch. Generally, the lumbar arteries and the superior mesenteric artery were ligated so that only the blood from the renal artery flowed to the adrenal gland.

Experimental Brain Research Series 16
© Springer-Verlag Berlin · Heidelberg 1987

Effluent blood from the adrenal gland was collected through the inferior vena cava. This technique allows perfusion of the gland without interruption of blood supply or trauma to the gland. Saline or the drugs to be tested were administered into the aortic pouch and the adrenal blood was sampled at various time points. Heparinized blood was collected in tubes and catecholamines were extracted as described by Neff and Costa (1968). Catecholamine concentration was measured using high performance liquid chromatography (HPLC) and electrochemical detection (Kataoka et al. 1986).

Results

In cultured bovine chromaffin cells approximately 40% of the cells displayed strong GAD immunoreactivity (Fig. 1E). In intact adrenal glands only few faintly stained chromaffin cells were observed (Fig. 1A). This reaction product was most intense in dog adrenal glands from all species studied. After splanchnic nerve transection the intensity of the reaction product and the number of GAD immunoreactive chromaffin cells was increased (Fig. 1F). In intact adrenals variable numbers of GAD-positive fibers were observed mainly in the medullary area (Fig. 1 A,B) and these fibers were often surrounding chromaffin cells. In dog adrenals the fibers formed a dense plexus at the cortico-medullary boundary (Fig. 1A). In rat adrenal gland GAD-positive fibers were located in the medulla but not in the cortex (Fig. 1B). In bovine adrenals some GAD-positive fibers were found also in the cortex (Fig. 1C).

In intact adrenals GABA immunoreactivity was identical to the distribution of GAD immunoreactivity in all species studied. In addition, in bovine and dog adrenals an intense punctate staining was detected in the medullary area. This punctate staining surrounded some chromaffin cells and blood vessels (Fig. 1D).

Denervation studies in rats and dogs indicated that two to four weeks after splanchnic nerve transection the amount of GAD-positive fibers were decreased but did not completely disappear. In addition, the number of GAD-immunoreactive chromaffin cells were increased after denervation in dog adrenals (Fig. 1F).

Perfusion studies of dog adrenal glands showed that GABA and GABA$_A$ receptor agonists such as 4,5,6,7-tetrahydroisoxazolopyridin-3 (THIP) or muscimol increased the catecholamine content in the adre-

Fig. 1. GAD (A,B,C,E,F) and GABA (D) immunoreactivity in adrenal ▶ glands (A,B,C,D,F) and cultured bovine chromaffin cells (E). A: In dog adrenal gland the GAD-immunoreactive fibers are mainly located in the boundary (arrowheads) between cortex (c) and medulla (m), some faintly stained chromaffin cells can be seen (arrows); bar = 10 um. B: In rat adrenal, the fibers (arrow) are surrounding groups of chromaffin cells in the medulla (m); bar = 10 um. C: In bovine adrenal, the fibers form often bundles before entering to the medulla (m); bar = 20 um. D: GABA immunoreactivity in bovine adrenal gland showing a high density of terminals surrounding chromaffin cells (open circles) and blood vessels (circles); bar = 20 um. E: 40% of cultured chromaffin cells display strong GAD immunoreactivity whereas some chromaffin cells (arrowhead), and fibroblasts (open arrows) completely devoid the staining; bar = 10 um. F: In denervated dog adrenal medulla, a group of chromaffin cells (arrow) display a strong immunoreactivity; a = artery, bar = 30 um

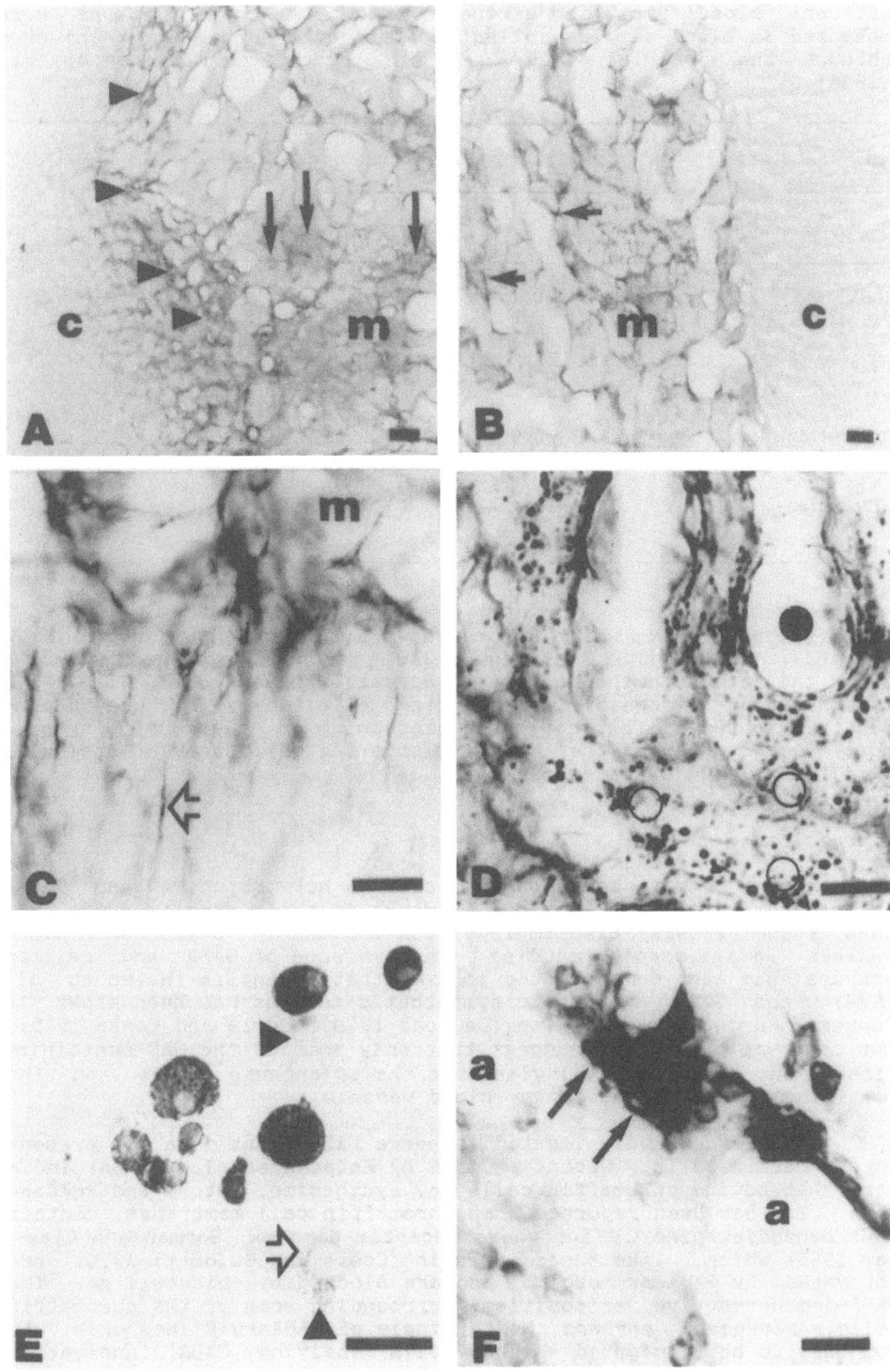

Table I. Effect of GABAergic drugs on catecholamine concentration in effluent blood from dog adrenal glands. CA concentrations were measured in blood samples collected 5 min after drugs were injected through the arterial pouch (2 ml/min). Data from Kataoka et al. (1986)

Drug (mmol/2ml)	Catecholamine Concentration (nmol/ml plasma)	
	Saline	Drug
GABA (19.4)	0.95 ± 0.1	53 ± 34
THIP (0.071)	1.6 ± 0.2	4.0 ± 0.4
Bicuculline Met. (0.05)	0.7 ± 0.3	0.4 ± 0.08
L-Baclofen (0.094)	1.6 ± 0.4	1.8 ± 0.8

nal effluent blood, while bicuculline, a GABA receptor antagonist, reduced it (Tab. I). Baclofen, a GABA receptor agonist, failed to alter the catecholamine content in adrenal effluent blood (Tab. I). The increased release of catecholamines elicited by THIP was prevented by bicuculline but not by hexamethonium or naloxone. Furthermore, denervation of the adrenal glands failed to prevent the THIP-elicited release of catecholamines.

Discussion

The present study documents that a dense network of GAD and GABA-immunoreactive fibers was found throughout the adrenal medulla and that these fibers also impinge upon chromaffin cells and blood vessels in all species studied. The presence of GABA and related enzymes has been demonstrated in sympathetic ganglia (Walsh et al. 1974), and GABA applied to sympathetic neurons has been shown to depress neuronal transmission (DeGroat 1970; Suria and Costa 1975). Our denervation studies suggest that only some of the GAD-containing fibers enter the adrenal gland via the splanchnic nerve and the remaining fibers may accompany blood vessels.

GAD is not exclusively located in nerve fibers but is also present in chromaffin cells. Recent studies by Kataoka et al. (1984) indicate that bovine chromaffin cells may synthesize, store and release GABA. It has been reported that chromaffin cell membranes contain GABA/benzodiazepine/Cl⁻ ionophore receptor complex (Borman and Clapham 1985) which, like those in brain (Costa and Guidotti 1978), are activated by GABA or muscimol and are blocked by bicuculline. The GABA-immunoreactive varicosities, surrounding some of the chromaffin cells, represent perhaps the terminals of GABAergic neurones. It remains to be elucidated whether cells receiving GABA innervation are epinephrine or norepinephrine storing cells. Hence, it may be inferred that only a specific population of chromaffin cells can be activated by GABA released from GABA—containing nerve terminals. Alternatively chromaffin cells might be activated by GABA released from GABA—containing adjacent chromaffin cells.

The dog adrenal perfusion studies in vivo showed that appropriate doses of GABA or other GABA—mimetic drugs release catecholamines into the circulation. The magnitude of this release is comparable to that obtained by injecting a pharmacological dose of carbachol or by maximal efficient electrical stimulation of the splanchnic nerve (Kataoka et al. 1986). These data suggest that GABA$_A$ receptor stimulation facilitates the release of catecholamines by depolarizing chromaffin cells in a fashion that is independent from nicotine receptor stimulation.

In conclusion, GABA may be involved in the physiological regulation of catecholamine secretion from the adrenal medulla. We believe that GABA should be included in the list of important neuromodulators (as several neuropeptides) that are located in chromaffin cells and control the responsiveness of chromaffin cells to incoming cholinergic stimuli.

References

Alho H, Fujimoto M, Guidotti A, Hanbauer I, Kataoka Y, Costa E (1986) Gamma—aminobutyric acid in the adrenal medulla: location, pharmacology and function. In: Panula P, Soinila S (eds.) Neurochemistry: modern methods and applications. Alan R Liss, New York, pp 453—464

Borman J, Clapham DE (1985) Gamma—aminobutyric acid receptor channels in adrenal chromaffin cells: a patch clamp study. Proc Natl Acad Sci USA 82: 2168—2172

Costa E, Guidotti A (1979) Molecular mechanism in the receptor action of benzodiazepines. Ann Rev Pharmacol Toxicol 19: 531—545

DeGroat WC (1970) The action of gamma—aminobutyric acid and related amino acids on mammalian autonomic ganglia. J Pharmacol Exp Ther 172: 384—396

Erdo SL, Bowery NG (eds) (1986) GABAergic mechanism in the mammalian periphery. Raven, New York

Hilton JG, Weaver DC, Muehlheims G, Galviano VV, Wegria R (1958) Perfusion of isolated adrenal in situ. Am J Physiol 192: 525—530

Kataoka Y, Fujimoto M, Alho H, Guidotti A, Geffard G, Kelly GD, Hanbauer I (1986) Intrinsic GABA receptors modulate the release of catecholamine from canine adrenal gland in situ. J Pharmacol Exp Ther 239: 584—590

Kataoka Y, Gutman Y, Guidotti A, Panula P, Wrobleski J, Cosenza-Murphy D, Wu JY, Costa E (1984) Intrinsic GABAergic system of adrenal chromaffin cells. Proc Natl Acad Sci USA 81: 3212—3222

Kitayama S, Morita K, Dohi T, Tsujimoto A (1984) The nature of the stimulatory action of gamma—aminobutyric acid in the isolated perfused dog adrenals. Naunyn Schmiedeberg's Arch Pharmacol 326: 106—110

Neff N, Costa E (1968) Application of steady state kinetics to the study of catecholamine turnover after monoamine oxidase inhibition or reserpine administration. J Pharmacol Exp Ther 160: 40–49

Oertel WH, Schmechel DE, Tappaz ML, Kopin IJ (1981) Production of specific antiserum to rat brain glutamic acid decarboxylase by injection of an antigen-antibody complex. Neuroscience 6: 2589–2700

Seguela P, Geffard M, Buijs R, Le-Moal M (1984) Antibodies against gamma-aminobutyric acid: specificity studies and immunocytochemistry results. Proc Natl Acad Sci USA 81: 3888–3892

Sternberger LA (1979) Immunocytohistochemistry, 2nd edn. Wiley, Chichester, pp 104–169

Suria A, Costa E (1975) Action of diazepam, dibutyryl cCMP and GABA on presynaptic nerve terminals in bullfrog sympathetic ganglia. Brain Res 87: 102–106

Walsh JM, Bowery NG, Brown NA, Clark JB (1974) Metabolism of gamma-aminobutyric acid (GABA) by peripheral nervous tissue. J Neurochem 22: 1145–1147

Immunohistochemistry of Opioid Peptides in Guinea Pig Paraganglia

M. Colombo, W. Kummer, and Ch. Heym

Anatomisches Institut, Universität Heidelberg, Im Neuenheimer Feld, 307, 6900 Heidelberg 1, FRG

Introduction

Adrenal medullary cells and paraganglionic cells in guinea pig autonomic ganglia are known to display opioid peptide immunoreactivity (Schultzberg et al. 1978; Watson et al. 1981; Vincent et al. 1984; Evans et al. 1985; Kobayashi et al. 1985; Kummer et al. 1986). Despite these extensive studies, only limited data are available concerning coincidence of different opioid peptides within single paraganglionic cells (see Evans et al. 1985; Kummer et al. 1986). In the present study, antisera against eight different opioid peptides were applied to serial sections of guinea pig adrenal medulla, autonomic ganglia and carotid body. Particular attention was given to the immunohistochemical coexistence of opioids derived from two different precursor molecules: pro-enkephalin (e.g., met-enkephalin-arg-phe = MEAP) and pro-dynorphin (e.g., dynorphin A = DYN; alpha-neo-endorphin = alpha-NEO).

Materials and Methods

Five um serial paraplast sections of formalin-fixed guinea pig tissues were processed for immunohistochemistry using biotinylated IgG as a second antibody and streptavidin-biotin-peroxidase complex as the detector agent (Bonnard et al. 1984). In order to correlate the aldehyde-induced catecholamine fluorescence with the immunohistochemistry of opioid peptides, additional adrenal medullae and carotid bodies were shock-frozen, fixed by paraformaldehyde vapour and embedded in araldite. After plastic removal, the sections were processed in the same manner as paraplast sections. Primary antisera are listed in Table 1. Specificity tests included tissue incubation with pre-immune serum, preabsorption tests (20–100 ug peptide/ml antiserum), and spot tests on nitrocellulose sheets (Scopsi and Larsson 1986). Leu-enkephalin (ENK) antisera could be preabsorbed with high concentrations of both alpha-NEO and DYN, but did not recognize 1 ug spots of both peptides. Met-ENK was not recognized by the leu-ENK antisera. The met-ENK antibody recognized beta-endorphin (END), while all other antisera showed positive reactions solely with the appropriate peptide.

Results

Catecholamine-storing cells in all investigated tissues exhibit immunoreactivity to cleavage products of pro-ENK and pro-DYN but not to cleavage products of pro-opiomelanocortin (e.g., beta-END and ACTH).

Experimental Brain Research Series 16
© Springer-Verlag Berlin · Heidelberg 1987

Table 1. List of applied antisera

Antiserum	Source	Lot	Dilution
leu-ENK	INC	35381	1:5000
leu-ENK	Dr. Lang	XVI 7	1:2000
met-ENK	INC	8416015	1:5000
met-ENK	INC	13391	1:5000
MEAP	Peninsula	000112	1:1500
DYN 1-8	Dr. Lang	-	1:5000
DYN 1-13	Dr. Lang	IV	1:10000
DYN 1-13	Dr. Lang	III	1:5000
DYN 1-17	Dr. Lang	-	1:4000
DYN 1-17	own	Paula	1:2000
alpha-NEO	Dr. Hoellt	Agathe	1:3000
alpha-NEO	own	Klara	1:2000
alpha-NEO	own	Lotte	1:2000
beta-END	INC	8410016	1:5000
beta-END	Dr. Lang	beta-E 4I	1:3000
ACTH	INC	11151	1:4000

Adrenal Medulla. Adrenaline-storing medullary cells ("ordinary" cells; Unsicker et al. 1978) display a moderate green formaldehyde-induced fluorescence, whereas scattered clusters of distinctly smaller noradrenergic cells are distinguished by their intense yellow fluorescence (small granule-containing "SGC"-cells; Unsicker et al. 1978; Fig. 1a). No difference in their immunoreaction to opioid peptides is observed between ordinary cells and SGC cells. Adrenal medullary catecholamine-storing cells display differential immunoreactivity to opioid antisera (Table 2). Most cells are observed to be alpha-NEO-immunoreactive (Fig. 2a). In many of these cells, leu-ENK immunoreactivity can be visualized (Fig. 1b), although there are some cells displaying exclusively either alpha-NEO- or leu-ENK immunoreactivity (Fig. 2a,d). DYN-immunoreactive cells are less numerous, but if present DYN immunoreactivity is often co-localized with alpha-NEO immunoreactivity (Figs. 2a-c), and in some cells also with leu-ENK immunoreactivity. DYN 1-17 immunoreactivity, DYN 1-13 immunoreactivity and DYN 1-8 immunoreactivity are not identically distributed in adrenal medullary cells; some cells are seen solely to be immunoreactive to antisera of one or two of the three DYN-A sequences (Figs. 2b,c). Scattered met-ENK-immunoreactive cells exhibit co-localization with all other investigated peptides or only with leu-ENK (Figs. 1b,c) and with alpha-NEO. Distribution patterns of met-ENK immunoreactivity and MEAP immunoreactivity appear to be identical although cells stain somewhat weaker to MEAP antiserum than to met-ENK antiserum. Few cells are non-reactive to any of the investigated peptides.

Carotid Body. Most glomus cells of the carotid body reveal immunoreactivity to alpha-NEO (Fig. 3a); less numerous cells are found to be immunoreactive to DYN 1-17 or to leu-ENK. The overall distribution of DYN 1-17 immunoreactivity resembles that of alpha-NEO immu-

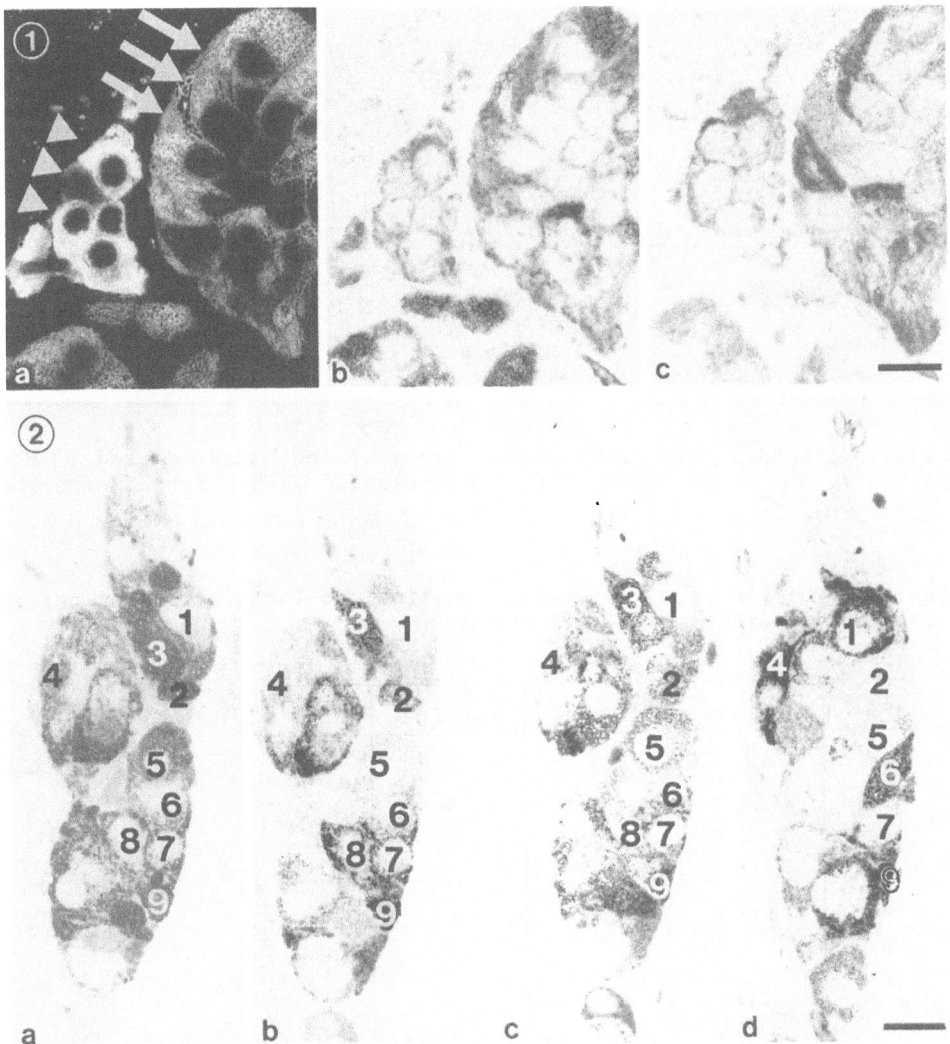

Fig. 1. Consecutive sections of adrenal medullary "ordinary" cells (arrow heads) and "SGC" cells (small arrows). Bar indicates 10 um. a: Formaldehyde-induced fluorescence. b: Leu-ENK immunoreaction. c: Met-ENK immunoreaction

Fig. 2. Consecutive semithin sections of immunoreactive adrenal medullar "ordinary" cells; letters indicate identical cells. Bar indicates 10 um. a: Alpha-NEO immunoreaction. b: DYN 1-8 immunoreaction. c: DYN 1-17 immunoreaction. d: Leu-ENK immunoreaction

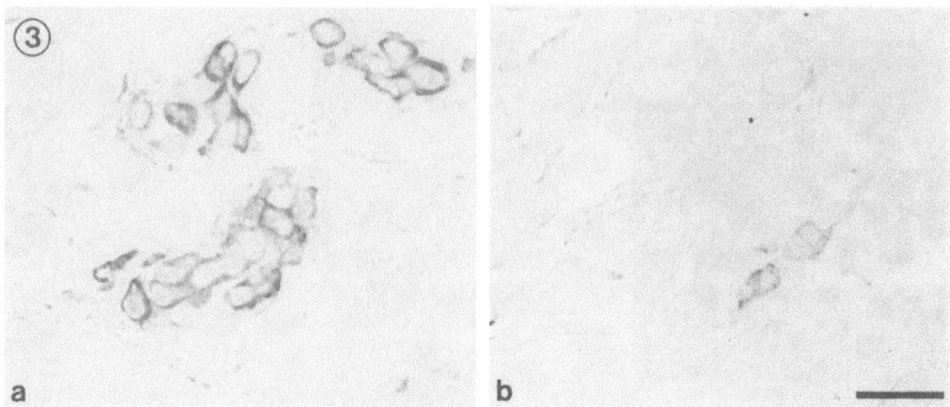

Fig. 3. Consecutive semithin sections of the carotid body. Bar indicates 30 um. a: alpha-NEO immunoreaction. b: Met-ENK-arg-phe immunoreaction

Table 2. Differential immunohistochemical co-localization of opioid peptides in adrenal medullary cells

α-NEO	DYN 1-17	DYN 1-8	leu-ENK	MEAP
+	+	+	+	+
+	+	−	+	+
+	+	+	−	−
+	+	−	+	−
+	−	−	+	+
+	−	−	+	−
+	−	−	−	−
−	+	−	−	−
−	−	−	+	−
−	−	−	−	−

noreactivity, but cells are stained in varying extent. Only a few cells are immunoreactive to met-ENK and to MEAP (Fig. 3b). These cells are also alpha-NEO-immunoreactive. DYN 1-8 immunoreactivity is not observed in glomus cells.

SIF-Cells. Immunoreactions of small intensely fluorescent (SIF) cells in the superior cervical ganglion (SCG) and in the celiac-superior mesenteric ganglion resemble that of glomus cells – with practically all cells showing alpha-NEO immunoreactivity and some cells being immunoreactive to DYN 1-17 or to leu-ENK. Met-ENK-immunoreactive and MEAP-immunoreactive SIF cells are rarely found. DYN 1-8 is absent.

Discussion

The precursor molecule pro-DYN includes one copy of each DYN A and alpha-NEO (Horikawa et al. 1983); complete processing of the precursor within all pro-DYN-containing cells should generate both these opioids in equimolar amounts. The present immunohistochemical study, however, revealed an independent distribution of DYN A- (1-8, 1-13, 1-17) and alpha-NEO-immunoreactive cells in the guinea pig adrenal medulla, indicating incomplete cleavage of pro-DYN in both cell types, i.e., ordinary and SCG cells (Unsicker et al. 1978). Accordingly, gel filtration showed the DYN-immunoreactive material of guinea pig adrenal extracts to be of a high molecular weight (M = 8,000 to 10,000; Evans et al. 1985). In view of the biochemical data it appears to be likely that in the adrenal medulla the various DYN A antisera used in our study recognized larger fragments than those against which they had been raised. Nevertheless, the incongruent, although overlapping immunohistochemical distribution patterns obtained by DYN A 1-8, 1-13 and 1-17 antisera, respectively, point to the existence of different molecular variants which are independently generated by individual adrenal medullary cells.

No correlation was found between the immunohistochemical demonstration of pro-DYN- and pro-ENK-related opioids; coexistence (cf. Evans et al. 1985) as well as exclusive occurrence was observed - indicating both simultaneous and separate processing of the two different precursors. Again, high molecular weight fragments are likely to cause the immunoreactions obtained by the MEAP antiserum (Evans et al. 1985).

The multiplicity of phenotypes recognized by serial incubations with different antisera may either reflect momentary functional states of only a few cell types or may point to adrenal medullary cell populations more heterogeneous than hitherto believed (cf. Shaw and Letourneau 1986). Accordingly, it cannot be decided from the present knowledge, whether the more uniform staining patterns of carotid body and SIF cells reflect functional synchronization or are due to the presence of a smaller number of cell types than in the adrenal medulla.

References

Bonnard C, Papermaster DS, Kraehenbuhl JP (1984) The streptavidin-biotin bridge technique: application in light and electron microscope immunocytochemistry. In: Polak JM, Varndell IM (eds) Immunolabelling for electron microscopy. Elsevier, Amsterdam, pp 95-112

Evans CJ, Erdelyi E, Hunter J, Barchas JD (1985) Co-localization and characterization of immunoreactive peptides derived from two opioid precursors in guinea pig adrenal glands. J Neurosci 5: 3423-3427

Horikawa S, Takai T, Toyosato M, Takahashi H, Noda M, Kakidani H, Kubo T, Hirose T, Inayama S, Hayashida H, Miyata T, Numa S (1983) Isolation and structural organization of the human prepro-enkephalin B gene. Nature 306: 611-614

Kobayashi S, Miyabayashi T, Uchida T, Yanaihara N (1985) Met-enkephalin-arg^6-gly^7-leu^8 in large-cored vesicles of splanchnic nerve terminals innervating guinea pig adrenal chromaffin cells. Neurosci Lett 53: 247-252

Kummer W, Heym Ch, Colombo M, Lang R (1986) Immunohistochemical evidence for extrinsic and intrinsic opioid systems in the guinea pig superior cervical ganglion. Anat Embryol 174: 401–405

Schultzberg M, Lundberg JM, Hökfelt T, Terenius L, Brandt J, Elde RP, Goldstein M (1978) Enkephalin-like immunoreactivity in gland cells and nerve terminals of the adrenal medulla. Neuroscience 3: 1169–1186

Scopsi L, Larsson L-I (1986) Increased sensitivity in peroxidase immunocytochemistry. A comparative study of a number of peroxidase visualization methods employing a model system. Histochemistry 84: 221–230

Shaw TJ, Letourneau PC (1986) Chromaffin cell heterogeneity of process formation and neuropeptide content under control and nerve growth factor-altered conditions in cultures of chick embryonic adrenal gland. J Neurosci Res 16: 337–355

Unsicker K, Habura-Flüh O, Zwarg U (1978) Different types of small granule-containing cells and neurons in the guinea pig adrenal medulla. Cell Tissue Res 189: 109–130

Vincent SR, Dalsgaard C-J, Schultzberg M, Hökfelt T, Christensson I, Terenius L (1984) Dynorphin-immunoreactive neurons in the autonomic nervous system. Neuroscience 11: 973–987

Watson SJ, Akil H, Ghazarossian VE, Goldstein A (1981) Dynorphin immunocytochemical localization in brain and peripheral nervous system: preliminary studies. Proc Natl Acad Sci USA 78: 1260–1263

Localization of Calcitonin Gene-Related Peptide (CGRP) in Chromaffin Cells and Nerve Terminals in Adrenal Medulla

M. Pelto-Huikko and T. Salminen

Department of Biomedical Sciences, University of Tampere, 33101 Tampere 10, Finland

Introduction

Several neuropeptides have been demonstrated in adrenal chromaffin cells and/or nerve terminals during the last decade. These include enkephalins, somatostatin, vasoactive intestinal polypeptide, substance P, neurotensin and neuropeptide Y (Schultzberg et al. 1978; Lundberg et al. 1979, 1986; Linnoila et al. 1980; Pelto-Huikko et al. 1985a,b). A recently characterized neuropeptide, calcitonin gene-related peptide (CGRP), has also been demonstrated in nerve fibers and chromaffin cells in rat adrenal gland (Rosenfeld et al. 1983). The present work was carried out to study the distribution of CGRP in adrenal gland of rat and man and to correlate its distribution with catecholamines and other neuropeptides.

Materials and Methods

Human adrenal glands were obtained at adrenalectomy performed in conjunction with renal surgery. Rat adrenals were excised under ether or pentobarbitone (30 mg/kg i.p.) anesthesia. Tissues were immediately sliced with a razor blade and fixed by immersion in 4% paraformaldehyde or in a mixture of 4% paraformaldehyde and 0.2% glutaraldehyde for six hours. Immunocytochemistry was carried out applying the pre-embedding modification of the PAP-method. Correlation of neuropeptides with each other or with catecholamine synthesizing enzymes was studied in consecutive sections or by double staining the same section in contrasting colours (Sternberger and Joseph 1979).

Experimental animals were treated neonatally with capsaicin (CAPS) (50 mg/kg, s.c.) and killed at three months of age or injected twice at three months of age (interval 24 h, survival time 24 h after second injection). A third group of animals (aged three months) was injected twice with 6-OHDA (50 mg/kg, i.v.) with one week interval and killed two days after the second injection. Control animals were injected with the vehicles only.

Antisera to CGRP were obtained from Amersham and Peninsula and the antiserum to phenylethanol-N-methyltransferase (PNMT) was purchased from Eugene Technics. Several antisera to enkephalins (ENK), neuropeptide Y (NPY), substance P (SP) and vasoactive intestinal polypeptide (VIP) from different commercial sources were used for this study.

Experimental Brain Research Series 16
© Springer-Verlag Berlin · Heidelberg 1987

Results

Rat Adrenal Gland. CGRP was localized in a few chromaffin cells (Fig. 1a). CGRP-immunoreactive cells were also NPY-immunoreactive. ENK-immunoreactive cells were seen in the areas of CGRP-immunoreactive cells, but the coexistence of the two peptides could not be demonstrated with certainty. At the ultrastructural level CGRP was localized in the chromaffin granules (diameter 200-250 nm) (Fig. 3).

CGRP-immunoreactive nerve fibers were seen in nerve bundles and around blood vessels in cortex and medulla. Usually only few fibers were seen among chromaffin cells but some chromaffin cell groups and ganglion cells were densely surrounded with immunoreactive terminals (Fig. 1a,b). CGRP- and SP-immunoreactive nerves had a very similar distribution. ENK-, NPY- and VIP-immunoreactive nerves had a different distribution than CGRP-immunoreactive fibers.

All CGRP- and NPY-immunoreactive and most of the ENK-immunoreactive cells were PNMT-immunoreactive. Most of the CGRP-immunoreactive nerves were seen among adrenaline (PNMT-immunoreactive) cells (Fig. 1b).

Experimental Animals. In rats treated with CAPS most of the CGRP- and SP-immunoreactive fibers disappeared. Treatment with 6-OHDA had no effect on CGRP-immunoreactive fibers. The number of ENK-, NPY- and VIP-immunoreactive nerves was not affected by these treatments. After treatment of adult rats with CAPS or 6-OHDA the number of

Fig. 1. Adrenal medulla of the normal rat. CGRP is localized in a few chromaffin cells (large arrows) and nerve fibers and terminals and in nerve bundles among chromaffin cells (small arrows) (a). A section double labelled with antibodies to PNMT to demonstrate adrenaline cells (a) (originally blue). CGRP-immunoreactive cells (stars) (originally brown) are seen in the areas of adrenaline cells. Arrows point to CGRP-immunoreactive nerve fibers (b). na = noradrenaline cells; bars indicate a: 100 um, b: 50 um

Fig. 2. Adrenal medulla of the adult rat treated with CAPS. The number of CGRP-immunoreactive chromaffin cells is clearly higher than in normal adrenal medulla. Immunoreactive nerve fibers are not present (a). A section double stained in the same way as the section in 1b. CGRP-immunoreactive cells (stars) are seen only in the areas of adrenaline cells (a) (b). na = noradrenaline cells; bars indicate a: 100 um b: 50 um

Fig. 3. Rat adrenal medulla. Part of the storage granules are heavily labelled in CGRP-immunoreactive adrenaline cell (arrows). Bar indicates 0.5 um

Fig. 4. Human adrenal medulla. Numerous CGRP-immunoreactive nerve terminals (arrows) are seen among chromaffin cells (a). A section double stained similarly as the section in 1b. CGRP-immunoreactive terminals (small arrows) are seen in close contact with PNMT-immunoreactive adrenaline cells (large arrows) and also nonlabelled noradrenaline cells (b). Bars indicate a: 100 um, b: 50 um

Fig. 5. Human adrenal medulla. CGRP-immunoreactive nerve terminal is in contact with chromaffin cell. CGRP immunoreactivity is localized mainly in large dense cored vesicles (arrows). Bar indicates 1 um

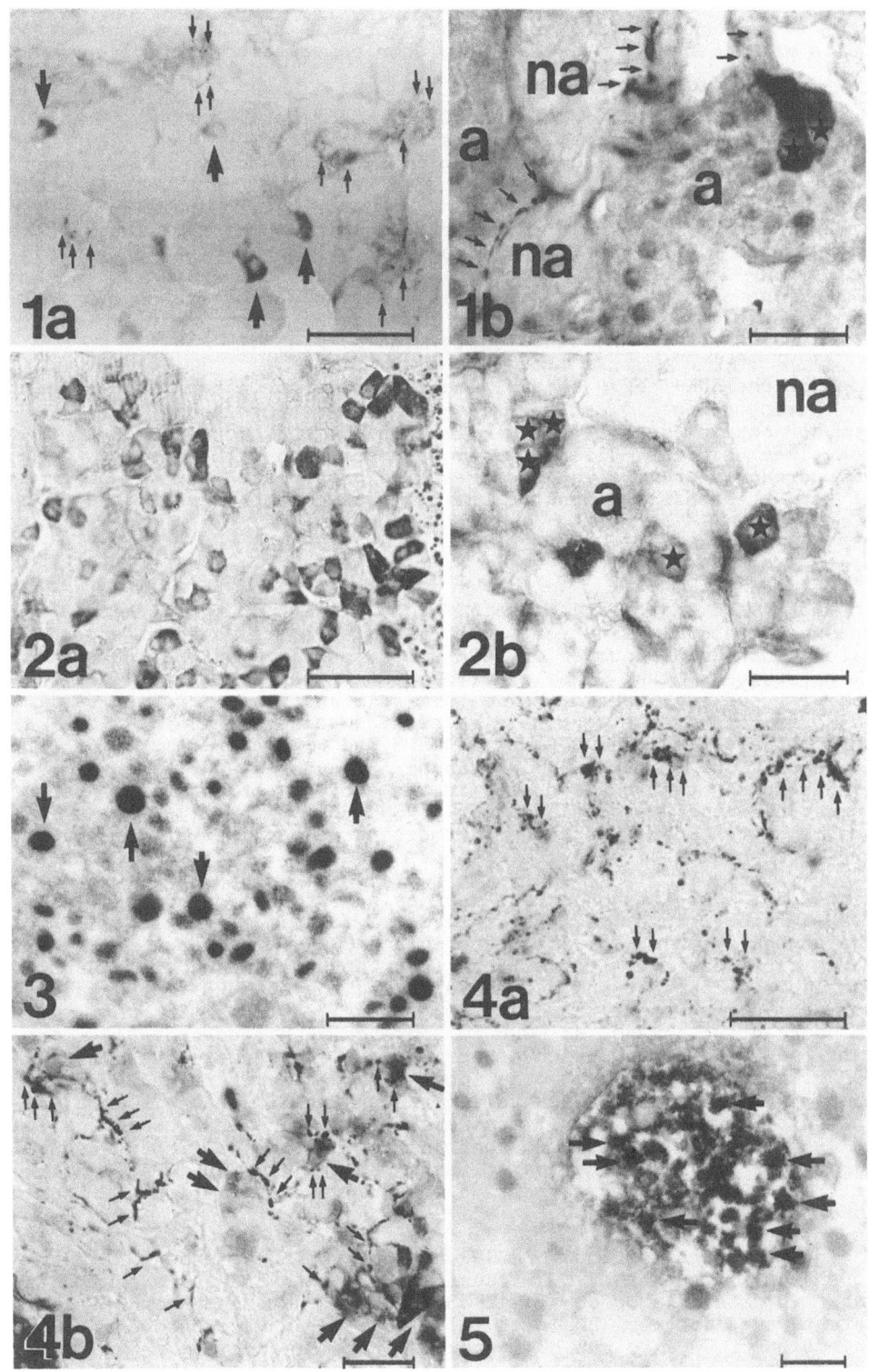

CGRP-immunoreactive adrenaline cells clearly increased (Figs. 2a,b). The number and intensity of staining of NPY- and ENK-immunoreactive cells was reduced in these animals when compared to control animals.

Human Adrenal Gland. Numerous CGRP-immunoreactive nerve terminals were seen among chromaffin cells and ganglion cells (Fig. 4a). Single chromaffin cells were also CGRP-immunoreactive. There were large variations in the number of CGRP-immunoreactive terminals between different regions of adrenal medulla but the number of CGRP-immunoreactive terminals among noradrenaline (PNMT-negative) cells usually was higher than among adrenaline (PNMT-positive) cells (Fig. 4b). CGRP-immunoreactive terminals were also seen among ENK- and NPY-immunoreactive chromaffin cells. Under electron microscope CGRP-immunoreactivity was localized in large dense cored vesicles (diameter 100–120 nm) in the labelled terminals (Fig. 5).

Most of the ENK-, NPY- and SP-immunoreactive nerve terminals present in human adrenal gland were seen in different regions than CGRP-immunoreactive terminals. VIP-immunoreactive nerve terminals had sometimes similar distributions when compared with that of CGRP-immunoreactive terminals. However, no crossreactivity was observed between the antisera when preabsorbed with synthetic peptides.

Discussion

In normal adrenal gland of rat CGRP is present in a few adrenaline cells. Stressful treatment with pharmacological agents caused a clear increase in number of CGRP-immunoreactive adrenaline cells. This suggests that the levels of CGRP are under control of the activity of splanchnic nerve terminals and the increase of the activity of these terminals raises the concentration of CGRP in chromaffin cells. The effects of CAPS and 6-OHDA treatments on the concentration of ENK and NPY appeared opposite and it has been demonstrated that the decrease in the activity of nicotinic receptors increases the concentration of ENK (Bohn et al. 1983). This suggests that concentrations of these three peptides are regulated by different mechanisms.

Since no CGRP-immunoreactive neurons were found in adrenal gland CGRP-immunoreactive terminals evidently originate from outside of the adrenal gland. In rat adrenal gland CGRP-immunoreactive nerves were CAPS-sensitive suggesting that they originate from sensory ganglia. In both rat and man the distribution of CGRP-immunoreactive nerves in most cases was different from VIP-, SP-, ENK- and NPY-immunoreactive nerves, so CGRP-immunoreactive fibers probably represent a different population of nerves from other peptidergic nerves. CGRP-immunoreactive and other peptidergic nerve terminals had no clear specificitiy to any cell type in human adrenal medulla although the number of CGRP-immunoreactive nerves among noradrenaline cells was higher. This is in contrast to the specific innervation of adrenaline and noradrenaline cells in rat and hamster (Pelto-Huikko et al. 1985a,b) and suggests a more diffuse regulation of human adrenal medullary cells.

The physiological effects of CGRP are not yet fully known. Intravenous administration of CGRP causes hypotension and increased heart rate in rats (Fisher et al. 1983) and CGRP has been shown to be a potent vasodilatator (Brain et al. 1985). Intracerebroventricular injection of CGRP to rat brain causes increase in blood pressure and heart rate and raise in the plasma concentration of noradrenalin

(Fisher et al. 1983). Presence of CGRP-immunoreactive nerve terminals and chromaffin cells in adrenal medulla suggests that CGRP has
a regulatory function in secretion of catecholamines and neuropeptides from adrenal medulla and that CGRP may also have endocrine
functions when cosecreted from adrenal medulla with other biologically active substances.

References

Bohn M, Kessler L, Golightly L, Black I (1983) Appearance of enke-
phalin-immunoreactivity in rat adrenal medulla following treatment
with nicotinic antagonist or reserpine. Cell Tissue Res 231: 469-479

Brain SD, Williams TJ, Tippins JR, Morris HR, McIntyre I (1985)
Calcitonin gene-related peptide is a potent vasodilatator. Nature
313: 54-56

Fisher LA, Kikkawa DO, Rivier JE, Amara SG, Evans RM, Rosenfeld MG,
Vale WW, Brown MR (1983) Stimulation of noradrenergic sympathetic
outflow by calcitonin gene-related peptide. Nature 305: 534-536

Hökfelt T, Lundberg J, Schultzberg M, Fahrenkrug J (1981) Immunohistochemical evidence for a local VIP-ergic neuron system in the
adrenal of the rat. Acta Physiol Scand 113: 575-576

Linnoila I, DiAugustine RP, Hervonen A, Miller R (1980) Distribution
of (leu5) and (met5) enkephalin-, VIP- and SP-like immunoreactivity
in human adrenal glands. Neuroscience 5: 2247-2259

Lundberg JM, Hamberger B, Schultzberg M, Hökfelt T, Granberg P-O,
Efendic S, Terenius L, Goldstein M, Luft R (1979) Enkephalin- and
somatostatin-like immunoreactivities in human adrenal medulla and
pheochromocytoma. Proc Natl Acad Sci USA 76: 4079-4083

Lundberg JM, Hökfelt T, Hemsen A, Theodorsson-Norheim E, Pernow J,
Hamberger B, Goldstein M (1986) Neuropeptide Y-like immunoreactivity
in adrenaline cells of adrenal medulla and in tumors and plasma of
pheochromocytoma patients. Regul Peptid 13: 169-182

Pelto-Huikko M, Salminen T, Hervonen A (1985a) Localization of
enkephalins in adrenaline cells and the nerves innervating adrenaline cells in rat adrenal medulla. Histochemistry 82: 377-383

Pelto-Huikko M, Salminen T, Toivanen M, Partanen M, Hervonen A
(1985b) Immunohistochemical localization of neurotensin in hamster
adrenal medulla. Anat Rec 211: 458-464

Rosenfeld MG, Mermod J-J, Amara SG, Swanson LW, Sawchenko JR, Rivier
J, Vale WW, Evans RM (1983) Production of a novel neuropeptide
encoded by the calcitonin gene via tissue-specific RNA processing.
Nature 304: 129-135

Schultzberg M, Lundberg JM, Hökfelt T, Terenius L, Brandt J, Elde R,
Goldstein M (1978) Enkephalin-like immunoreactivity in gland cells
and nerve terminals of the adrenal medulla. Neuroscience 3: 1169-
1186

Sternberger LA, Joseph SA (1979) The unlabeled antibody method.
Contrasting color staining of paired pituitary hormones without
antibody removal. J Histochem Cytochem 27: 1424-1429

Substance P in the Carotid Body

R. D. Yates and I. Chen

Department of Anatomy, Tulane Medical School, 1430 Tulane Avenue, New Orleans, LA 70112, USA

Introduction

Several lines of evidence suggest that substance P (SP) may be a neurotransmitter/neuromodulator in baroreceptor and chemoreceptor afferent neurons. These include: (1) localization of SP in the neuronal perikarya of the petrosal and nodose ganglia, in some nerve fibers of the sinus and aortic nerves (Lundberg et al. 1978; Helke et al. 1980; Katz and Karten 1980) and in some axon terminals in the nucleus of the tractus solitarius (NTS) (Pickel et al. 1979; Voorn and Buijs 1983; Kalia et al. 1984; Maley 1985); (2) modulatory effects of SP after the application of the neuropeptide into the region of the NTS where central afferent processes of the glossopharyngeal and vagus terminate (Haeusler and Osterwalder 1980; Hedner et al. 1981; Henry and Sessle 1985); and, (3) depletion of SP in the NTS following denervation of the 9th and 10th cranial nerves (Gillis et al. 1980; Helke et al. 1980). These data, however, do not provide direct evidence that SP is present in the primary chemo- and baroreceptor afferent neurons. The presence of a few SP-immunoreactive (SP-I) nerve fibers in the carotid body have been demonstrated (Lundberg et al. 1979; Jacobowitz and Helke 1980; Cuello and McQueen 1980; Chen et al. 1984) but synaptic connections between SP-fibers and glomus cells which are apparently essential for chemoreception (Verna et al. 1975; Zapata et al. 1977; Eyzaguirre and Fidone 1980; Nishi et al. 1981; Monti-Bloch et al. 1983) have not been identified. Additionally, the functional roles of SP in the chemoreflex are still controversial (Reis et al. 1981; Furness et al. 1982; Mueller et al. 1982; Carter and Lightman 1983, 1985; Lorez et al. 1983). We have undertaken research to visualize SP-I structures in the carotid bodies of rats and cats at both the light and ultrastructural levels.

Materials and Methods

Adult rats (Sprague-Dawley) and adult cats which included both sexes were anesthetized with sodium pentobarbital (50 mg/kg body weight) and were perfused through the left ventricle with 4% paraformaldehyde in 0.1 M phosphate buffer (pH 7.2-7.3) for 30 min at room temperature. Carotid bodies were excised and fixed an additional 3 to 16 h in the perfusate at 4°C before being transferred to a cold phosphate buffer containing 8% sucrose (for 1 h) and 30% sucrose (overnight). The tissues were then frozen with acetone which had been precooled with dry ice, and sectioned in a cryostat at 40 um or hand-sliced following 2 cycles of freezing and thawing. Sections were preincubated in 10% normal goat serum (NGS) in phosphate buffered saline (PBS) with or without 0.1% Triton X-100 for 1 h at

Experimental Brain Research Series 16
© Springer-Verlag Berlin · Heidelberg 1987

25°C and then were immunocytochemically processed for SP using peroxidase-antiperoxidasee (PAP) or avidin-biotin-peroxidase complex (ABC). The rabbit anti-SP serum (Immunonuclear Corp., Stillwater, Minnesota) was diluted to 1:2000 (for PAP technique) or 1:5000 (for ABC technique) times with 1% NGS in PBS (staining times, 16 h at 25°C and 60 h at 4°C). For control staining either the normal rabbit serum or anti-SP serum preabsorbed with synthetic SP (100 ug/ml of original anti-SP serum; Sigma Chem. Co., St. Louis, Montana) were used in place of the anti-SP serum at the same dilutions. Following diaminobenzidine (DAB)-H_2O_2 incubation some of the sections were mounted on microscopic slides with a glycerine-PBS mixture and photographed. Other tissue slices were postosmicated and processed for electron microscopy. Sections either uncontrasted with a heavy metal or contrasted with lead citrate were examined in a JEOL 100B, JEOL 100CX or a Siemens 101 electron microscope.

Results

In 40 um frozen sections treated with the anti-SP serum, nerve fibers showing SP-like immunoreactivity occurred singly in the interstitial connective tissue of both rat and cat carotid bodies. Most of the SP-I fibers were associated with blood vessels. However, some appeared to lie around the periphery of groups of parenchymal cells. The number of SP-I fibers was much smaller than DBH-I ones in the carotid body (Chen et al. 1985). In cat carotid bodies some parenchymal cells showed moderate to strong SP-I. Intensely SP-I cells occurred singly or in small groups and often were provided with long, single processes. In rat carotid bodies, we were not able to detect SP-I consistently in parenchymal cells.

In control sections treated with the normal rabbit serum the follo-wing structures showed non-specific peroxidase activity: some blood cells, cytoplasmic inclusions, presumably lysosomes, and mast cells (only following ABC techniques; Bussolati and Gugliotta 1983). Pre-absorption of the antiserum with SP abolished staining of varicose nerve fibers.

Electron microscopy revealed that SP-I fibers were small unmyelina-ted ones. One or two profiles of SP-I fibers, if present, versus several SP-negative fibers in a common Schwann sheath was the most frequent arrangement. Occasionally, in both rat and cat carotid bodies, a myelinated fiber was seen accompanied by a group of unmye-linated ones containing an SP-I axon within a common perineurial sheath (Fig. 1). Small profiles of SP-I fibers contained mitochon-dria, microtubules and dense-cored vesicles of 85-140 nm in diame-ter. Large profiles of SP-I fibers often exhibited small clear synaptic vesicles (Fig. 1). The latter profiles probably represented varicose parts of SP-I fibers seen in light microscopy and often were incompletely surrounded by Schwann sheaths. The frequency of occurrence of dense-cored vesicles in SP-I fibers was higher in the cat than in the rat carotid bodies. None of the nerve terminals making synaptic contacts with glomus cells exhibited SP-I, although a few SP-I fibers were observed to be ensheathed in processes of supporting cells at the periphery of parenchymal cell groups in both rat and cat carotid bodies.

In cat carotid bodies the parenchymal cells which exhibited SP-I were glomus cells (Fig. 2). In general, those glomus cells showing intense staining were provided with numerous large dense-cored ve-sicles of around 140 nm in diameter and long processes containing similar vesicles. Immunoreactivity primarily resided in the dense-

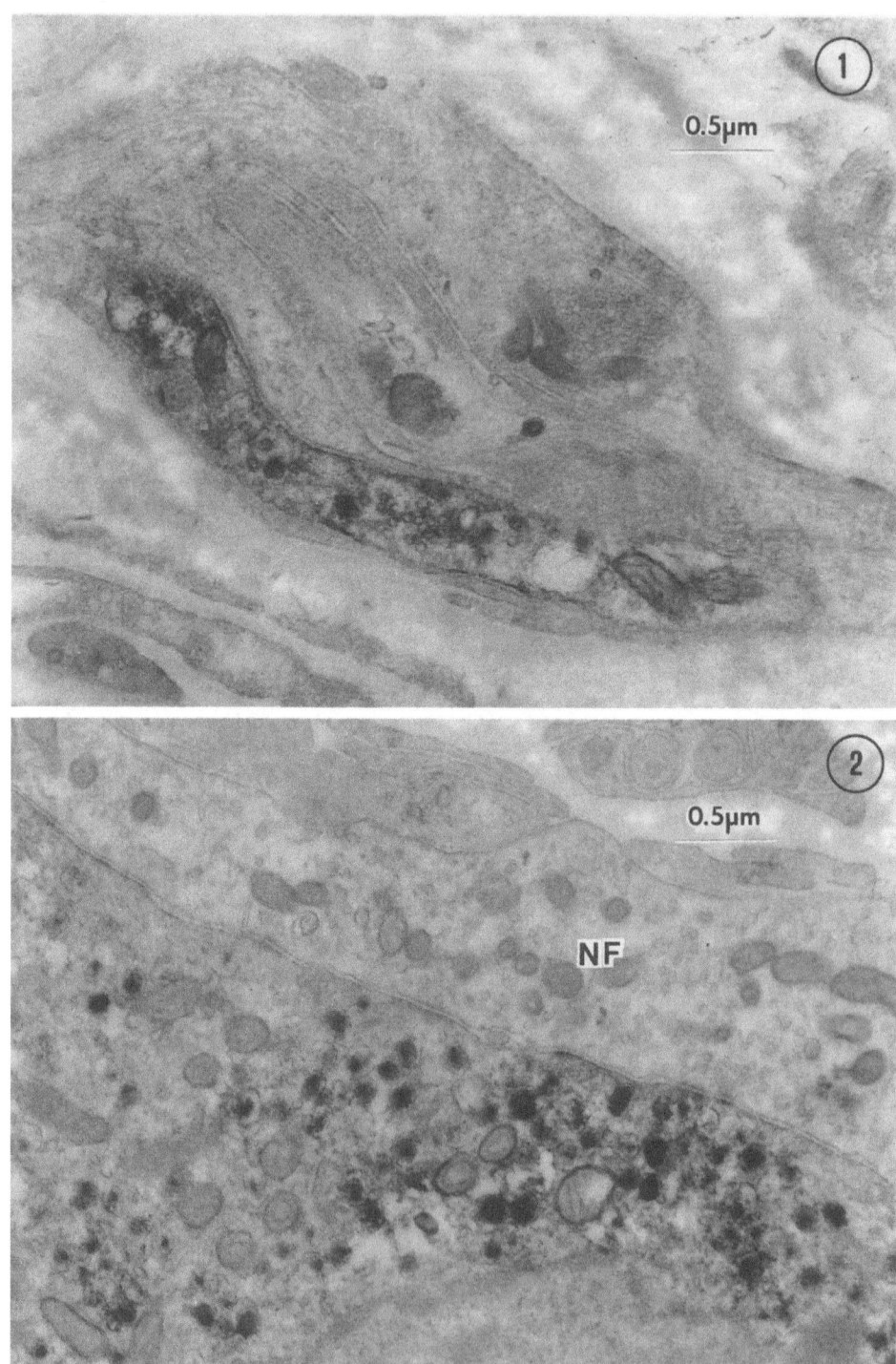

cores of vesicles and also varied from undetectable to strong even in the same glomus cells. No SP-I was detected in glomus cells which contained many dense-cored vesicles whose diameter was larger than 200 nm. SP-I glomus cells were often seen to make synaptic contact with nerve terminals containing numerous small clear vesicles and accumulations of such structures as well as SP-I vesicles on the glomus cell side at active zones of synapses were often noted (Fig. 2).

Discussion

Our data do not provide positive support for the idea of a neuro-transmitter or neuromodulator role of SP in chemoreceptor primary neurons although we do not completely rule out the possibility that the amount of SP in these afferents is below the sensitivity for detection with the techniques employed or, that SP synthesized in the perikarya is mainly transported toward the CNS.

The origin and function of small myelinated SP-I fibers in the carotid body are not clear at the present time. Using combined retrograde transport of macromolecules and SP immunocytochemistry, Hess (1981) concluded that SP-I fibers in the sinus nerve arise from a few aberrant wandering ganglion cells in the glossopharyngeal nerve. Another nerve supply to the carotid body is the superior cervical sympathetic source of SP-I fibers via the glomerular branch of the superior cervical sympathetic ganglion to the carotid body but ganglionectomy does not result in a significant reduction of the number of SP-I fibers in the carotid body (Hess 1981).

Our electron microscope studies confirm that all SP-I parenchymal cells are glomus cells in the cat carotid body. It is not clear if differences in SP-I represent different types of glomus cells or the same type under different physiological conditions. Although there are considerable differences in the size of vesicles, possibly largely attributable to different fixation and tissue processing, the intensely SP-I glomus cells in our studies fit into the type II category of major glomus cells. The dense-cored vesicles in these glomus cells also display intense DBH-immunoreactivity (Chen et al. 1985). Frequent occurrence of synaptic contacts between SP-I glomus cells and nerve terminals as well as accumulation of SP-I vesicles at active sites of synapses support neuromodulator roles of SP in chemoreceptor transduction in the cat glomus cells. Failure of consistent localization of SP-I in rat glomus cells in which the cell types and the mode of innervation are more thoroughly under-stood (McDonald and Mitchell 1975; Chen et al. 1984) than in cats, casts some doubt about the neuromodulator role of SP in the carotid body in general. Perhaps the SP content of the glomus cells under normal conditions is barely at or below the sensitivity of the immunocytochemical techniques employed. Perhaps hypoxia or hyperoxia will induce changes in the neuromodulator content of glomus cells and thereby assist us in determining the role of SP in carotid body activity.

◀ **Fig. 1.** Electron micrograph illustrating substance P immunoreactivi-ty in the terminal portion of a nerve fiber. Note dense-cored and clear vesicles

Fig. 2. Electron micrograph illustrating substance P immunoreactivi-ty in a glomus cell. Note substance P negative nerve fibers (NF) apposing the glomus cell

References

Bussolati G, Gugliotta P (1983) Nonspecific staining of mast cells by avidin-biotin peroxidase complexes (ABC). J Histochem Cytochem 12: 1419-1421

Carter DA, Lightman SL (1983) Substance P microinjections into the nucleus tractus solitarius elicit a pressor response in capsaicin-treated rats. Neurosci Lett 43: 253-257

Carter D, Lightman SL (1985) Cardio-respiratory actions of substance P, TRH and t-HT in the nucleus tractus solitarius of rats: evidence for functional interaction of neuropeptides and amine neurotransmitters. Neuropeptides 6: 425-436

Chen I, Hansen JT, Yates RD (1985) Dopamine-beta-hydroxylase-like immunoreactivity in the rat and cat carotid body: a light and electron microscopic study. J Neurocytol 14: 131-144

Chen I, Yates RD, Hansen JT (1984) Localization of substance P and dopamine-beta-hydroxylase-like immunoreactivities (SP-I and DBH-I) in the rat and cat carotid bodies. Anat Rec 208: 29A-30A

Cuello AC, McQueen DS (1980) Substance P: a carotid body peptide. Neurosci Lett 17: 215-219

Eyzaguirre C, Fidone SJ (1980) Transduction mechanisms in the carotid body: glomus cells, putative neurotransmitters and nerve endings. Amer J Physiol 239: C135-C152

Furness JB, Papka RE, Della NG, Costa M, Eskay RL (1982) Substance P-like immunoreactivity in nerves associated with the vascular system of guinea pigs. Neuroscience 7: 447-459

Gillis RA, Helke CJ, Hamilton BL, Norman WP, Jacobowitz DM (1980) Evidence that substance P is a neurotransmitter of baro- and chemoreceptor afferents in nucleus tractus solitarius. Brain Res 181: 476-481

Haeusler G, Osterwalder R (1980) Evidence suggesting a transmitter or neuromodulatory role for substance P at the first synapse of the baroreceptor reflex. Naunyn Schmiedeberg's Arch Pharmacol 314: 111-121

Hedner T, Hedner J, Jonason J, Lundberg D (1981) Evidence suggesting a role for substancee P in central respiratory regulation in the rat. Acta Physiol Scand 112: 487-489

Helke CJ, O'Donohue TL, Jacobowitz DM (1980) Substance P as a baro- and chemoreceptor afferent neurotransmitter: immunocytochemical and neurochemical evidence in the rat. Peptides 1: 1-9

Henry JL, Sessle BJ (1985) Effects of glutamate, substance P and eledoisin-related peptide on solitary tract neurons involved in respiration and respiratory reflexes. Neuroscience 14: 863

Hess A (1981) On the origin of substance P-positive fibers in the rat carotid body. Anat Rec 199: 114A

Kalia M, Fuxe K, Hökfelt T, Johansson O, Lang R, Ganten D, Cuello C, Terenius L (1984) Distribution of neuropeptide immunoreactive nerve terminals within the subnuclei of the nucleus of the tractus solitarius of the rat. J Comp Neurol 222: 409-444

Katz DM, Karten HJ (1980) Substance P in the vagal sensory ganglia: localization in cell bodies and pericellular arborizations. J Comp Neurol 193: 549-564

Lorez HP, Haeusler G, Aeppli L (1983) Substance P neurons in medullary baroreflex area and baroreflex function of capsaicin-treated rats. Comparison with other primary afferent systems. Neuroscience 8: 507-523

Lundberg JM, Hökfelt T, Nilsson G, Terenius L, Rehfeld J, Alde R, Said SP (1978) Peptide neurons in the vagus, splanchnic and sciatic nerves. Acta Physiol Scand 104: 499-501

Lundberg JM, Hökfelt T, Fahrenkrug J, Nilsson G, Terenius L (1979) Peptides in the cat carotid body (glomus caroticum): VIP-enkephalin and substance P-like immunoreactivity. Acta Physiol Scand 107: 279-281

Maley BE (1985) The ultrastructural localization of enkephalin and substance P immunoreactivities in the nucleus tractus solitarii of the cat. J Comp Neurol 233: 490-496

McDonald DM, Mitchell RA (1975) The innervation of glomus cells, ganglion cells and blood vessels in the rat carotid body: a quantitative ultrastructural analysis. J Neurocytol 4: 177-230

Monti-Bloch L, Stensaas LJ, Eyzaguirre C (1983) Effects of ischemia on the function and structure of the cat carotid body. Brain Res 270: 63-76

Mueller RA, Lundberg DBA, Breese GR, Hedner T, Jonason J (1982) The neuropharmacology of respiratory control. Pharmacol Rec 34: 255-285

Nishi K, Okajima Y, Ito H, Sugahara K (1981) Alteration of chemoreceptor responses and structural features of ischemic carotid body of the cat. Jap J Physiol 31: 677-694

Pickel VM, Joh TH, Reis DJ, Leeman SE, Miller RJ (1979) Electron micrsocopic localization of substance P and enkephalin in axon terminals related to dendrites of catecholaminergic neurons. Brain Res 160: 387-400

Reis DJ, Granata AR, Perrone MH, Talman WT (1981) Evidence that glutamic acid is the neurotransmitter of baroreceptor afferents terminating in the nucleus tractus solitarius (NTS). J Auton Nerv Syst 3: 321-334

Verna A, Roumy M, Leitner L-M (1975) Loss of chemoreceptor properties of the rabbit carotid body after destruction of the glomus cells. Brain Res 100: 13-23

Voorn P, Buijs RM (1983) An immuno-electron microscopical study comparing vasopressin, oxytocin, substance P and enkephalin containing nerve terminals in the nucleus of the solitary tract of the rat. Brain Res 270: 169-173

Zapata P, Stensaas LJ, Eyzaguirre C (1977) Recovery of chemosensory function of regenerating carotid nerve fibers. In: Acker H, Fidone S, Pallot D, Eyzaguirre C, Lübberg DW, Torrance RW (eds) Chemoreceptor in the carotid body. Springer, New York, pp 44-50

Calcitonin Gene-Related Peptide-Immunoreactive Nerve Fibers in Carotid Body and in Carotid Sinus

W. Kummer

Anatomisches Institut, Universität Heidelberg, Im Neuenheimer Feld 307, 6900 Heidelberg 1, FRG

Introduction

Calcitonin gene-related peptide (CGRP) has been found to be widely distributed throughout the central and peripheral nervous system (Rosenfeld et al. 1983; Lee et al. 1985; Skofitsch and Jacobowitz 1985). Radioimmunoassays revealed high CGRP concentrations at the carotid bifurcation compared with other vascular segments (Mulderry et al. 1985). The identification of these CGRP-containing elements was addressed in the present immunohistochemical study. In view of the frequent co-localization of substance P (SP) and CGRP immunoreactivity in peripheral neurones (Lee et al. 1985), parallel incubations were carried out using SP antisera.

Materials and Methods

Carotid bodies were obtained from untreated rats (n = 5), guinea pigs (n = 5), rabbits (n = 2), cats (n = 3) and dogs (beagle, n = 2). The origin of CGRP-immunoreactive (-IR) nerve fibers in the carotid sinus and carotid body was investigated in guinea pigs subjected to unilateral 1) transection of the cervical sympathetic trunk (n = 3), 2) removal of the superior cervical ganglion (n = 7), 3) transection of the carotid sinus nerve (CSN) close to its origin from the glossopharyngeal nerve (n = 8), 4) transection of the CSN in conjunction with removal of the superior cervical ganglion (n = 3), and 5) CSN transection in conjunction with removal of the nodose ganglion (n = 3). Survival times were varied from 7 to 13 days.

Sham-operated contralateral carotid bifurcations served as controls. All animals were anaesthetized and perfusion-fixed with acid-free Bouins fluid, and the specimens obtained were routinely embedded in paraplast and sectioned serially at 7 um. Primary antibodies were raised against synthetic rat CGRP (A: Peninsula Lab., UK; lot 000114; working dilution 1:4000; B: Amersham Buchler GmbH, FRG; 1:150), against human CGRP (Peninsula Lab., UK; code "neat"; 1:4000) and against SP (polyclonal antibody obtained from Merseyside Lab., UK; lot 201; 1:4000; and monoclonal antibody from Sera-Lab., UK; clone NC1/34 HL, lot B6E 35; 1:2000). Preabsorption tests with 20 ug of either rat CGRP or SP (synthetic peptides from Sigma, FRG) per ml diluted antiserum revealed specificity of the observed immunoreactions. Biotinylated second antibodies, streptavidin-biotin-peroxidase complex (both Amersham Buchler GmbH) and 3,3'-diaminobenzidine were used as detection system.

Experimental Brain Research Series 16
© Springer-Verlag Berlin · Heidelberg 1987

Results

The three CGRP antisera used gave identical results; no difference in the staining patterns obtained by the polyclonal and by the monoclonal SP antibody was observed.

Carotid Body

CGRP-IR fibers were found in carotid bodies of all species investigated, being particularly numerous in rabbit, cat and guinea pig. Many CGRP-IR varicosities showed a close association to small arteries and arterioles, whereas others encircled clusters of glomus cells (Fig. 1b). SP-IR fibers exhibited a similar distribution pattern, but generally appeared to be slightly less numerous. In contrast to CGRP, SP-IR was observed also in glomus cells of cat, and occasionally, of guinea pig.

Neither transection of the cervical sympathetic trunk nor removal of the superior cervical ganglion altered the amount and distribution of CGRP- and SP-IR fibers in the guinea pig carotid body (Fig. 1c). Inconsistent results were obtained following CSN transection: in most cases, persisting fibers can be seen only occasionally, but their amount showed a great interindividual variability and one case revealed an apparently normal innervation density. Additional excision of the superior cervical ganglion still left a population of CGRP- and SP-IR fibers in the guinea pig carotid body unaffected. Combined resection of the nodose ganglion and CSN transection, however, resulted in a complete loss of both CGRP- and SP-IR nerve fibers in the carotid body (Fig. 1d).

Carotid Sinus

In the guinea pig, the carotid sinus and adjacent arteries were enmashed by a dense adventitial plexus of CGRP-IR fibers (Fig. 1a). Occasionally, delicate fibers and varicosities penetrated the media. SP-IR fibers were restricted to the adventitial layer and were not as numerous as CGRP-IR fibers. Following denervation experiments, the density of SP- and CGRP-IR nerve fibers in the carotid sinus paralleled the innervation density of the carotid body, resulting in a lack of both SP- and CGRP-IR vascular nerve fibers in response to CSN transection plus excision of the nodose ganglion. Immunoreactive fibers along the proximal segments of the common carotid artery were not affected by any of the denervation procedures.

Discussion

The present observations on the distribution of SP-IR elements in the carotid body and carotid sinus are in agreement with previous reports (Cuello and McQueen 1980; Jacobowitz and Helke 1980; Gorgas et al. 1983). The frequent co-localization of SP- and CGRP immunoreactivities in sensory neurones (Lee et al. 1985) suggests coexistence of these peptides in nerve fibers of the carotid body and carotid sinus, however, this hypothesis needs ultrastructural confirmation.

As concluded from the loss of immunoreactive fibers induced by denervation, both CGRP- and SP-IR fibers in the guinea pig carotid body and carotid sinus approach their targets via the CSN, and, to an interindividually variable extent, via vagal branches. Tracer and

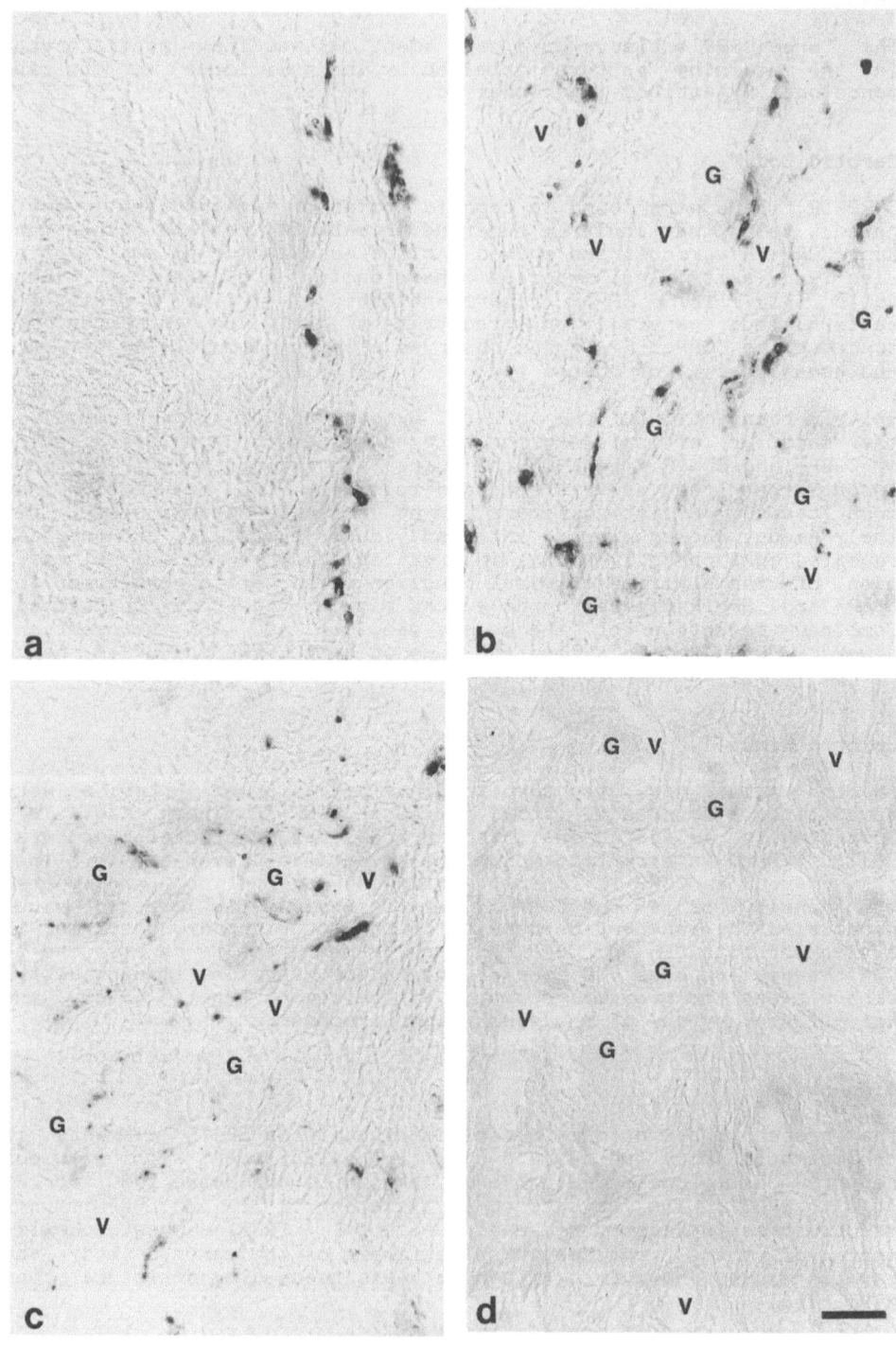

degeneration experiments have shown that these nerves in all mammalian species so far investigated, contain primary afferent fibers as
well as a minor population of efferent fibers being preganglionic to
multipolar neurones in the periphery of the carotid body (e.g.,
McDonald and Mitchell 1975; Smith and Mills 1976). Contacts of
either SP- or CGRP-IR fibers with neurones of the carotid body were
never observed in this study; such neurones were absent in the
guinea pig. Thus, it can be inferred that CGRP- and SP-IR fibers of
the guinea pig carotid body and carotid sinus originate from glossopharyngeal and vagal primary afferent neurones.

There is experimental evidence for an influence of peripherally
released SP on carotid chemoreceptor discharge (Prabhakar et al.
1984), but the sensory modality (e.g., nociception, presso- or chemoreception) transduced by carotid SP-IR afferents is still a
matter of dispute (see Helke et al. 1980; Furness et al. 1982).
Specific effects of CGRP on carotid presso- and chemoreception as
yet have not been investigated. Its involvement in pressoreception
is questionable, because the delicate appearance of CGRP-IR fibers
in the tunica media does not correlate with the coarse shape of
ultrastructurally characterized pressoreceptors (see Gorgas et al.
1983). Considering the potent vasodilatatory properties of CGRP
(Marshall et al. 1986) and the frequent association of CGRP-IR nerve
fibers with arterial blood vessels of the carotid body, this peptide
may play an important role in controlling blood flow through this
arterial chemoreceptor organ.

Acknowledgements. The skilful technical and editorial assistance of
Ms. B. Schoell is greatly acknowledged. This study was supported by
the DFG (grant He 919/6-1).

References

Cuello AC, McQueen DS (1980) Substance P: a carotid body peptide.
Neurosci Lett 17: 215-219

Furness JB, Elliott JM, Murphy R, Costa M, Chalmers JP (1982) Baroreceptor reflexes in conscious guinea-pigs are unaffected by depletion of cardiovascular substance P nerves. Neurosci Lett 32: 285-290

Gorgas K, Reinecke M, Weihe E, Forssmann WG (1983) Neurotensin and
substance P immunoreactive nerve endings in the guinea pig carotid
sinus and their ultrastructural counterparts. Anat Embryol 167: 347-
354

Helke CJ, O'Donohue TL, Jacobowitz DM (1980) Substance P as a baro-
and chemoreceptor afferent neurotransmitter: immunocytochemical and
neurochemical evidence in the rat. Peptides 1: 1-9

Jacobowitz DM, Helke CJ (1980) Localization of substance P immunoreactive nerves in the carotid body. Brain Res Bull 5: 195-197

◄ **Fig. 1.** CGRP-IR fibers in the adventitial layer of the guinea pig
carotid sinus (a) and in the guinea pig carotid body (b-d). CGRP-IR
fibers approach blood vessels (V) as well as glomus cell clusters
(G) in both control (b) and sympathectomized (c) carotid bodies, but
are absent following CSN transection plus excision of the nodose
ganglion (d). Bar = 20 um

Lee Y, Takami K, Kawai Y, Girgis S, Hillyard CJ, MacIntyre I, Emson PC, Tohyama M (1985) Distribution of calcitonin gene-related peptide in the rat peripheral nervous system with reference to its coexistence with substance P. Neuroscience 15: 1227-1237

Marshall I, Al-Kazwini S, Holman JJ, Craig RK (1986) Human and rat alpha-CGRP but not calcitonin cause mesenteric vasodilatation in rats. Eur J Pharmacol 123: 217-222

McDonald DM, Mitchell RA (1975) The innervation of glomus cells, ganglion cells and blood vessels in the rat carotid body: a quantitative ultrastructural analysis. J Neurocytol 4: 177-230

Mulderry PK, Ghatei MA, Rodrigo J, Allen JM, Rosenfeld MG, Polak JM, Bloom SR (1985) Calcitonin gene-related peptide in cariovascular tissues of the rat. Neuroscience 14: 947-954

Prabhakar NR, Runold M, Yamamoto Y, Lagercrantz H, von Euler C (1984) Effect of substance P antagonist on the hypoxia-induced carotid chemoreceptor activity. Acta Physiol Scand 121: 301-303

Rosenfeld MG, Mermod JJ, Amara SG, Swanson LW, Sawchenko PE, Rivier J, Vale WW, Evans RM (1983) Production of a novel neuropeptide encoded by the calcitonin gene via tissue specific RNA processing. Nature (Lond) 304: 129-135

Skofitsch G, Jacobowitz DM (1985) Calcitonin gene-related peptide: detailed immunohistochemical distribution in the central nervous system. Peptides 6: 721-745

Smith PG, Mills E (1976) Autoradiographic identification of the terminations of petrosal ganglion neurons in the cat carotid body. Brain Res 113: 174-178

Fluorescence Histochemical and Microspectrofluorometric Study of Three Different Catecholamines in Rat Uterus and Paracervical Ganglion

Ch. Owman[1], O. Alm[2], N.-O. Sjöberg[3], and M. Stjernquist[3]

[1] Department of Histology, University of Lund, Biskopsgatan 5, 22362 Lund, Sweden
[2] Department of Pathology, University of Lund, Sölvegatan 25, 22362 Lund, Sweden
[3] Department of Obstetrics and Gynecology, University Hospital, 22185 Lund, Sweden

Introduction

In a recent fluorometric investigation of catecholamines in the rat uterus (Sjöberg et al. 1987) it has been established that the uterine proper contains noradrenaline which, like in other animals (Owman et al. 1986a), agrees with the adrenergic innervation of the organ. Utero-vaginal tissue, in addition, stores substantial quantities of dopamine. Significant amounts of adrenaline have also been reported to be present in the rat uterus (see Sjöberg et al. 1987 for references).

The functional role of the endogenous catecholamines in the rat uterus have been difficult to evaluate because their storage sites have been unclear. The uterus of this animal is a frequently used organ in studies on the importance of adrenergic mechanisms in reproductive functions. The present investigation was therefore undertaken in an attempt to define the localization of the three catecholamines present in the rat uterus and its paracervical ganglion formations using a combination of formaldehyde and glyoxylic acid histochemistry in combination with cytospectrofluorometry.

Material and Methods

The study included adult female Sprague-Dawley rats, weighing 200–250 g and maintained on a standard pellet diet and tap water ad lib. All animals were killed by decapitation under light ether anesthesia. Vaginal smears were stained for determination of the cyclic phase.

Ten untreated animals from each of the proestrus, estrus, and diestrus stage were killed within 30 min after vaginal smears had been taken. Another group of 5 animals (irrespective of cyclic phase) received 100 mg/kg i.p. of L-3,4-dihydroxyphenylalanine (L-DOPA) 30 min before sacrifice and a further 5 animals (irrespective of phase) were given 40 mg/kg i.v. of DL-dihydroxyphenylserine (DL-DOPS) 1 h before sacrifice.

Half of the untreated animals and all of those receiving amine precursors were used for formaldehyde histochemistry (Björklund et al. 1970b) applied to pieces from several parts of the uterine horns, and a large preparation from the utero-vaginal junction including adjacent connective tissue. After freeze-drying, the specimens were exposed to formaldehyde gas at +80°C for 1 h (specimens from injected animals) and for either 1 or 3 h (specimens from untreated animals) for histochemical demonstration of catechola-

Experimental Brain Research Series 16
© Springer-Verlag Berlin · Heidelberg 1987

mines. After embedding in paraffin, 6 u thick sections were prepared and mounted in either Entellan (Merck) containing a small amount of xylene, or in liquid paraffin, for fluorescence microscopic examination. The other half of the untreated animals was processed for glyoxylic acid histochemistry (Lindvall and Björklund 1974).

Preparations from the utero-vaginal junction with adjacent connective tissue including paracervical ganglion formations were taken from 5 untreated animals irrespective of cyclic phase. The specimens were freeze-drieed and prepared according to the formaldehyde method (reaction at 1 h, +80°C), embedded and sectioned, all as above. The sections were mounted on coverslips and deparaffinized in xylene. Cytospectrofluorometric analysis was carried out with a modified Leitz microspectrograph (Björklund et al. 1968). The sections were analysed before and after stepwise acidification in HCl fumes in a closed vessel for 30 sec and 3 min (Björklund et al. 1968).

Results

Fluorescence Histochemistry

The most prominent feature of the rat uterus processed according to the Falck-Hillarp formaldehyde technique was a rich vascular supply of adrenergic nerve terminals. Only fairly few adrenergic nerve terminals could be observed in the smooth muscle (Fig. 1). The fibres followed the course of the muscle cells. Isolated nerve terminals could also be seen to run for a short distance into the mucosal layer, accompanying small blood vessels (Fig. 1). The pattern of innervation was similar in the uterine horns and in the cervix.

Fig. 1. Fluorescence photomicrograph of rat uterine horn, transverse ▶ section. Large number of green-fluorescent adrenergic nerves around blood vessels running between the outer longitudinal and inner circular smooth muscle layers. Few delicate varicose terminals (arrows) in the circular smooth muscle (M) and fibres (arrow) accompanying a blood vessel in the outer part of the mucosa (m). Bar indicates 50 um

Fig. 2. Large clusters of small, polymorph cells (SIF-cells) with an intense fluorescence located in the otherwise non-fluorescent utero-vaginal ganglia. Bar indicates 50 um

Fig. 3. (a) Animal pretreated with amine precursor (L-DOPA 100 mg/kg, 30 min). Groups of nerve cells in the utero-vaginal ganglion. Part of the ganglion cells are non-fluorescent, whereas others exhibit a moderate green fluorescence; in untreated animals, all ganglion cells are non-fluorescent. Clusters of intensely fluorescent, chromaffin-like cells (SIF-cells) are seen. Bar indicates 50 um. (b) The same picture is obtained after administration of another amine precursor (DL-DOPS 40 mg/kg, 1 h). Higher magnification of a group of ganglion cells. Bar indicates 50 um

Fig. 4. Fluorescence photomicrograph of transversally sectioned rat uterus processed according to the glyoxylic acid method. The myometrium proper contains a larger number of fluorescent adrenergic nerve terminals than visible with the formaldehyde reaction (cf. Fig. 1). Also perivascular nerve fibres are seen. Bar indicates 50 um

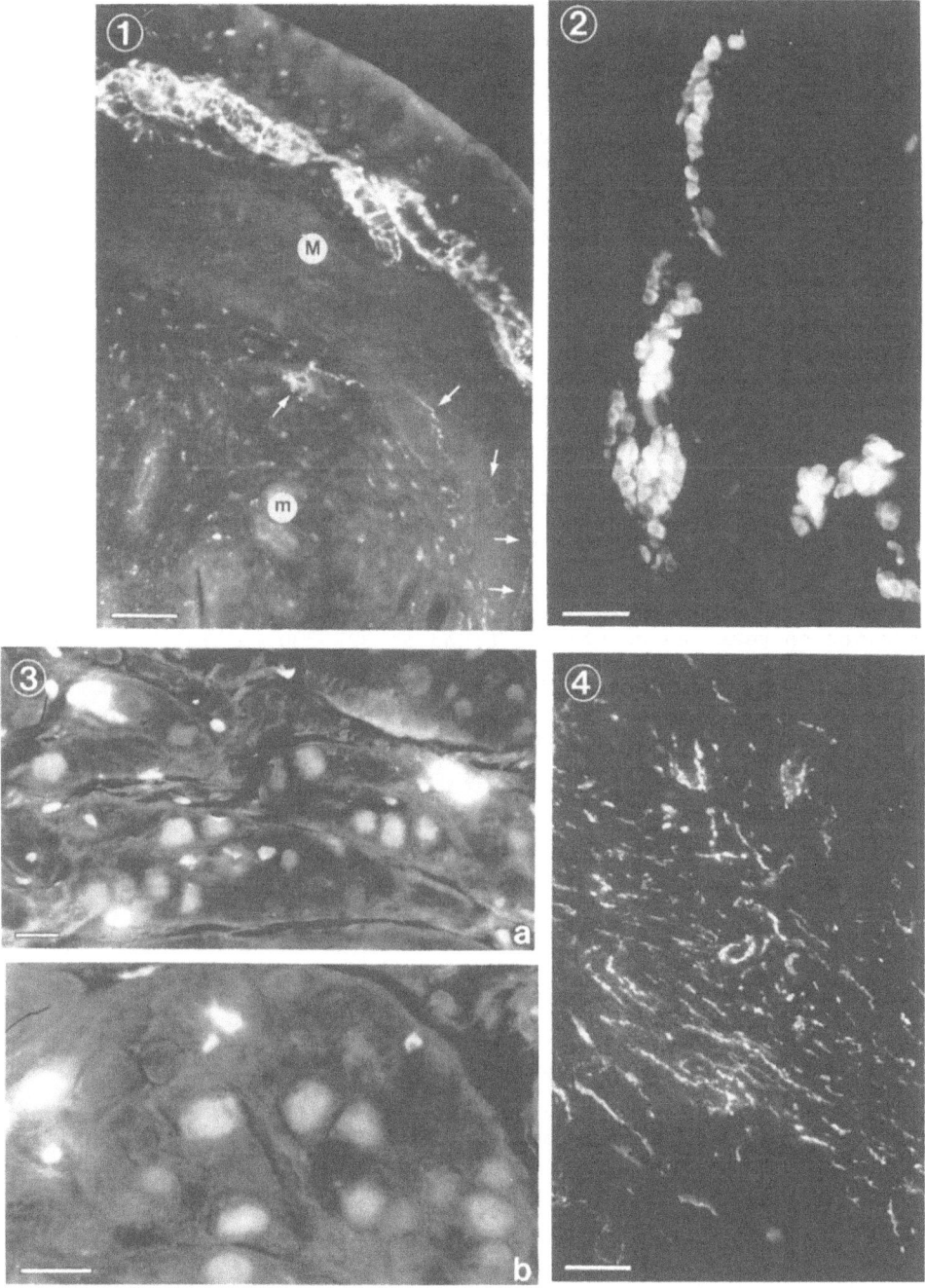

No overt difference was found in the number or intensity of the nerves at various cyclic phases of the animal. Prolonged treatment with formaldehyde gas did not alter the intensity of the fluorophore. It was the same whether the sections were mounted in Entellan or in liquid paraffin.

Large ganglion formations were located laterally and posterolaterally in the adventitial tissue at the utero-vaginal junction. The ganglion cells exhibited no formaldehyde-induced fluorescence; they were visible only through their slight autofluorescence. The ganglia also contained a large number of small, branching, polymorph cells ("SIF-cells"; Eränkö et al. 1980) emitting an intense green to green-yellow, formaldehyde-induced fluorescence (Fig. 2 and 3).

After the administration of DOPA or DOPS, the number and intensity of uterine nerve terminals was somewhat increased. A similar number of fluorescent nerve fibres were visible in (untreated) animals processed according to the glyoxylic acid method. The increased number of fibres was particularly evident in the myometrium proper, where they ran in the same direction as the smooth muscle cells (Fig. 4). About half of the previously non-fluorescent ganglion cells attained a green cytoplasmic fluorescence of moderate intensity following treatment with the amine precursors (Fig. 3). The fluorescence of the SIF-cells appeared unchanged.

The excitation/emission maxima of the paracervical SIF-cells as well as of the uterine nerves were at 410/475 mu. A tendency to a second excitation peak was constantly found at 325-330 mu (Fig. 5). After exposure to HCl vapours for 30 sec, the excitation maximum in all structures measured was shifted to 370 mu with unchanged fluorescence intensity. After further acidification, up to 3 min, the initially developed excitation peak at 370 mu was markedly reduced, concomitant with a decrease in the fluorescence intensity. Simultaneously, an increase in the excitation curve at about 330 mu occurred, both in the nerves (Fig. 5a) and in the majority of SIF-cells (Fig. 5b). In some of these cells, however, the intensity of the emitted light was essentially unchanged upon prolonged acidification, and the excitation maximum remained at 370 mu (Fig. 5c).

Discussion

Adrenergic nerves in the rat uterus could be visualized through their fluorescence induced by either formaldehyde or glyoxylic acid. The nerve terminals showing up in the former reaction were distributed mainly around blood vessels, whereas with glyoxylic acid more fibres became visible in the myometrium proper. The fluorophore had excitation and emission spectra characteristic for catecholamines, behaving like a primary catecholamine (Björklund et al. 1970b). This is in agreement with the recent confirmation that noradrenaline is the only catecholamine present in the uterus proper (Sjöberg et al. 1987). Accordingly, the fluorescent nerves showed the typical microspectrofluorometric features of noradrenaline upon acidification (Björklund et al. 1968). The cyclic variations in the number and fluorescence intensity of these nerves described by Adham and Schenk (1969) could not be confirmed. In line with those findings uterine noradrenaline in rat is uninfluenced by oophorectomy or treatment with sex steroids, which markedly affect the levels in the guinea-pig myometrium (Falck et al. 1974).

It seems likely that nerves storing catecholamines in amounts small enough to escape microscopic detection with the formaldehyde reac-

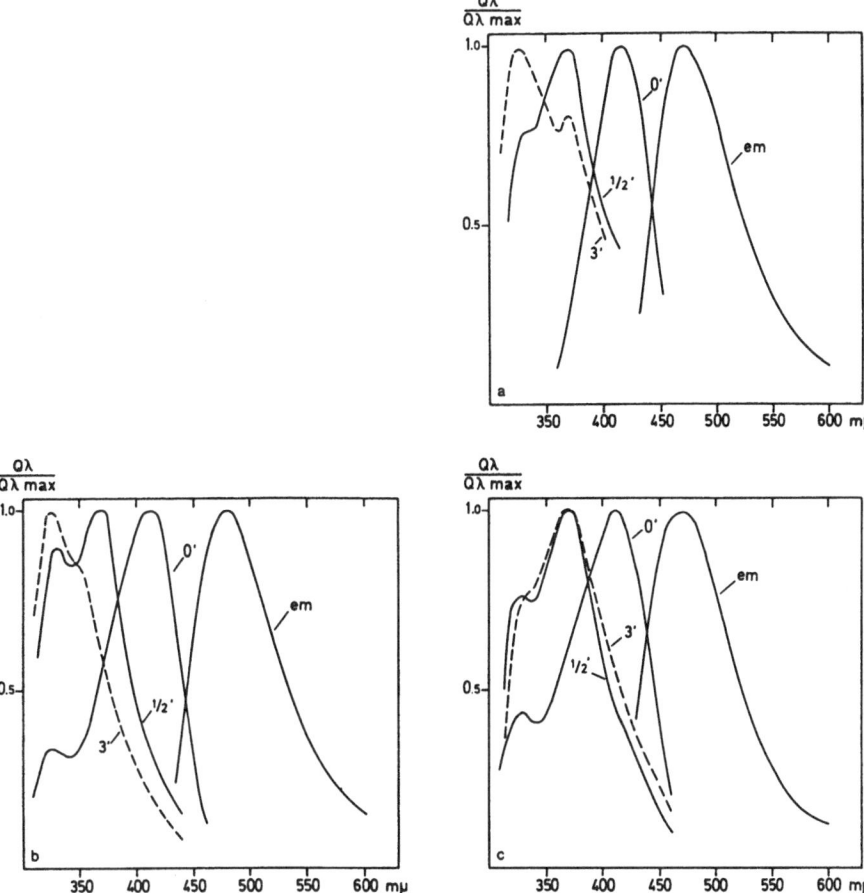

Fig. 5. Fluorescence emission (right curve) and excitation spectra after formaldehyde treatment only (0'), after combined exposure to HCl vapour at room temperature for 1/2 min, and after an additional acidification for 2 1/2 min. Recordings from (a) uterine cervical nerve terminals, (b) chromaffin—like cells (SIF—cells) with a yellow—green fluorescence, and (c) SIF—cells with a more reddish fluorescence.

tion was present in the myometrium. Thus, administration of DOPA or DOPS increased the number of fluorescent nerves, corresponding to the density of the nerve plexus visible with the more sensitive glyoxylic acid method.

Ganglion formations simultaneously containing both adrenergic and non-adrenergic nerve cells have constantly been found in the utero-vaginal junction of various species (for references, see Owman et al. 1983). There is much evidence that these ganglia are the site of origin for the short adrenergic neurons innervating the female reproductive tract (Owman et al. 1986a). Adrenergic nerve cells were not visible with the formaldehyde method in the corresponding ganglion formations in the rat. However, administration of the catechol-amine precursors, DOPA and DOPS, resulted in the appearance of a population of nerve cells in the ganglia with a green fluorescence.

It is well known that the catecholamine concentration in adrenergic nerve cell bodies is only about one thousandth to one hundreth of the concentration in the terminal varicosities (Dahlström 1966). It is thus conceivable that part of the nerve cells also in the rat utero-vaginal ganglia are adrenergic but normally contain too low concentrations of the amine to allow for its histochemical detection.

Small, polymorph branching cells with an intense formaldehyde-induced fluorescence (SIF-cells) have repeatedly been found at different sites, particularly in relation to autonomic ganglia, in various animal species. In the cells with a more reddish fluorescence it constantly showed a shift in the excitation maximum from 410 mu to 370 mu upon treatment with HCl, a maximum that remained after further acidification. This shift is characteristic for the form-aldehyde-induced dopamine fluorescence (Björklund et al. 1968) and due to an acid-induced change of the fluorophore (6,7-dihydroxy-3,4-dihydroisoquinoline) from its quinoidal into its tautomeric non-quinoidal form (Corrodi and Hillarp 1964; Jonsson 1966). We thus believe that the dopamine chemically found in the utero-vaginal junction (Sjöberg et al. 1987) is stored in the SIF-cells which exhibit a strong formaldehyde-induced fluorescence of reddish colour under the fluorescence microscopic conditions used. Similar dopamine-containing cells have been found in the sympathetic chain of rats, cats, and pigs (Björklund et al. 1970a).

The noradrenaline fluorophore will, in contrast to the situation for the dopamine fluorophore, only transiently occur in its non-quinoidal form (excitation maximum 370 mu) and is rapidly converted upon prolonged HCl exposure to the fully aromatic isoquinoline (excitation maximum 330 mu) (Björklund et al. 1968). Since a labile hydroxy group in 4-position is also present in the adrenaline fluorophore, it can be expected that a similar acid-induced shift will occur with this fluorophore, too. This was observed in the fluorescence of the more yellow-green population of SIF-cells in the utero-vaginal ganglion.

It is difficult to establish whether the fluorophore in this population of SIF-cells represents noradrenaline or adrenaline - the presence of major quantities of dopamine seems to be excluded. The noradrenaline chemically measured in the utero-vaginal junction (Sjöberg et al. 1987) can be fully accounted for by the adrenergic nerves present in this region. Histochemical and chemical studies (Owman and Sjöstrand 1965; Sjöstrand 1965) have indicated that the SIF-cells in the male genital tract of certain species store large amounts of adrenaline. It is conceivable that the population of chromaffin-like cells with a yellow-green fluorescence found in the ganglia contain adrenaline.

Acknowledgement. Supported by grant No. 14X-5680 from the Swedish Medical Research Foundation.

References

Adham N, Schenk EA (1969) Autonomic innervation of the rat vagina, cervix and uterus and its cyclic variation. Am J Obstet Gynecol 104: 508-516

Björklund A, Ehinger B, Falck B (1968) A method for differentiating dopamine from noradrenaline in tissue sections by microspectrofluorometry. J Histochem Cytochem 16: 263-270

Björklund A, Cegrell L, Falck B, Ritzèn M, Rosengren E (1970a) Dopamine-containing cells in sympathetic ganglia. Acta Physiol Scand 78: 334-338

Björklund A, Falck B, Owman Ch (1970b) Fluorescence microscopic and microspectrofluorometric techniques for the cellular localization and characterization of biogenic amines. In: Berson S, Yalow R, Dorfman R, Rall E, Kopin IJ (eds) Methods in investigative and diagnostic endocrinology. Elsevier, Amsterdam, pp

Corrodi H, Hillarp N-A (1964) Fluoreszenzmethoden zur histochemischen Sichtbarmachung von Monoaminen. 2. Identifizierung des fluoreszierenden Produktes aus Dopamin und Formaldehyd. Helv Chim Acta 47: 911-918

Dahlström A (1966) The intraneuronal distribution of noradrenaline and the transport and life-span of amine storage granules in the sympathetic adrenergic neuron. A histochemical and biochemical study. Ivar Haeggström, Stockholm

Eränkö O, Soinila S, Päivärinta H (eds) (1980) Histochemistry and cell biology of autonomic neurons, SIF cells and paraneurons. Raven, New York

Falck B, Gardmark S, Nybell G, Owman Ch, Rosengren E, Sjöberg N-O (1974) Ovarian influence on the content of norepinephrine transmitter in guinea pig and rat uterus. Endocrinology 94: 1475-1479

Jonsson G (1966) Fluorescence studies on some 6,7-substituted 3,4-dihydroisoquinolines formed 3-hydroxytyramine (dopamine) and formaldehyde. Acta Chem Scand 20: 2755-2762

Lindvall O, Björklund A (1974) The glyoxylic acid fluorescence histochemical method: a detailed account of the methodology for the visualization of central catecholamine neurons. Histochemistry 39: 97-127

Owman Ch, Alm P, Sjöberg N-O (1983) Pelvic autonomic ganglia: structure, transmitters, function and steroid influence. In: Elfvin L-G (ed) Autonomic ganglia. Wiley, Chichester, pp 125-143

Owman Ch, Alm P, Sjöberg N-O, Stjernquist M (1986a) Structural, biochemical, and pharmacological aspects of the uterine autonomic innervation and its remodeling during pregnancy. In: Huszar G (ed) The physiology and biochemistry of the uterus in pregnancy and labor. CRC Press, Boca Raton, Florida, pp 5-23

Owman Ch, Sjöstrand NO (1965) Short adrenergic neurons and catecholamine-containing cells in the vas deferens and accessory male genital glands of different mammals. Z Zellforsch 66: 300-320

Sjöberg N-O, Owman Ch, Rosengren E, Stjernquist M, Swedin G (1987) Chemical analysis of catecholamines in uterine tissue of rat: evidence for the presence of noradrenaline as well as dopamine and adrenaline. Exp Brain Res (in press)

Sjöstrand NO (1965) The adrenergic innervation of the vas deferens and the accessory male genital glands. Acta Physiol Scand 65, Suppl 257: 1-82

Chemical Analysis of Catecholamines in Uterine Tissue of Rat: Evidence for the Presence of Noradrenaline as well as Dopamine and Adrenaline

N.-O. Sjöberg[1], Ch. Owman[2], E. Rosengren[3], M. Stjernquist[1], and G. Swedin[4]

[1] Department of Obstetrics and Gynecology, University Hospital, 22185 Lund, Sweden
[2] Department of Histology, University of Lund, Biskopsgatan 5, 22362 Lund, Sweden
[3] Department of Pharmacology, University of Lund, Sölvegatan 10, 22362 Lund, Sweden
[4] Department of Physiology I, Karolinska Institutet, P.O. Box 60400, 10401 Stockholm, Sweden

Introduction

The uterus of several animal species, including man, receives a well-developed adrenergic nerve supply (see Owman et al. 1986a for references). Fluorometric determinations have shown that the transmitter substance in the uterine adrenergic nerves is noradrenaline; neither adrenaline nor dopamine has been found in measurable amounts in the tissues (Sjöberg 1967). From denervation experiments it has been established that the adrenergic innervation of the myometrium belongs to the system of short adrenergic neurons – in part or entirely, depending on species (Owman et al. 1966; Rosengren and Sjöberg 1967; Thorbert et al. 1977), originating in paracervical ganglia located in the utero-vaginal junction (Owman et al. 1983).

The rat uterus appears to be an exception in that the smooth muscle layers are less well innervated by adrenergic nerve terminals, most of which supply the vascular system (Adham and Schenk 1969; Falck et al. 1974). Besides noradrenaline, significant quantities of adrenaline (29–229 mug/g tissue or 17–90 mug/organ) have been reported to be present in the uterus (Rudzik and Miller 1962; Wurtman et al. 1963a,b; Cha et al. 1965; Klingman 1965; Spratto and Miller 1968). Finally, it has been shown that the utero-vaginal tissue contains small amounts of dopamine (Swedin and Brundin 1968).

The present study was undertaken to provide extensive re-evaluation of the distribution and identity of catecholamines in rat uterine tissue. Fluorometric determinations suggest that noradrenaline is the only catecholamine present in the uterus proper, but that both dopamine and adrenaline may be stored in utero-vaginal tissue in which the paracervical ganglia are located. The further detailed localization of uterine catecholamines has been followed up in a separate fluorescence histochemical and fluorometric investigation (Owman et al. 1987).

Material and Methods

The study included adult female Sprague-Dawley rats, weighing 200–250 g and maintained on a standard pellet diet and tap water ad lib. All animals were killed by decapitation under light ether anesthesia. Uterine catecholamines were determined by two techniques in two separate laboratories.

(1) The method of Bertler et al. (1958) was used with the modifications of Häggendal (1963). The cyclic phase was determined in vaginal smears and 5 animals were killed in each of the proestrus,

estrus, and diestrus stage within 30 min after the smears were taken. The entire uterus, carefully freed from connective tissue, was taken for determination (1 organ per determination). Uteri (dissected as previously described) from another 6 animals in each of the three cyclic stages mentioned were pooled for separate determination in each stage. Tissues consisting of the utero—vaginal junction including adjacent connective tissue from 5 animals (irrespective of cyclic phase) were pooled for one determination. The supernatant of a centrifuged tissue homogenate was neutralized with K_2CO_3 to pH 6.0 and the precipitate of $KClO_4$ spun down. The resulting supernatant was then put on a column containing Dowex 50, and the catecholamines were eluted with 1.0 N HCl. After neutralization of the eluate the amines were oxidized with $K_3Fe(CN)_6$ to their aminochromes, which were then transformed into their highly fluorescent lutins by means of 10 N NaOH containing BAL to prevent further oxidation. The fluorescence was read in an Aminco—Bowman spectrophotofluorometer.

(2) Twenty animals irrespective of cyclic phase were used in the second series of determinations. The uterine horns from the bifurcation to 5 mm from the tubo—uterine junctions (8 animals) were pooled for 2 determinations. Samples from the uterine horns were pooled together (2 animals) for 2 recovery studies. The utero—vaginal junctions from 20 animals (cyclic phase not determined) were pooled together for one determination. The supernatants of homogenized and centrifuged tissues were adjusted to pH 3.8 with 2.0 N KOH and recentrifuged. After addition of 3 ml of 1 M Tris—buffer (pH 8.5) the supernatants were passed through alumina columns (0.6 g of Al_2O_3). The catecholamines were eluted with 10 ml of 0.25 N acetic acid. Oxidation was performed with iodine according to Chang (1964) and the fluorescence was read in an Aminco—Bowman spectrophotofluorometer.

Results and Discussion

Neither of the flurometric methods used revealed the presence of any adrenaline or dopamine in the uterus, even though organs from 6—8 animals were pooled together. The amount of noradrenaline present is evident from Tables 1 and 2. The level of this catecholamine did not significantly change with the cyclic phase of the animal (Table 1).

Table 1. Concentration of catecholamines in rat uterus at various stages during the estrus cycle. Neither adrenaline nor dopamine occurred in measurable amounts. Differences between mean noradrenaline levels were evaluated by Student's t—test. Because of the uterine weight changes occurring during the estrus cycle, only comparison between the total noradrenaline contents are relevant

| Phase | Single organs from 5 animals | | | Pooled organs from 6 animals | |
	Weight (g)	Total content (µg)	Concentration (µg/g)	Total content (µg)	Concentration (µg/g)
Proestrus	0.31 ± 0.03	0.05 ± 0.01	0.18 ± 0.02	0.51	0.19
Estrus	0.35 ± 0.03	0.06 ± 0.01	0.17 ± 0.02	0.43	0.19
Diestrus	0.26 ± 0.04	0.07 ± 0.01	0.26 ± 0.04	0.54	0.27

Table 2. Concentration and recovery of adrenaline (A) and noradrenaline (NA) in uterine horns taken from rats irrespective of cyclic phase

No. of determinations	Weight of uterine horns (g)	Added catecholamine	Adrenaline			Noradrenaline		
			Total (µg)	µg/g	Recovery (%)	Total (µg)	µg/g	Recovery (%)
2	0.63	0.20 µg NA	0	0	-	0.25	-	95
	0.63	-	0	0	-	0.06	0.18	-
2	0.59	0.20 µg A	0.16	-	80	0.07	0.23	-
	0.59	-	0	0	-	0.07	0.23	-
8	2.15	-	0	0	-	0.37	0.17	-
8	2.10	-	0	0	-	0.49	0.23	-

Tissues pooled from the utero-vaginal junction contained considerable amounts of both dopamine (0.27 ug/g) and noradrenaline (0.24 ug/g), whereas only little adrenaline (0.09 ug/g) was found.

There is, hence, no evidence that catecholamines other than noradrenaline (see Owman et al. 1987) is stored in the uterine adrenergic nerves of the rat. The variation during the estrous cycle in the number and fluorescence intensity of these nerves reported by Adham and Schenk (1969) could not be corroborated. Such variation has been observed in other species of animals with a more well-developed myometrial adrenergic nerve supply (Owman et al. 1986a). In accordance with these observations, uterine noradrenaline in rat is uninfluenced by oophorectomy or treatment with sex steroids, which markedly affect the levels in the myometrium of the guinea-pig (Falck et al. 1974).

Small, polymorph branching cells with an intense formaldehyde-induced fluorescence (SIF-cells; Eränkö et al. 1980) have repeatedly been found at different sites, particularly in relation to autonomic ganglia, in various animal species. In microspectrofluorometric analyses (Owman et al. 1987) the flurophore has turned out to be typical for that of catecholamines in tissues. In one population of SIF cells with a more reddish formaldehyde-induced fluorescence there is reason to believe (Björklund et al. 1968) that the catecholamine present is dopamine, corresponding to the dopamine measured chemically in the utero-vaginal junction. The existence of similar dopamine-containing cells has been demonstrated in the sympathetic chain of rats, cats and pigs (Björklund et al. 1970).

In fluorometric analyses of rat's uterine nerves (Owman et al. 1987) the formaldehyde-induced fluorescence was reduced in intensity upon prolonged acidification because a new excitation maximum developed at lower wavelengths (about 330 mu). The noradrenaline fluorophore will therefore, in contrast to the situation for the dopamine fluorophore, only transiently occur in its non-quinoidal form (excitation maximum 370 mu) and is rapidly converted to the fully aromatic isoquinoline (excitation maximum 330 mu) upon further acidification (Björklund et al. 1968). Because a labile hydroxy group in 4-position is also present in the adrenaline fluorophore, it can be expected that a similar acid-induced shift will occur with this fluorophore, too. Such a shift was observed in the fluorescence of a more yellow-green population of SIF-cells in the utero-vaginal ganglion (Owman et al. 1987).

Obviously, it was difficult to conclusively state in the study by Owman et al. (1987) whether the fluorophore in this population of SIF-cells was due to noradrenaline or adrenaline - major quantities of dopamine seemed to be excluded. The noradrenaline presently measured in the utero-vaginal junction can be fully accounted for by the adrenergic nerves running in this region. There is evidence from histochemical and chemical studies (Owman and Sjöstrand 1965; Sjöstrand 1965) that the SIF-cells in the male genital tract of certain species store large amounts of adrenaline. In view of the above reasoning and on the basis of the small quantities of adrenaline found in the utero-vaginal junctional tissue, it can be assumed that the population of chromaffin-like cells with a yellow-green fluorescence found in the ganglia contain adrenaline.

As mentioned in the introduction, several investigators have reported the presence of chemically measurable amounts of adrenaline in the rat uterus. There is no evidence that uterine adrenaline has a neuronal localization. The present and certain other reports, however, suggest two aspects of the chemically measured adrenaline in rat uterine tissue. One is that endogenous adrenaline is localized in a large number of chromaffin cells in the utero-vaginal junction and that this adrenaline contaminates the uterine tissue, unless the organ is carefully freed from connective tissues when taken for analysis. The other aspect is that the majority of adrenaline is not synthesized locally (Wurtman et al. 1963a) but is taken up from the circulation (Wurtman et al. 1963b, 1964; Green and Miller 1966a) and then stored at some extra-neuronal, extra-chromaffin site, e.g., diffusely in the smooth musculature (Wurtman et al. 1964). Since plasma adrenaline concentrations (Green and Miller 1966b) and the uterine binding of the amine (Wurtman et al. 1963b, 1964) changes during the estrus cycle as well as with pregnancy, an explanation is offered for the considerably varying levels of adrenaline previously reported for the rat uterus.

Acknowledgement. Supported by grant No. 14X-5680 from the Swedish Medical Research Foundation.

References

Adham N, Schenk EA (1969) Autonomic innervation of the rat vagina, cervix and uterus and its cyclic variation. Am J Obstet Gynecol 104: 508-516

Bertler A, Carlsson A, Rosengren E (1958) A method for the fluorimetric determination of adrenaline and noradrenaline in tissue. Acta Physiol Scand 44: 273-292

Björklund A, Ehinger B, Falck B (1968) A method for differentiating dopamine from noradrenaline in tissue sections by microspectrofluorometry. J Histochem Cytochem 16: 263-270

Björklund A, Cegrell L, Falck B, Ritzén M, Rosengren E (1970) Dopamine-containing cells in sympathetic ganglia. Acta Physiol Scand 78: 334-338

Cha K-S, Lee W-C, Rudzik A, Miller JW (1965) A comparison of the catecholamine concentration of uteri from several species and the alterations which occur during pregnancy. J Pharmacol exp Ther 148: 9-13

Chang CC (1964) A sensitive method for spectrophotofluorometric assay of catecholamines. Int J Neuropharmacol 3: 643-649

Eränkö O, Soinila S, Päivärinta H (eds) (1980) Histochemistry and cell biology of autonomic neurons, SIF cells and paraneurons. Raven, New York

Falck B, Gardmark S, Nybell G, Owman Ch, Rosengren E, Sjöberg N-O (1974) Ovarian influence on the content of norepinephrine transmitter in guinea pig and rat uterus. Endocrinology 94: 1475-1479

Green RD, Miller JW (1966a) Evidence for the active transport of epinephrine and norepinephrine by the uterus of the rat. J Pharmacol exp Ther 152: 42-50

Green RD, Miller JW (1966b) Catecholamine concentrations: changes in plasma of rats during estrous cycle and pregnancy. Science 151: 825-826

Häggendal J (1963) An improved method for fluorimetric determination of small amounts of adrenaline and noradrenaline in plasma and tissues. Acta Physiol Scand 59: 242-254

Klingman GI (1965) Catecholamine levels and DOPA-decarboxylase activity in peripheral organs and adrenergic tissues in the rat after immunosympathectomy. J Pharmacol exp Ther 148: 14-21

Owman Ch, Alm P, Sjöberg N-O (1983) Pelvic autonomic ganglia: structure, transmitters, function and steroid influence. In: Elfvin L-G (ed) Autonomic ganglia. Wiley, Chichester, pp 125-143

Owman Ch, Rosengren E, Sjöberg N-O (1966) Origin of the adrenergic innervation to the female genital tract of the rabbit. Life Sci 5: 1389-1396

Owman Ch, Alm P, Sjöberg N-O, Stjernquist M (1986a) Structural, biochemical, and pharmacological aspects of the uterine autonomic innervation and its remodeling during pregnancy. In: Huszar G (ed)

The physiology and biochemistry of the uterus in pregnancy and labor. CRC Press, Boca Raton, Florida, pp 5-23

Owman Ch, Alm P, Sjöberg N-O, Stjernquist M (1987) Fluorescence histochemical and microspectrofluorometric study of three different catecholamines in rat uterus and para-cervical ganglion. Exp Brain Res (in press)

Owman Ch, Sjöstrand NO (1965) Short adrenergic neurons and catechol-amine-containing cells in the vas deferens and accessory male genital glands of different mammals. Z Zellforsch 66: 300-320

Rosengren E, Sjöberg NO (1967) The adrenergic nerve supply to the female reproductive tract of the cat. Amer J Anat 121: 271-284

Rudzik AD, Miller JW (1962) The effect of altering the catecholamine content of the uterus on the rat of contractions and the sensitivity of the myometrium to relaxin. J Pharmacol exp Ther 1: 88-95

Sjöberg N-O (1967) The adrenergic transmitter of the female reproductive tract: distribution and functional changes. Acta Physiol Scand Suppl 305: 1-32

Sjöstrand NO (1965) The adrenergic innervation of the vas deferens and the accessory male genital glands. Acta Physiol Scand 65, Suppl 257: 1-82

Spratto GR, Miller JW (1968) The effect of various estrogens on the weight, catecholamine content and rate of contractions of rat uteri. J Pharmacol exp Ther 161: 1-6

Swedin G, Brundin J (1968) Distribution of noradrenaline in the genital organs of the female rat with a remark on dopamine in the cervix and vagina. Experientia 24: 1015-1016

Thorbert G, Alm P, Owman Ch, Sjöberg N-O (1977) Regional distribution of autonomic nerves in guinea-pig uterus. Am J Physiol 233: C25-C34

Wurtman RJ, Axelrod J, Kopin IJ (1963a) Uterine epinephrine and blood flow in pregnant and postparturient rats. Endocrinology 73: 501-503

Wurtman RJ, Axelrod J, Potter LT (1964) The disposition of catecholamines in the rat uterus and the effect of drugs and hormones. J Pharmacol exp Ther 2: 150-155

Wurtman RJ, Chu EW, Axelrod J (1963b) Relation between the oestrus cycle and the binding of catecholamines in the rat uterus. Nature 198: 547-548

The Distribution of Vasoactive Intestinal Polypeptide (VIP) in the Opossum Esophagus in Relation to Function

J. Christensen[1], T. H. Williams[1], J. Jew[1], and T. M. O'Dorisio[2]

[1] Department of Medicine and Anatomy, University of Iowa, College of Medicine, Iowa City, IA 52242, USA
[2] Department of Medicine, Ohio State University, Columbus, OH, USA

Introduction

Vasoactive inhibitory polypeptide (VIP) has been discussed as a candidate for the transmitter of the inhibitory nerves that regulate function of the circular muscle layer of the esophagus (Rattan and Goyal 1974; Christensen 1975, 1983; Rattan et al. 1977; Goyal and Rattan 1978; Goyal et al. 1980; Daniel et al. 1983; Christensen and Percy 1984). VIP-mediated nerve transmission seems not to be involved in the control of the longitudinal layer of the muscularis propria or muscularis mucosae (Christensen 1975; Christensen and Percy 1984). The evidence comes mostly from physiological studies of the North American opossum, D. virginiana, used as an animal model because marsupials possess a long distal segment of the esophagus composed of smooth muscle like primates. The distribution of VIP has not been examined in this organ. We studied VIP-immunostained sections throughout the opossum esophagus by light microscopy to seek evidence that the distribution of VIP-immunoreactive nerves might support the idea that such nerves regulate the function of the circular layer of esophageal smooth muscle in this species.

Materials and Methods

Immunostaining. We used seven mature opossums weighing 2–4 kg and anesthetized with 50 mg/kg sodium pentobarbital. After intravascular perfusion with a modified Zamboni's solution (Zamboni and de Martino 1967) the esophagus was removed, further immersed in the fixative solution and then immersed in 10% sucrose in phosphate buffer overnight. The organ was divided into serial segments 1–1.5 cm long which were cut on a freezing microtome in the plane of the flattened segments into 35 um thick serial sections. 15–25 sections spanned the whole thickness from the mucosal to the serosal surfaces. Immunostaining for VIP (1:1000) was performed by the avidin-biotin method (Hsu et al. 1981; Jew et al. 1986). Control sections were run for specificity of binding using normal rabbit serum and preadsorbed VIP antiserum instead of the primary VIP antiserum.

VIP Content of the Esophagus. The esophagus from each of 3 mature opossums was divided into 9 segments of equal length spanning the whole distance from the pharynx to the stomach. In each segment the mucosa was stripped from the muscularis to give two fragments. The 54 specimens (9 segments x 2 fragments x 3 animals) were identified by code, frozen and assayed separately for VIP content by a radioimmunoassay method (Gaginella et al. 1978) that detects VIP at 5 pg/ml

Experimental Brain Research Series 16
© Springer-Verlag Berlin · Heidelberg 1987

with coefficients of reproducibility in buffers and in tissues for intra- and inter-assay of 5% and 18% respectively. The protein content of each piece was measured (Lowry et al. 1951). The code was broken after the assays were done and VIP content was calculated for mucosal and muscularis fragments for each of the 9 levels of the esophagus.

Results

Many structures were stained immunocytochemically for VIP. Fine beaded linearities, identified as terminal nerves, were present in all 3 layers of smooth muscle, the muscularis mucosae and both layers of the muscularis propria. In the muscularis mucosae and the longitudinal layer of the muscularis propria, they lay as single fibers or small fiber bundles aligned with the muscle bundles and crossing them obliquely (Fig. 1A). The pattern was different in the circular layer where a more arborized and irregular pattern of branching of single fibers occurred (Fig. 1B). The muscle of the lower esophageal sphincter could not be distinguished from that of the esophageal body in the density or distribution of VIP-immuno-reactive terminal nerves. Immunoreactive nerves in the part of the esophagus composed wholly of striated muscle (the proximal one-third) occurred in relation to single cells scattered at wide inter-vals (Fig. 1C), not associated with motor end-plates. Terminal nerves outlined blood vessels and supplied submucosal glands which are prominent in this organ. Ganglion cells in both the submucosal

Table 1. VIP content of muscularis propria and mucosa at 9 levels of the opossum esophagus

Level of Esophagus (% of Length)	Nature of Muscularis Propria	VIP Content of Muscularis Propria (pg/mg protein)	VIP Content of Mucosa (pg/mg protein)
1 - 11%	striated	22.1 ± 5.7*	75.3 ± 17.4
12 - 22%	striated	†† 27.4 ± 3.3	† 120.4 ± 7.3
23 - 33%	striated	50.8 ± 3.2	224.0 ± 114.7
34 - 44%	striated/smooth	76.3 ± 25.3	227.3 ± 108.5
45 - 55%	striated/smooth	71.3 ± 22.4	132.1 ± 11.9
56 - 66%	smooth	89.5 ± 26.5	196.7 ± 79.3
67 - 77%	smooth	† 80.8 ± 7.3	† 162.8 ± 93.5
78 - 88%	smooth	63.5 ± 12.9	96.6 ± 10.2
89 - 99%	smooth (including sphincter)	†† 29.3 ± 1.0	†† 56.6 ± 22.0

*   for each value, x ± SE, n = 3
+,++   the average of values marked with + exceeds (p≤ 0.05) the average of those marked with ++, as tested by ANOVA

and myenteric plexuses were less immunoreactive and varicosities were stained in ganglia. Immunoreactivity was also detected in abundant small oval cells with a single (occasionally double) round clear nucleus that were dispersed diffusely in all muscle layers, both smooth and striated muscle, and were concentrated especially in the planes of the submucous and myenteric plexuses (Fig. 1D).

The assay of VIP content in the esophagus showed that the content was lower (p $\leq$ 0.05) in the striated muscle regions and in the level of the lower esophageal sphincter than it was in the smooth muscle parts of the esophageal body (Table 1). Similarly, the VIP content in the mucosa was lower in the segment containing the lower eso-phageal sphincter than it was at other levels of the esophagus (Table 1).

Discussion

Our observations differ in some respects from and augment those in other species (Uddman et al. 1978; Alumets et al. 1979; Chayvialle et al. 1980). All smooth muscle layers are well-endowed with VIP-immunoreactive nerves. Reactive nerves occur in the striated muscle layers and supply glands and vessels. There is a considerable extra-neuronal localization of VIP in small oval cells distributed abun-dantly throughout all layers of the esophageal wall in all regions. VIP is identifiable by radioimmunoassay throughout the esophagus, with a greater concentration (p $\leq$ 0.05) in the smooth muscle of the esophageal body than in the striated muscle and the lower esophageal sphincter.

The results suggest that VIP-immunoreactivity may characterize more than one functional class of nerves so that immunoreactivity does not directly correlate with VIP-mediated neurohumoral transmission. Alternatively, VIP may support or modulate many functions without being the precipitating or sole cause of any.

References

Alumets J, Fahrenkrug J, Hakanson R, Schaffalitzky de Muckadell O, Sundler F, Uddman R (1979) A rich VIP nerve supply is characteristic of sphincters. Nature 280: 155-156

Chayvialle J-A, Miyata M, Rayford PL, Thompson JC (1980) Immunoreac-tive somatostatin and vasoactive intestinal peptide in the digestive tract of cats. Gastroenterology 79: 837-843

Christensen J (1975) Pharmacology of the esophageal motor function. Ann Rev Pharmac 25: 243-258

Christensen J (1983) The oesophagus. In: Christensen J, Wingate DL (eds) A guide to gastrointestinal motility. John Wright and Sons, Bristol, pp 75-100

Christensen J, Percy WH (1984) A pharmacologic study of oesophageal muscularis mucosae from the cat, dog and American opossum (Didelphis virginiana). Brit J Pharmacol 83: 329-336

Daniel EE, Helmy-Elkoly A, Hager LP, Kannan MS (1983) Neither a purine nor VIP is the mediator of inhibitory nerves of opossum oesophageal smooth muscle. J Physiol (Lond) 336: 243-260

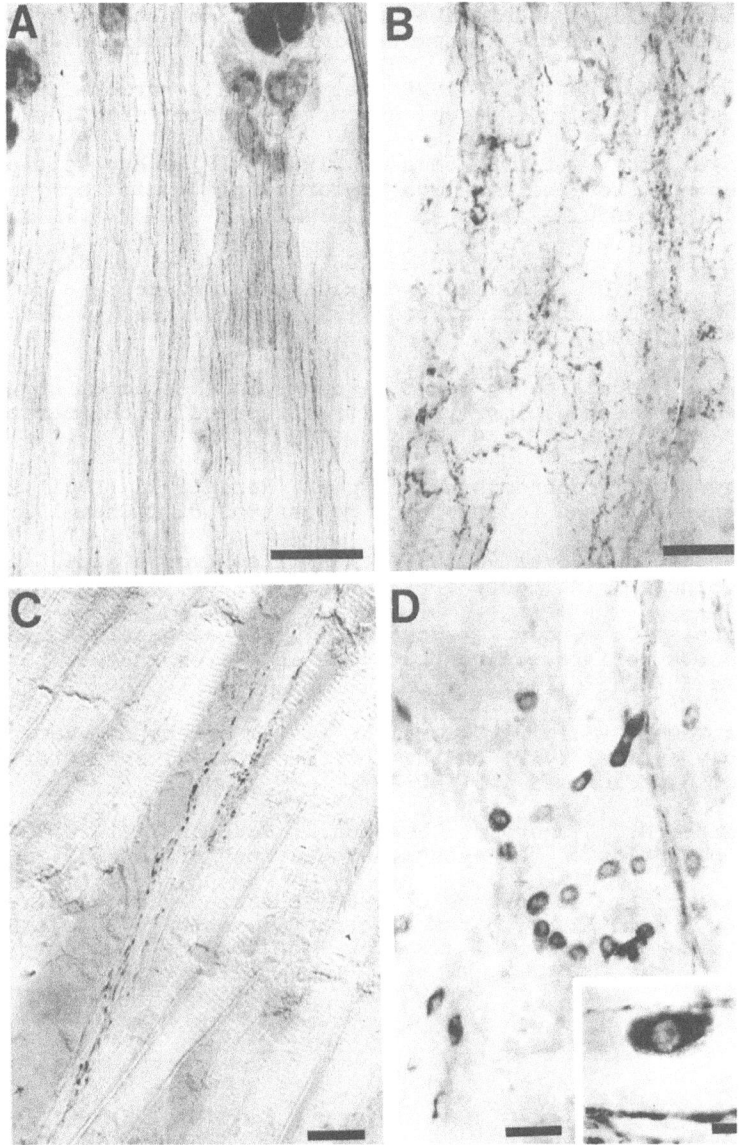

Fig. 1. VIP-immunoreactive structures in the opossum esophagus. **A.** VIP-immunoreactivity in a section of muscularis mucosae with a submucosal gland. Terminal nerves lie in small bundles oriented in line with the muscle bundles. The pattern is the same in the longitudinal layer of the muscularis propria. Bar indicates 200 um. **B.** VIP-immunoreactivity in a section of the circular layer of muscularis propria. Terminal nerves interlace the muscle bundles randomy, in contrast to the pattern in the muscularis mucosae and longitudinal layer of muscularis propria. Bar indicates 50 um. **C.** VIP-immunoreactivity in terminal nerves in relation to striated muscle. Bar indicates 20 um. **D.** VIP-immunoreactive oval cells. Bar indicates 10 um. **Inset**: a single oval cell under oil immersion. Bar indicates 2 um

Gaginella TS, Mekhjian HS, O'Dorisio TM (1978) Vasoactive intestinal polypeptide: quantification by radioimmunoassay in isolated cells, mucosa and muscle of hamster intestine. Gastroenterology 74: 718–721

Goyal RK, Rattan S (1978) Neurohumoral, hormonal and drug receptors for the lower esophageal sphincter. Gastroenterology 74: 598–619

Goyal RK, Rattan S, Said SI (1980) VIP as a possible neurotransmitter of non-cholinergic non-adrenergic inhibitory neurons. Nature 288: 378–380

Hsu SM, Raine L, Fanger H (1981) The use of avidin-biotin peroxidase complex (ABC) in immunoperoxidase techniques: a comparison between ABC and unlabeled antibody (PAP) procedures. J Histochem Cytochem 29: 577–580

Jew J, Fountaine N, Jew E (1986) Immunohistochemistry of peptides in neural tissues. Improving ultrastructural localization. Peptides (in press)

Lowry OH, Rosebrough NJ, Farr AL, Randall RJ (1951) Protein measurements with the folin phenol reagent. J Biol Chem 193: 265–275

Rattan S, Goyal RK (1974) Neural control of the lower esophageal sphincter. Influence of the vagus nerves. J Clin Invest 54: 899–906

Rattan S, Grady M, Goyal RK (1982) Vasoactive intestinal peptide causes peristaltic contractions in the esophageal body. Life Sciences 30: 1557–1563

Rattan S, Said SI, Goyal RK (1977) Effect of vasoactive intestinal polypeptide (VIP) on the lower esophageal sphincter pressure. Proc Soc Exp Biol Med 155: 40–43

Uddman R, Alumets J, Hakanson R, Sundler F (1978) Peptidergic (VIP) innervation of the esophagus. Gastroenterology 75: 5–8

Zamboni L, de Martino C (1967) Buffered picric acid-formaldehyde: a new, rapid fixative for electron microscopy. J Cell Biol 35: 148A

Peptidergic and Catecholaminergic Innervation in Rat and Human Brown Adipose Tissue

J. D. Lever[1], R. T. Jung[2], D. Norman[1], D. Symons[1], P. J. Leslie[2], and J. O. Nnodim[3]

[1] Department of Anatomy, University College, P.O. Box 78, Cardiff CF1 1XL, Great Britain
[2] Department of Medicine, University of Dundee, Dundee, Great Britain
[3] Department of Anatomy, University of Benen, Benen, Nigeria

Introduction

An important role for the sympathetic nervous system in the adaptational responses of brown adipose tissue (BAT) in rodents is indicated by an increased noradrenaline turnover in the tissue (Cottle et al. 1976) and by the demonstration of heightened catecholamine levels in nerve plexuses related to brown adipocytes coincidentally with cold acclimation (Cottle and Cottle 1970).

Following demonstration of thermogenesis in rat and marmot BAT (Smith and Roberts 1964) and recognition of the trophic role of sympathetic nerves in non-shivering thermogenesis in the rat (Foster and Frydman 1978), there has been sustained interest in the distribution, blood supply and innervation of this tissue. Further, a connection between the lipolytic and thermogenic potential of BAT and the regulation of body weight has been postulated by Rothwell and Stock (1979) as a result of studies which demonstrated a large increase in the metabolic sensitivity of rat BAT to exogenous noradrenaline concurrent with hyperphagia (luxuskonsumption), during which animal body weight did not significantly rise. The distribution of BAT in man has been described in light microscopic studies (Heaton 1972); and recently Cunningham et al. (1985) have determined its presence in human perinephric fat depots over a wide age range (extending to the sixth decade) by visual inspection, electron microscopy and the demonstration by nucleotide binding of the tissue specific uncoupling protein 32KDa, the recognised prerequisite for thermogenesis. Pertinent to this work and to the finding of Mory (1984) that noradrenaline controls the concentration of 32KDa in (rodent) BAT, is the recent histochemical demonstration (Lever et al. 1985a, 1986) of catecholaminergic nerve plexuses in human perirenal BAT.

Research into the exact nature and distribution of the nerves in BAT is not only indicated in the context of thermogenesis per se, but also in connection with related vasomotor phenomena within the tissue. While Selye and Timiras (1949) observed an increased blood flow in rat BAT during cold stress, Brück and Wünnenberg (1965) later claimed that this hyperaemia was inhibited by beta-blockers. However, observations by Derry et al. (1969) that vasomotor and parenchymal nerves in rat BAT responded differently to reserpine treatment and to immuno-sympathectomy prompted them to postulate a dual innervation for BAT.

As has been pointed out by Hökfelt et al. (1977), the peripheral autonomic nervous system is assembled in a more complex manner than has hitherto been assumed. In addition to the classic cholinergic

Experimental Brain Research Series 16
© Springer-Verlag Berlin · Heidelberg 1987

and adrenergic neurons it is now necessary to consider the arrangement of peptidergic neurons within the system.

In the present study catecholaminergic and substance P-immunoreactive elements are demonstrated within autonomic nervous plexuses in
interscapular brown adipose tissue (ISBAT) from adult rats and in
perirenal brown adipose tissue (PRBAT) from human donors.

Materials and Methods

ISBAT was obtained from adult rats as was PRBAT from human donors
aged 5, 23, 27, 39 and 47 years at operation for benign renal
conditions. Material was treated as follows: (1) **For transmission
electron microscopy** (TEM) by 3% glutaraldehyde fixation, buffer wash
and subsequent post-fixation in 1% osmium tetroxide before processing to araldite. (2) **For catecholamine histochemistry,** material was
either treated by a standard formaldehyde induction method or fresh-
frozen and sectioned by cryostat at 12 um: the sections being thawed
onto glass slides and processed by the sucrose-potassium phosphate
glyoxylic acid (SPG) technique: five 1 sec dips in SPG solution : 5
min drying in a warm airstream : 5 min incubation at 80°C : after
liquid paraffin mounting sections were viewed in a Leitz Laborlux 12
microscope with Pleomopak 2.5 Vertical Fluorescence Illuminator
(filters BP 340-380 nm excitor: LP430 nm suppressor). (3) **For immu-
nofluroescent histochemistry** by fixation in 0.4% parabenzoquinone
prior to cryostat sectioning : sections reacted successively with
rabbit anti-substance P, donkey biotinylated anti-rabbit IgG F(ab)₂
fragment and fluorescein-streptavidin conjugate were viewed in a
fluorescence microscope embodying 450-490 nm (excitor) and 515 nm
(suppressor) filters : methodological control was ensured by parallel section incubations omitting primary antibody, while specificity
was indicated by parallel section treatment with antigen (substance
P)-absorbed primary antibody.

Results

Both in ISBAT (rat) and in human PRBAT, adrenergic nerves can be
visualised by SPG histochemistry as a beaded fluoresent plexus upon

Fig. 1. Fluorescence micrograph of rat BAT showing catecholaminergic ▶
nerve fibre distribution as a vasomotor plexus around a supply
artery (⬆) and as a parenchymal plexus (↑) among brown adipocytes:
scale bar = 50 um. (Taken from Fig. 19 in J.O. Nnodim and J.D. Lever
(1985) Anat. Embryol. 173: 215-223)

Figs. 2 and 3. Both electron micrographs of human perirenal BAT from
a 5 year old donor showing nerve terminals: at the medio-adventitial
junction of a small supply artery (Fig. 2), in an intercellular
interval in the parenchymal field (Fig. 3). Note large diameter
(1500 Å (⬆) and small diameter (500 Å (⇑) dense cored vesicles.
medial muscle (M); brown adipocyte mitochondria (B); scale bar =
5,500 A

Figs. 4 and 5. Both fluorescence micrographs of small supply arteries to rat (Fig. 4) and human (perirenal) BAT (Fig. 5) showing
substance P-like immunoreactive putative nerves upon their walls
(↑); see Materials and Methods for procedural details. Scale bar =
30 um

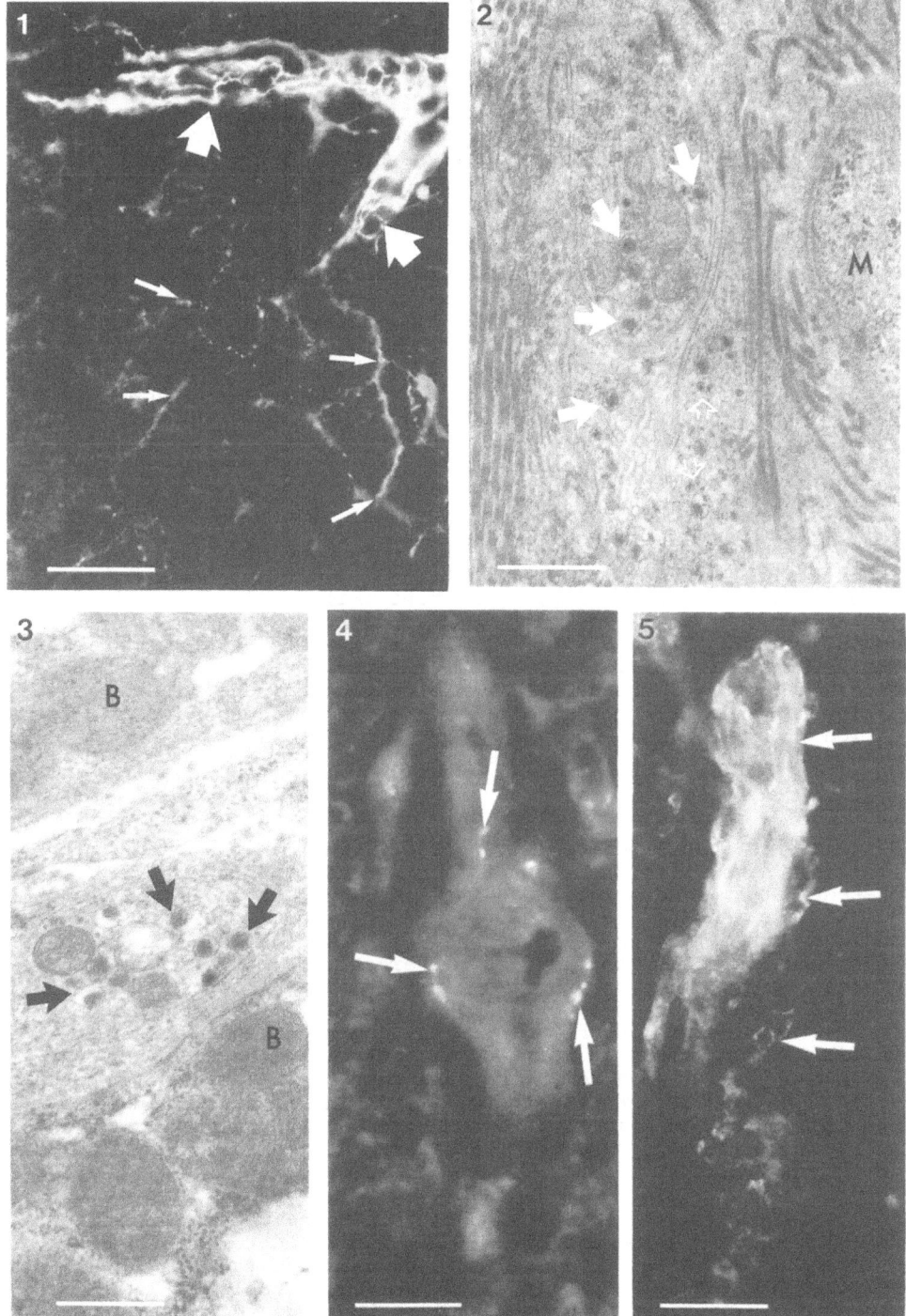

the walls of inter- and intralobular arteries and extending to the parenchymal domain into direct relationship with individual brown adipocytes (Fig. 1; see Lever et al. 1986). In a review of SPG preparations from all the human donors it was clear that while the extent and fluorescence intensity of the paravascular catecholaminergic nerve plexuses remained essentially comparable, extent and intensity of parenchymal plexuses were diminished with age especially in the 39 and 47 year donors (see Figs. 2-4 in Lever et al. 1986).

Unmyelinated nerve bundles can readily be found by TEM in rat ISBAT and human PRBAT. Nerve distribution is both vasomotor and parenchymal. At the medio-adventitial interface of small supply arteries, nerve terminals of differing fine structural appearance may be found in both rat and human BAT: terminals containing numbers of small (500 Å d) dense cored vesicles (dcvs), putative adrenergic terminals, may be seen near other terminals characterised by clusters of large (1300-1500 Å d) dcvs (Fig. 2). Furthermore, in human PRBAT, axons with terminal features including clusters of larger diameter dcvs have been observed between parenchymal cell groups (Fig. 3) while more peripherally terminals can be found juxtaposed to, or indenting the surface of, individual adipocytes in rat ISBAT and in human PRBAT.

In both human PRBAT and rat ISBAT, we have consistently found substance P-like immuno-reactive beaded elements (putative nerves) upon the walls of small supply arteries (Figs. 4 and 5). It would appear that this peptidergic plexus does not extend into the parenchymal domain, but in human PRBAT, substance P-positive elements have been observed around thin-walled intralobular vessels. Results obtained (not illustrated) in the methodological and antibody specificity controls were confirmatory of the bona fides of the investigation.

Discussion

Evidence has been produced (Lever et al. 1986) that a tissue morphologically similar to the brown fat of rodents and other mammals, is present at perirenal sites in man, not only in childhood but in adult life and that this tissue has a catecholaminergic vasomotor and parenchymal innervation and is well vascularised. Since human PRBAT contains 32KDa and does have energetic potential, it is appropriate to speculate on the competence of human brown fat in similar lipolytic and thermogenic roles to those already attributed to BAT in rodents (Foster and Frydman 1978) and indeed to surmise at its possible involvement in body weight control. In the latter context Cawthorne (1983) has demonstrated a direct relationship between catecholamine turnover and thermogenesis in BAT and body weight in relation to excess caloric intake in the rat: in such circumstances increased levels of lipolysis (and thermogenesis) mitigate against weight increase.

The significance of substance P-containing nerve terminals on the walls of supply vessels to ISBAT in the rat and to human PRBAT is conjectural. Substance P is known to be a smooth muscle relaxant and possible vasodilator and is of course demonstrable in nociceptor neurones in dorsal root ganglia and in their central and peripheral processes (Henry 1977). In the context of rat ISBAT Sidman and Fawcett (1954) observed that cold-induced hyperaemia in the fat pad did not occur if its nerve supply was surgically divided prior to cold exposure: this segmentally derived nerve traverses the pad en route for an extensive cutaneous distribution on the body wall. Pursuing such a line of thought, the present findings (at any rate in the rat) might suggest a mechanism behind cold-induced hyperaemia in ISBAT if the substance P-positive nerve terminals on the walls of its supply arteries (Fig. 4) are collaterals of nerves conveying sensory information from the integument, which collaterals might be vasodilator to the BAT vessels, that it might be functioning as the efferent limb in an axon reflex pathway.

Pluralism in the innervation pattern of BAT is not only indicated by the present histochemical revelation of both catecholaminergic and peptidergic elements within its nerve plexuses, but is also suggested by fine structural observation. Thus in nerves on the walls of small supply arteries to BAT in rat and man, which by their position could be regarded as vasomotor, terminal features show variation (Fig. 2) which might be indicative of plurality of type. Also germane to this question is the earlier finding (Lever et al. 1984) that after 6-hydroxydopamine administration in the rat, some but not all terminals are stigmatised in vasomotor plexuses on intralobular supply arteries in ISBAT. The identity of the unstigmatised (therefore non-adrenergic) terminals in these vasomotor plexuses is unknown. It is conceivable that they may be peptidergic, an hypothesis which could be tested by gold label electron immunocytochemical means employing in the first instance substance P antibody.

If the adrenergic innervation to BAT is both trophic to the metabolic activity of the adipocytes and vasoconstrictive of the intralobular supply arteries, perhaps the substance P-immunoreactive elements visualised in the present study are vasodilator nerve fibers. In his review entitled "Sympathetic Vasodilator Fibres" Burn (1938) quoted Dale's belief that in some locations vasodilatation is exerted through "axon reflexes at the termination of sensory nerves". Whatever is the significance of these words, a caveat has to be made in the specific context of rat ISBAT because of the recent demonstration (Lever et al. 1985b) of arteriovenous anastomoses at perilobular sites in ISBAT. In terms of capillary haemodynamics, if these anastomoses were closed by a vasoconstrictive (viz. noradrenergic) stimulus, more blood would _ipso_ _facto_ flow along capillary routes through the tissue.

References

Brück K, Wünnenberg B (1965) Untersuchungen über die Bedeutung des multilokularen Fettgewebes für die Thermogenese des neugeborenen Meerschweinchens. Pflügers Arch Ges Physiol 283: 1–16

Burn JH (1938) Sympathetic vasodilator fibres. Physiol Rev 18: 137–153

Cawthorne MA (1983) Brown adipose tissue hibernation and obesity. J Soc Med 76: 173–174

Cottle MKW, Cottle WH (1970) Adrenergic fibres in brown fat of cold-acclimated rats. J Histochem Cytochem 18: 116–119

Cottle WH, Nash CW, Ueress AT, Fergusson BA (1967) Release of noradrenaline from brown fat of cold-acclimated rats. Life Sci 6: 2267–2271

Cunningham S, Leslie P, Hopwood D, Illingworth P, Jung RT, Nicholls DG, Peden N, Rafael J, Rial E (1985) The characterisation and energetic potential of brown adipose tissue in man. Clin Sci 69: 343–348

Derry DM, Schonbaum E, Steiner G (1969) Two sympathetic nerve supplies to brown adipose tissue of the rat. Can J Physiol Pharmacol 47: 57–63

Foster DO, Frydman ML (1978) Non–shivering thermogenesis in the rat. II. Measurements of blood flow with microspheres point to brown adipose tissue as the dominant site of the calorigenesis induced by noradrenaline. Can J Physiol Pharmacol 56: 110–122

Heaton JM (1972) The distribution of brown adipose tissue in the human. J Anat 112: 35–39

Henry JL (1977) Substance P and pain: a possible relation in afferent transmission. In: von Euler US, Pernow B (eds) Substance P. Raven, New York, pp 231–240

Hökfelt T, Elfvin LG, Schultzberg M, Goldstein M, Nilsson G (1977) On the occurrence of substance P–containing fibers in sympathetic ganglia: immunohistochemical evidence. Brain Res 129: 29–43

Lever JD, Nnodim JO, Symons D (1984) Observations on the innervation of brown adipose tissue in the adult rat: an ultrastructural and fluorescence histochemical study after 5–hydroxy– and 6–hydroxydopamine treatment. J Anat 138: 565

Lever JD, Jung RT, Nnodim JO, Leslie P, Symons D (1985a) Adrenergic innervation of brown adipose tissue at adult rat and human perirenal sites: fluorescence histochemical and histological study. Proc XII Internat Anat Cong London, p 402

Lever JD, Nnodim JO, Symons D (1985b) Arteriovenous anastomoses in interscapular brown adipose tissue in the rat. J Anat 143: 207–210

Lever JD, Jung RT, Nnodim JO, Leslie PJ, Symons DS (1986) Demonstration of a catecholaminergic innervation in human perirenal brown adipose tissue at various ages in the adult. Anat Rec 215: 251–255

Mory G, Bouilland F, Combes–George M, Ricquier D (1984) Noradrenaline controls the concentration of the uncoupling protein in brown adipose tissue. FEBS letters 166: 393–396

Rothwell NJ, Stock MJ (1979) A role for brown adipose tissue in diet–induced thermogenesis. Nature 281: 31–35

Selye H, Timiras PS (1949) Participation of brown fat in the alarm reaction. Nature 164: 745–746

Sidman RL, Fawcett DW (1954) The effect of peripheral nerve section on some metabolic responses of brown adipose tissue in mice. Anat Rec 118: 487–507

Smith RE, Roberts JC (1964) Thermogenesis of brown adipose tissue in cold–acclimated rats. Amer J Physiol 206: 143–148

Enzymes and Proteins

Molecular Biology of Catecholamine Synthesis, Storage and Release

T. H. Joh and O. Hwang

Laboratory of Molecular Neurobiology, Cornell University Medical College, 1300 York Avenue, New York, NY 10021, USA

Introduction

Biosynthesis of catecholamines (CA) involves four enzymatic steps: (1) tyrosine hydroxylase (TH) catalyzes the conversion of tyrosine to DOPA; (2) DOPA is converted to dopamine (DA) by aromatic L—amino acid decarboxylase (DOPA—decarboxylase, AADC); (3) dopamine beta-hydroxylase (DBH) catalyzes the enzymatic reaction of DA to norepinephrine (NE); and (4) epinephrine (EP) is synthesized from NE by phenylethanolamine N—methyltransferase (PNMT).

An interesting feature exists in this biosynthetic pathway: The tissue specificity observed in the gene expression of each enzyme. In DA neurons, TH and AADC genes are expressed, while those of DBH and PNMT are silent. In NE neurons, PNMT gene is not expressed. The genes for all four enzymes are expressed only in adrenergic neurons and chromaffin cells. Thus, the phenotype of each CA neuron, namely dopaminergic, noradrenergic and adrenergic neuron, is determined by the presence of the phenotype specific enzyme(s). In addition, another distinctive marker for the neuronal phenotype is the presence of high—affinity uptake site(s) for the specific neurotransmitter at its nerve endings. This implies that specific protein(s) at the nerve endings can also determine the phenotype.

CA neurotransmitters are stored in their specific storage vesicles and are released by exocytosis. It is not clearly established whether different neurotransmitters are stored in different storage granules, and whether the membrane structures of different neurotransmitters storage vesicles are different. We do know, however, that when the neurotransmitter is released upon exocytosis, the membrane bound proteins inside the vesicles are exposed to outer-membrane of nerve endings. For instance, upon exocytosis, vesicular membrane form of DBH (mDBH) (Smith and Kirshner 1967; Winkler 1976; Klein et al. 1977) is exposed to outer membrane of noradrenergic nerve endings (Winkler et al. 1974), while soluble form of DBH, stored in the vesicular interior, is secreted with NE (Vivero et al. 1968; Pollard et al. 1979). Thus, the association of DBH, noradrenergic marker, with nerve endings provides the structural characteristics of this phenotype. It is of interest to investigate whether DA neurons also express their specific marker protein, TH, in their nerve endings. Furthermore, other neurons, such as cholinergic, serotonergic and GABA—ergic, may posess similar characteristics.

Complement Mediated Immunolysis of Striatal Synaptosomes

Docherty et al. (1985) reported that, in the presence of complement, antibodies directed against specific neurotransmitter synthesizing enzymes are capable of selectively lysing specific synaptosomal subpopulations.

Incubation of striatal synaptosomes with antibodies to TH and complement apparently caused the antibody dose-dependent selective lysis of the dopaminergic but not GABAergic or cholinergic subpopulation of synaptosomes. The lysis was tissue specific, as the noradrenergic synaptosomes of cortex were not affected. Similar experiments with antibodies to glutamate dehydrogenase (GAD) and choline acetyltransferase (ChAT) caused selective lyses of the GABAergic and cholinergic subpopulations of the synaptosomes, respectively. More recent results (personal communication with Dr. M. Docherty) showed that antibodies against tryptophan hydroxylase, the serotonine biosynthetic enzyme, caused the selective lysis of serotonergic synaptosomes.

The results indicate that antibodies to a neurotransmitter-synthesizing enzyme selectively recognize a certain membrane protein(s) present in the specific subpopulation of synaptosomes which contain that enzyme, and suggest that the biosynthetic enzyme and a certain nerve ending protein(s) share similar protein domains. This specific membrane protein(s) may be a full-length or partial length protein(s) of the neurotransmitter synthesizing enzyme.

Evidence for the Presence of TH in Synaptosomal Membranes of Rat Striatum

Tyrosine hydroxylase is believed to be localized in the cytosol of catecholamine synthesizing cells. Although TH activity is also associated with the particulate fraction, this had been attributed to aggregated soluble TH (Wurzberger and Mussacchio 1974). However, Kuczenski and Mandell (1972) reported that TH exists in two forms in rat striatum, which is enriched in dopaminergic nerve endings: a soluble and a membrane-bound form. The membrane bound form was observed to be brain region specific and mainly present in striatum. As mentioned earlier, TH-immunoreactive protein is present in the synaptic membrane of dopaminergic nerve endings. We utilized the method of proteolytic digestion of solubilized and intact striatal synaptosomal plasma membrane proteins to further investigate the identity of these membrane antigens. We found that a membrane associated TH-immunoreactive protein, which is undistinguishable from the soluble TH in its size and trypsin digestion pattern as detected by immunoblotting, exists in striatum.

Figure 1 shows the immunoblot of trypsin-treated synaptosomal membrane proteins with TH antibodies at various time points. The proteins were treated with trypsin either in the presence of intact membrane (panel 1) or after the solubilization by removal of the plasma membrane (panel II). At time zero (before the addition of trypsin) a TH-immunoreactive protein of 60 kd, the molecular weight of TH, is evident in both panels. In the presence of the intact membrane (panel I), the proteolytic digestion yields two populations of TH-immunoreactive proteins: a digested and an undigested form. The digested protein is sequentially cleaved to 56 kd, 37 kd, and 34 kd molecules whereas the band corresponding to the undigested form appears to remain unchanged in this size and intensity during the 60 min period. On the other hand, upon removal of the membrane by Triton X-100 (panel II), only the digested form is observed.

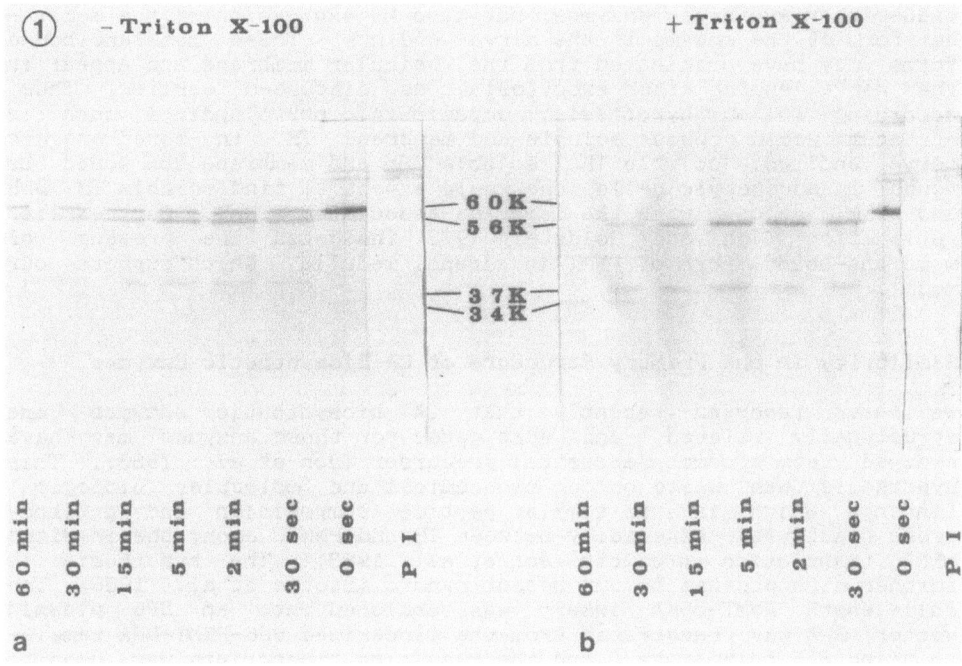

Fig. 1. Immunoblot of trypsin-treated synaptosomal membrane proteins with TH antibodies at various time points. a) Intact plasma membrane. b) After removal of the plasma membrane

The digested protein in panel I most likely represents the soluble TH loosely bound to the membrane, as TH is known to be associated with the particulate fraction (Petrack et al. 1968; Shiman et al. 1971) and digested into a 34 kd molecule by trypsin (Mussacchio et al. 1971). This indicates that even after repeated washing and rehomogenization in high salt and high pH, a large amount of the aggregated soluble TH still remains in the particulate fraction. The trypsin resistant form, on the other hand, probably represents the TH-immunoreactive membrane protein observed in our previous immunolysis studies (Docherty et al. 1985). It is interesting to note that this protein is identical in size to the soluble TH. As seen in panel II, however, only the digested form is evident upon removal of the membrane. This suggests that the membrane protein is, when solubilized, similar to the soluble TH in its molecular weight and proteolytic digestion pattern. Before the Triton X-100 treatment, this protein is presumably associated with the membrane in such a way that its trypsin-sensitive sites are protected. Thus, using the method of proteolytic digestion and immunoblot detection, we present evidence that the TH-immunoreactive protein in striatum synaptosomal membrane is indistinguishable from the soluble TH by its size and digestion pattern.

Above results suggest that dopaminergic cells express TH in the membrane of their nerve endings in addition to the soluble TH and AADC. The presence of the membrane and soluble forms of DBH in noradrenergic cells are well documented as described earlier. Together with our previous results of specific immunolysis of synaptosomes, we speculate that the phenotype of CA synthesizing cell may be defined not only by the presence of the specific, soluble neuro-transmitter synthetic enzymes, but also by expression of the membrane form of the enzyme in the nerve endings. These membrane-bound forms may have originated from the vesicular membrane and appear in the nerve ending after exocytosis, as discussed earlier. Thus, according to our hypothesis, dopaminergic nerve endings such as striatum would possess soluble and membrane TH. In noradrenergic nerve endings, soluble TH, soluble DBH and membrane DBH would be found. In adrenergic cells, one would expect to find soluble TH, DBH and PNMT, along with the membrane-associated PNMT. Our earlier publication (Joh and Goldstein 1973) indicated the presence of membrane-bound form of PNMT in adrenal medulla, which support our model.

Similarity in the Primary Structure of CA Biosynthetic Enzymes

We have reported recently that CA biosynthetic enzymes are structurally related, and that genes for these enzymes may have evolved from a common ancestral precursor (Joh et al. 1983). This hypothesis was based on our biochemical and molecular biological findings which include similar peptide composition and antibody cross reactivity, especially between TH and PNMT, among the in vitro mRNA translation products (Joh et al. 1983). The hypothesis is further strengthened by our recent report (Baetge et al. 1986). The full-length PNMT-cDNA insert was subcloned into an SP6 plasmid vector, RNA was transcribed from the linearized SP6-PNMT-DNA template using SP6 polymerase, and the resulting transcripts were used to program an in vitro translation system. Both antibodies to PNMT and TH recognized the sole protein product transcribed from PNMT-cDNA and translated in vitro. Furthermore, the analysis of amino acid sequences between bovine PNMT and rat TH indicated the existence of partial amino acid sequence homology between these two enzymes (Baetge et al. 1986).

It is important to stress that antibody cross reactivity is found only with in vitro translation products and not with the native proteins. Others (Ledley et al. 1985) have shown that rat phenylalanine hydroxylase and TH share more than 70% amino acid sequence homology, and yet that the antibodies cross react only with in vitro mRNA translation products. Thus, although the primary structures of CA-synthesizing enzymes are related, the native forms must be significantly different as a result of posttranslational modification and different folding of the proteins, allowing the highly specific antibodies to cross react only with the newly synthesized proteins. Similarly, these antibodies will react only with the specific synaptosomal membrane protein, as shown earlier, despite the homology in the primary structures of the different CA biosynthetic enzymes.

Conclusion

The above results imply that the phenotype of catecholamine synthesizing cell can be defined not only by the presence of the specific neurotransmitter synthesizing enzymes, but also by expression of the membrane form of the enzyme in the nerve ending, which probably have originated from the neurotransmitter containing vesicular membrane.

References

Baetge EE, Suh YH, Joh TH (1986) Complete nucleotide and deduced amino acid sequence of bovine phenylethanolamine N-methyltransferase: partial amino acid homology with rat tyrosine hydroxylase. Proc Natl Acad Sci USA 83: 5454–5458

Docherty M, Bradford HF, Wu J-W, Joh TH, Reis DJ (1985) Evidence for specific immunolysis of nerve terminals using antisera against choline acetyltransferase, glutamate decarboxylase and tyrosine hydroxylase. Brain Res 339: 105–113

Joh TH, Goldstein M (1973) Isolation and characterization of multiple forms of phenylethanolamine N-methyltransferase. Mol Pharmacol 9: 117–129

Joh TH, Baetge EE, Ross ME, Reis DJ (1983) Evidence for the existence of homologous gene coding regions for the catecholamine biosynthetic enzymes. Cold Spring Harbor Symp. Quant. Biol. 48: 327–335

Klein RL, Kirksey DE, Rush RA, Goldstein M (1977) Preliminary estimates of the dopamine beta-hydroxylase content and activity in purified noradrenergic vesicles. J Neurochem 28: 81–86

Kuczenski R, Mandell AJ (1972) Regulatory properties of soluble and particulate rat brain tyrosine hydroxylase. J Biol Chem 247: 3114–3122

Ledley FD, DiLella AG, Kwok SCM, Woo SLC (1985) Homology between phenylalanine and tyrosine hydroxylases reveals common structural and functional domains. Biochemistry 24: 3389–3394

Mussacchio JM, Wurtzberger RJ, D'Angelo GL (1971) Different molecular forms of bovine adrenal tyrosine hydroxylase. Mol Pharmacol 7: 136–147

Petrack B, Sheppy F, Fetzer V (1968) Studies on tyrosine hydroxylase from bovine adrenal medulla. J Biol Chem 243: 743–748

Pollard H, Pazoles CJ, Creuts CE, Zindler O (1979) The chromaffin granule and possible mechanisms of exocytosis. Intl Rev Cytol 58: 159–197

Shiman R, Akino M, Kaufman S (1971) Solubilization and partial purification of tyrosine hydroxylase from bovine adrenal medulla. J Biol Chem 246: 1330–1340

Smith WJ, Kirshner N (1967) A specific soluble protein from the catecholamine storage vesicles of bovine adrenal medulla. I. Purification and chemical characterization. Mol Pharmacol 3: 52–62

Vivero OH, Arguero L, Connet RJ, Kirshner N (1968) Mechanism of secretion from the adrenal medulla. III. Studies of dopamine beta-hydroxylase as a marker for catecholamine storage vesicle membranes in rabbit adrenal glands. Mol Pharmacol 5: 60–68

Winkler H (1976) The composition of adrenal chromaffin granules. An assessment of controversial results. Neuroscience 1: 65–80

Winkler H, Schneider FH, Rufener G, Nakane PK, Hortnagl H (1974) Membranes of adrenal medulla: their role in exocytosis. Adv Cytopharmacol 2: 127–139

Wurzberger RM, Mussacchio JM (1974) Subcellular distribution and aggregation of bovine adrenal tyrosine hydroxylase. J Pharmacol Exp Ther 177: 155–168

Chromaffin Cells: A Novel Source for Neuronotrophic Factors

K. Unsicker, R. Lietzke, D. Gehrke, and F. Stögbauer

Abteilung für Anatomie und Zellbiologie, Universität Marburg, Robert-Koch-Strasse 6, 3550 Marburg, FRG

Introduction

During the past three decades the adrenal medulla and its endocrine chromaffin cells have become established model systems for studying a large variety of neuronal functions (cf. Unsicker 1983). Work described in this article adds a novel aspect to the model functions of chromaffin cells emphasizing their position as a target neuron within a defined neuronal circuitry (Fig. 1) and their putative neuronotrophic contributions for the development and maintenance of three neurons which innervate the adrenal medulla. Our data support the notion that neurons may store and secrete neuronotrophic proteins.

The Concept of Target Organ–Regulated Neuronal Cell Death: Neurons as a Putative Source of Trophic Agents

Naturally occurring losses of neurons are a normal feature of neurogenesis (Oppenheim 1985). Although many fundamental issues of this phenomenon are still unresolved, competition of neurons for target cell–derived trophic agents may be an important principle for understanding the regulation of neuronal survival and death. Evidence from work on nerve growth factor (NGF) and muscle–derived trophic agents has considerably substantiated this view, but may also have biased our perception of neuronal targets towards muscle, skin and exocrine glands.

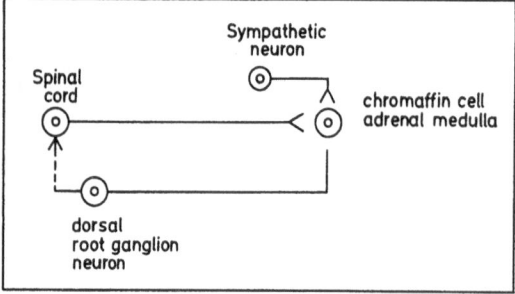

Fig. 1. Schematic drawing of principal neuronal connections to chromaffin cells and the adrenal medulla. The existence of sensory fibers has been documented in recent studies by Parker and Coupland (personal communication)

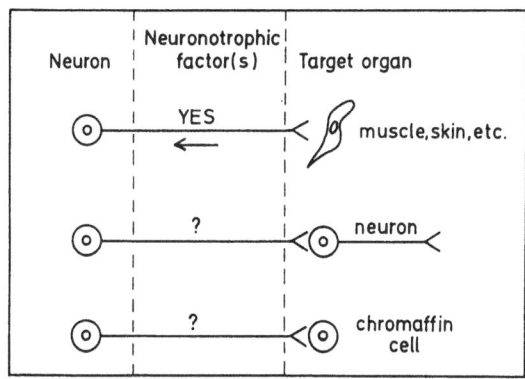

Fig. 2. Target organ-derived neuronotrophic factors: analogy of neurons and chromaffin cells with target cells of mesenchymal origin?

Within the nervous system neurons are the common targets of neurons. This raises the question whether neurons may have retrograde or anterograde regulative influences (beyond synaptic transmission) on neuronal survival by providing trophic factors (Fig. 2). A convincing demonstration of neuron-derived neuronotrophic factors requires elimination of putative contributions from non-neuronal cells, a large and sufficiently homogeneous population of neurons to enable biochemical studies and, in addition, sensitive test systems for monitoring neuronotrophic activity. The first set of criteria is met by cultured purified chromaffin cells (Unsicker and Müller 1981), which makes them particularly attractive as a tool to study this issue. Neuronal cells to be tested for a response to putative chromaffin cell-derived neuronotrophic factors would have to include those, which have chromaffin cells or the adrenal medulla, respectively, as their targets: preganglionic spinal cord, sensory dorsal root ganglion and sympathetic neurons (Fig. 1).

Neuronotrophic Activities in Cell Culture Media Conditioned by Purified Bovine Chromaffin Cells

Bovine chromaffin cells were freed from adrenal medullary non-chromaffin cells to better than 95% purity and cultured in fully defined serum-free media for 48 h. The conditioned medium carried neuronotrophic survival supporting activity (cf. Fig. 3) for all peripheral neurons tested (embryonic chick ciliary, dorsal root and sympathetic ganglia; cCG_8, $cDRG_8$, cSG_{11}; for abbreviations see Table 1; for rationale and details of assays, cf. Varon et al. 1983), but not for hippocampal neurons from E18 rats, which served as the CNS neuron model in our early studies. The activity(ies) was sensitive to trypsin and heat and not inhibited by (mouse) anti-NGF antibodies (which cross-react with bovine NGF) suggesting that we were dealing with one or several proteinaceous substances immunologically unrelated to NGF.

100 uM carbachol added to the cultured chromaffin cells for 2 h caused a 4- to 10-fold increase in the trophic units (TU) contained in the conditioned medium. A concomitant determination of chromogranin (by ELISA using a monoclonal antibody to chromogranin A) and catecholamines (by hplc and amperometric detection) revealed increa-

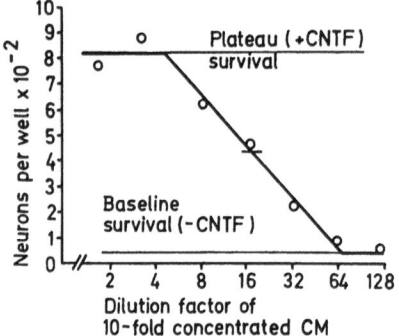

Fig. 3. Dose-response curve of culture medium conditioned by purified chromaffin cells (CM) towards 8-day chick embryo ciliary ganglion (CG) neurons. Saturating concentrations of CM supported the
survival of CG neurons at plateau values identical to those achieved
with CNTF. Half-maximal survival occurred at a 16-fold dilution of
the 10-fold concentrated CM, i.e. the original material contained
1.6 trophic units (TU) per ml

Table 1. Neuronotrophic activities in bovine chromaffin vesicles

	chick rat	Age	Target neuron	Abbreviation	TU/mg protein
Peripheral Nervous System	c	E 8	ciliary ganglion	cCG_8	~ 100
	c	E 8	dorsal root ganglion	$cDRG_8$	~ 100
	c	E 10	dorsal root ganglion	$cDRG_{10}$	~ 50
	c	E 11	lumbar sympathetic ganglia	cSG_{11}	~ 200
	r	new-born	superior cervical ganglion	$rSCG_n$	~ 70
	r	new-born	nodose ganglion	$rNOD_n$	none
	r	P 8	adrenal chromaffin cells	rAD_8	activity present, titer to be determined
CNS	c	E 6	spinal cord	cSC_6	~ 50
	r	E 18	hippocampal	$rHIPP_{18}$	~ 50

ses of these storage granule components by a factor of 10 and 14, respectively. Although these data did not suggest a strictly stoichiometrical release of neuronotrophic activities together with chromogranin A and catecholamines, they prompted us to further investigate the possibility that neuronotrophic activities were stored in chromaffin granules.

Neuronotrophic Activities in Isolated Chromaffin Granules

Chromaffin granule proteins isolated by established methods and free of lysosomal marker enzymes were tested for neuronotrophic, survival promoting activity on the peripheral and central neurons listed in Table 1. Depending on the type of neuron employed 5-20 ug of vesicle protein per ml were required to support half-maximal survival. The material was sensitive to heat (60°C, 1 h) and protease (trypsin) treatment.

Established Components of Chromaffin Granules and Their Putative Role as Neuronotrophic Agents

The low titers of neuronotrophic activities determined in the above experiments suggested the purification of active molecules to be an extremely difficult task. We therefore started with an attempt to identify established constituents of chromaffin granules in terms of putative neuronotrophic functions.

Chromogranins are the major catecholamine storage vesicle soluble proteins. Their functions are still enigmatic. In order to test for a putative neuronotrophic function polyclonal antibodies against chromogranin A, B and C and a monoclonal antibody against chromogranin A were added in excess to chromaffin vesicle proteins in bioassays using cCG$_s$ and cDRG$_s$ neurons. The antibodies failed to inhibit the survival promoting activity of the soluble vesicular proteins. It remains to be seen, however, whether the antibodies may immunoprecipitate the activity(ies). This experiment will be crucial since we have found that affinity-purified chromogranin A carries neuronotrophic activity for sensory, but not ciliary ganglion neurons. The activity was low in terms of TU/mg protein suggesting that only a minor breakdown product or a chromogranin-associated contaminant was responsible for the effect.

Neuropeptides are important constituents of chromaffin granules. A large number of peptides were tested on cCG$_s$, cDRG$_s$, and cSG$_{11}$ neurons. Positive results were obtained for met- and leu-enkephalin and somatostatin which fully supported the survival of cSG$_{11}$ neurons, but had only little effect on DRG and none on CG neurons. Half-maximal effects were obtained with doses of 5×10^{-6} M. Further experiments will have to clarify, whether the effects of the peptides can be blocked by specific antagonists and what fraction of the initially applied neuropeptide is actually present in the cultures after 24 h.

Towards Purification of Neuronotrophic Factors from Bovine Chromaffin Granules

Fractionation of chromaffin vesicle proteins by hplc using the molecular sieve columns SW 2000 and 3000 revealed that a CNTF-like activity (in terms of neuronal target cells being supported: cCG$_s$, cDRG$_{10}$, cSG$_{11}$, but not cDRG$_s$) resided in fractions covering relative molecular masses of 20 to 36 kd. The M$_r$ value for mammalian CNTF is 24 kd (Manthorpe et al. 1986). Activity addressing DRG, but not CG neurons was present in several fractions with peaks in the range of M$_r$ 40 to 50 as well as 70 to 80 kd. Studies are in progress to ascertain that CNTF is contained in chromaffin vesicles. We have recently shown that CNTF is stored in neuroblastoma cells, which are closely related to chromaffin cells (Heymanns and Unsicker 1987). Other vesicle constituents have not appeared so far to be promising candidates in our search for neuronotrophic factors. Ultrafiltration and removal of molecules smaller than 10 kd increased rather than reduced the survival supporting activity of the vesicle contents for cCG$_s$ and cDRG$_s$ neurons suggesting that molecules like catecholamines and ATP had not to be tested for trophic effects.

Conclusions

Experiments described in this article have provided evidence for the occurrence of neuronotrophic factors in the transmitter storing organelle of chromaffin cells. Our in vitro studies are only a first step towards the recognition of molecules possibly involved in trophic interactions of neurons. Purification of these molecules, production of antibodies and their in vivo administration will be required for clarifying the physiological relevance of these agents.

References

Heymanns J, Unsicker K (1987) Neuroblastoma cells contain a ciliary neuronotrophic factor. (submitted)

Manthorpe M, Skaper SD, Williams LR, Varon S (1986) Purification of adult rat sciatic nerve ciliary neuronotrophic factor. Brain Res 367: 282–286

Oppenheim RW (1985) Naturally occurring cell death during neural development. TINS 8: 487–493

Unsicker K (1983) Cell and tissue culture studies on the sympatho-adrenal system. In: Elfvin L-G (ed) Autonomic ganglia. Wiley, Chichester, pp 475–505

Unsicker K, Müller TH (1981) Purification of bovine adrenal chromaffin cells by differential plating. J Neurosci Meth 4: 227–241

Varon S, Adler R, Manthorpe M, Skaper SD (1983) Culture strategies for trophic and other factors directed to neurons. In: Pfeiffer SE (ed) Neuroscience approached through cell culture, Vol 2. CRC Press, Boca Raton, Florida, pp 53–57

Localization of Lysosomal Enzymes in Chromaffin Cells of the Rat Carotid Body

P. Böck

Abteilung für Mikromarophologie und elektronische Mikroskopie, Universität Wien, 1090 Wien, Austria

Introduction

The first report on strong cytoplasmic reactivity for nonspecific
acid phosphatase activity (AcPase) in adrenomedullary cells was
given by Eränkö (1952). Subsequently, comparative studies revealed
granular reaction product for AcPase in both adrenaline- and noradrenaline-storing cells from various species, although in varying
intensities (Coupland 1965). Using electron histochemical methods,
AcPase was shown in lysosomal organelles, within Golgi lamellae, and
in chromaffin granules (Greenberg et al. 1966; Holtzmann and Dominitz 1968; Coupland et al. 1971; Mastrolia and Manelli 1979). However, AcPase was found only in chromaffin granules of adrenaline-
storing cells, but was never observed in specific granules of noradrenaline-storing cells (Coupland et al. 1971). The present paper
refers to dopamine-storing chromaffin cells in the rat carotid body.
The localization of AcPase and inorganic trimetaphosphatase (TMPase)
will be described with particular reference to chromaffin granules
and tubular lysosomes.

Material and Methods

Adult Sprague-Dawley rats were anesthetized with ether and perfusion
fixation was performed with 2.5% glutaraldehyde in 0.1 M cacodylate
buffer, pH 7.4, via the left ventricle. Perfusion with the cooled
(4°C) fixative lasted for 10 min only. Thereafter the carotid bifurcation was removed and the carotid body was dissected using a microscope. During this time the tissues were saturated with the fixative. The carotid bodies were cut longitudinally into three or four
slices (with razor blades), for use for incubation. The entire
preparation time, from the beginning of perfusion fixation, was 20
min.

After fixation the specimens were washed in 0.1 M cacodylate buffer,
pH 7.4, to remove the aldehyde. Incubation for AcPase was performed
according to Barka (1964) in a pH 5.0 beta-glycerophosphate medium,
at 37°C for 50 min. Staining for TMPase was done as described by
Oliver (1980) in a pH 3.9 trimetaphosphate medium, at 37°C for 35
min.

After incubation the specimens were washed in 0.5% acetic acid (3
min) and in distilled water, and postfixed in 1% osmium tetroxide in
veronalacetate buffer. Dehydration and embedding in Epon 812 was
done as usual. Thin sections were cut on a Reichert OmU3 ultramicro-
tome and studied in a Zeiss EM9 electron microscope, either
unstained or after short staining with alkaline lead citrate.

Experimental Brain Research Series 16
© Springer-Verlag Berlin · Heidelberg 1987

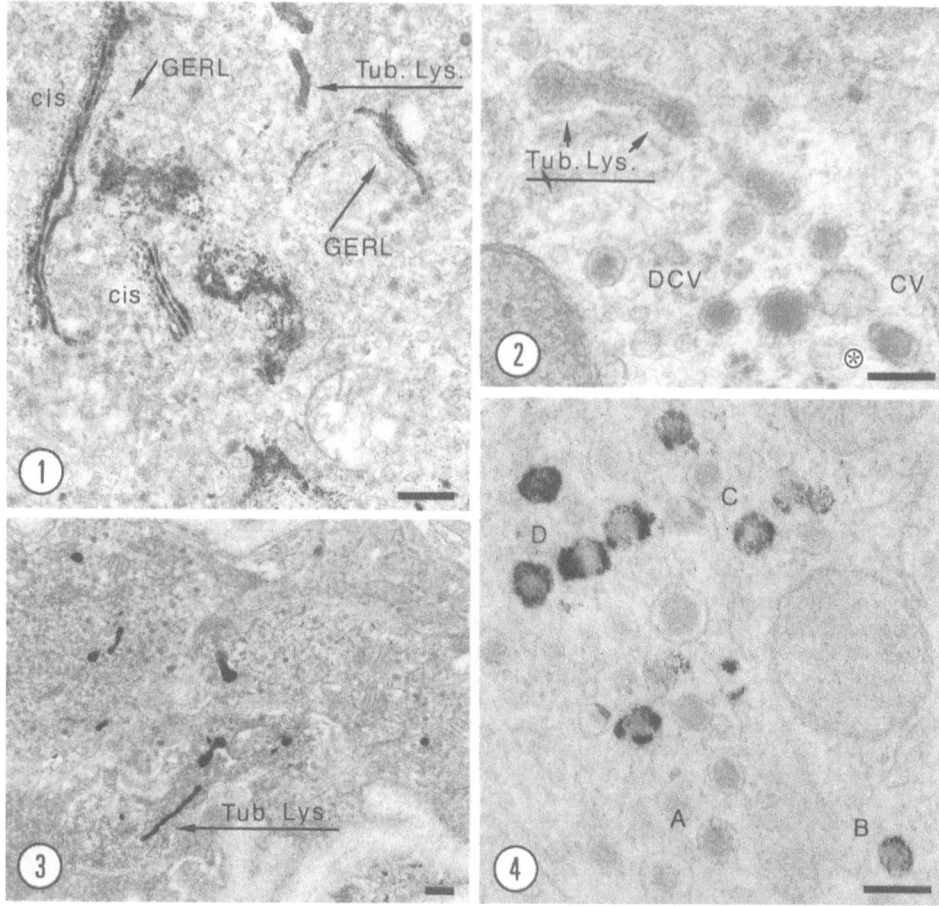

Fig. 1. Carotid body type I cell incubated for AcPase (50 min at
37°C). Reaction product is seen in all cisternae of the Golgi
stacks, the reaction intensity clearly being weaker in the cis
compartments as compared to the trans compartments. GERL elements,
by contrast, remain unstained. Elongated tubular lysosomes (Tub.
Lys) are reactive for AcPase. Slight staining with lead citrate. Bar
indicates 500 um

Fig. 2. Morphology of an elongated tubular lysosome (Tub. Lys) in
glutaraldehyde/osmium tetroxide fixed type I cell. The diameter of
the organelle varies within the same range as the diameters of
dense-cored vesicles (DCV). Asterisk indicates a profile that may
represeent a cross-sectioned tubular lysosome. CV: Coated vesicle.
Unstained. Bar indicates 200 um

Fig. 3. Carotid body incubated for TMPase (35 min at 37°C). Lead
deposits mark several lysosomal organelles, dense-cored vesicles do
not react. A longitudinally sectioned tubular lysosome (Tub. Lys.)
is easily discerned. Slight staining with lead citrate. Bar indicates 500 um

Fig. 4. Carotid body type I cell incubated for AcPase (50 min at
37°C). Some of the dense-cored vesicles show AcPase activity. Letters
A,B,C,D indicate a series of vesicles with increasing amounts of
lead deposits. There is no correlation between diameter or electron
density of the specific granules and its reactivity for AcPase.
Unstained. Bar indicates 200 um

Results

AcPase staining yields reaction product in all cisternae of the Golgi stacks in carotid body type I cells. Incubation for 50 min is short enough to visualize differing reaction intensities within the individual cisternae of a given stack. Reactivity for AcPase is lower in the cis constituent of the stack as compared to its trans constituent, the penultimate cisterna regularly being the most reactive one. GERL elements, by contrast, are clearly less reactive as compared to spatially related Golgi cisternae (Fig. 1).

Reaction product for AcPase is also found in lysosomal dense bodies, in multivesicular bodies, and in residual bodies and lipofuscin deposits. In addition, slender membrane-bound tubules, very similar to the "tubular lysosomes" described in albuminous gland cells (Oliver 1980), are seen to stain for AcPase (Fig. 1). After having been identified in AcPase preparations, these tubular lysosomes were readily found in glutaraldehyde/osmium preparations of carotid body tissue (Fig. 2). The organelles are characterized by size and shape, bounded by a thick unit membrane of the lysosomal type (measuring about 9.5 nm), and contain material of moderate electron density, which is separated from the limiting membrane by a clear space.

Reaction product for AcPase is also seen in dense-core vesicles of the type I cells. The amount of reaction product varies over a wide range. While some of the vesicles are free of lead deposits, others are strongly reactive. The reaction product is precipitated in the space between the dense core and the limiting membrane of the vesicle, where it is attached primarily to the inner surface of the membrane (Fig. 4).

Incubation for TMPase yields reaction product in lysosomal dense bodies, residual bodies, and multivesicular bodies. Elongated tubular lysosomes are regularly stained (Fig. 3). In a few cases Golgi cisternae were seen to stain. Dense-cored vesicles and GERL remained unreactive.

Discussion

The significance of AcPase reaction product in all cisternae of the Golgi stacks has already been discussed in detail by Böck (1986). The observation casts doubt on the reliability of the AcPase reaction as a marker of GERL elements and to distinguish between GERL and trans Golgi cisternae (as proposed by Novikoff and Novikoff 1977). Faint AcPase reaction in GERL but strong reactivity in Golgi cisternae nearby suggests that lysosomal elements are derived from Golgi cisternae rather than from GERL elements.

Tubular lysosomes were identified in carotid body type I cells using morphological criteria and by means of TMPase staining. The organelles also stain for AcPase. Tubular (elongated) lysosomes originally were thought to stain exclusively for TMPase (Oliver 1980). However, AcPase staining of tubular lysosomes has been reported in normal as well as in adenomatous parathyroid gland cells (Shannon and Roth 1974; Hörandner et al. 1984) and in carotid body type I cells in the present study.

A certain percentage of the dense-cored vesicles in the carotid body chromaffin cells stain for AcPase. This finding may indicate that these vesicles are derived from the Golgi cisternae, where the AcPase activity might be involved in degradation of nucleoside phosphates rather than acting as a lysosomal enzyme (Farquhar and Palade 1981). This interpretation is in agreement with the observed absence of TMPase activity in the dense-cored vesicles, which is thought to be a true lysosomal hydrolase (Doty et al. 1977). As a consequence, the AcPase positive vesicles should be considered as the younger ones. The interpretation that AcPase positive vesicles already had fused with a primary lysosome and therefore are old vesicles to be degraded is unlikely because of the absence of TMPase staining, an acknowledged lysosomal characteristic.

The presence of AcPase both in Golgi cisternae and in some of the dense-cored vesicles, most probably in younger ones, may indicate that final modification of the contents (proteolytic processing, glycosylation) is not achieved until the vesicles have left the Golgi region.

References

Barka T (1964) Electron histochemical localization of acid phosphatase activity in the small intestine of mouse. J Histochem Cytochem 12: 229-238

Böck P (1986) Histochemistry of the carotid body: fine-structural localization of acid phosphatase activities. Neurol Neurobiol 16: 617-628

Coupland RE (1965) The natural history of the chromaffin cell. Longmans, London

Coupland RE, Mastrolia L, Weakley BS (1971) Localization of acid phosphatase in the adrenal medulla of the albino rat. Prog Brain Res 34: 455-464

Doty SB, Smith CE, Hand AR, Oliver C (1977) Inorganic trimetaphosphatase as a histochemical marker for lysosomes in light and electron microscopy. J Histochem Cytochem 25: 1381-1384

Eränkö O (1952) On the histochemistry of the adrenal medulla of the rat, with special reference to acid phosphatase. Acta Anat (Basel) 16, Suppl 17: 1-60

Fraquhar MG, Palade GE (1981) The Golgi apparatus (complex) - (1954-1981) - from artifact to center stage. J Cell Biol 91: 77s-103s

Greenberg R, Falk GA, Kolen CA (1966) Localization of acid phosphatase in the stimulated adrenal medullary cell of the rat. J Cell Biol 31: 42A

Holtzman E, Dominitz R (1968) Cytochemical studies of lysosomes, Golgi apparatus and endoplasmic reticulum in secretion and protein uptake by adrenal medulla cells of the rat. J Histochem Cytochem 16: 320-336

Hörandner H, Ellinger A, Pavelka M, Klaushofer K (1984) A population of acid phosphatase and trimetaphosphatase positive tubular organelles in human parathyroid glands. Changing structure with different Ca^{++} concentrations in vitro. Acta Endocrinol 105, Suppl 264: 2-3

Mastrolia L, Manelli H (1979) Localization of acid phosphatase in the suprarenal gland of Rana esculenta. Cell Mol Biol 25: 11-15

Novikoff AB, Novikoff PM (1977) Cytochemical contributions to differentiating GERL from the Golgi apparatus. Histochem J 9: 525-551

Oliver C (1980) Cytochemical localization of acid phosphatase and trimetaphosphatase activities in exocrine acinar cells. J Histochem Cytochem 28: 78-81

Shannon WA, Roth SI (1974) An ultrastructural study of acid phosphatase activity in normal, adenomatous and hyperplastic (chief cell type) human parathyroid glands. Am J Pathol 77: 493-506

Distribution of the Ca^{++} Binding Protein Chromagranin A in the Pancreatic Islets

M. Gratzl

Abteilung für Klinische Morphologie, Universität Ulm, 7900 Ulm, FRG

Introduction

More than 20 years ago, experimental data suggested that proteins
are released together with catecholamines from the adrenal medulla
upon stimulation (Banks and Helle 1965). Co-storage and co-release
of these proteins, named chromgranins, with catecholamines is now
well established (cf. Winkler et al. 1986). Chromogranin A is the
most abundant member of this acidic group of proteins, which has
also been detected in many other endocrine cells secreting by exocytosis (cf. Winkler et al. 1986). Previous investigations differed in
the precise cellular location of the antigen in the pancreatic islet
(O'Connor et al. 1983; Cohn et al. 1984; Wilson and Lloyd 1984;
Varndell et al. 1985). The present contribution summarizes recent
data on the cellular and subcellular distributions of chromogranin A
in the bovine pancreatic islet (Ehrhart et al. 1986) and relates
them to the Ca^{++} binding function of chromogranin A (Reiffen and
Gratzl 1986a,b).

Immunocytochemical Observations

Four peptides constitute the established islet hormones within the
mammalian endocrine pancreas: insulin, glucagon, somatostatin and
pancreatic polypeptide. Appropriately diluted antisera against these
hormones were used to identify the different endocrine cells (PAP-
technique). A polyclonal antibody directed against bovine chromogranin A was applied to consecutive sections. It was observed by this
procedure that the cells which reacted with antibodies against
insulin, glucagon or somatostatin also displayed immunoreactivity
against chromogranin A.

At the ultrastructural level (protein A–gold technique), chromogranin A was confined exclusively to the secretory vesicles of the
pancreatic endocrine cells. As an example, immunostaining of the
insulin containing secretory vesicles of pancreatic B cells is shown
in Fig. 1. The topography of the gold particles after staining with
the antibody against insulin was different from that observed with
the anti-chromogranin A antibody. Approximately 70% of the insulin
immunoreactivity was located in the center of the electron dense
core of the vesicles. Whereas 64% of the gold particles indicating
chromogranin A immunoreactivity were observed overlying the clear
halo and the outer layer of the electron dense core. Similar results
were obtained in the glucagon containing vesicles. This subvesicular
distribution of chromogranin A is very much reminiscent of the
distribution of Ca^{++} in the vesicles as revealed by the pyroantimo-
nate technique (cf. Lenzen and Klöppel 1984).

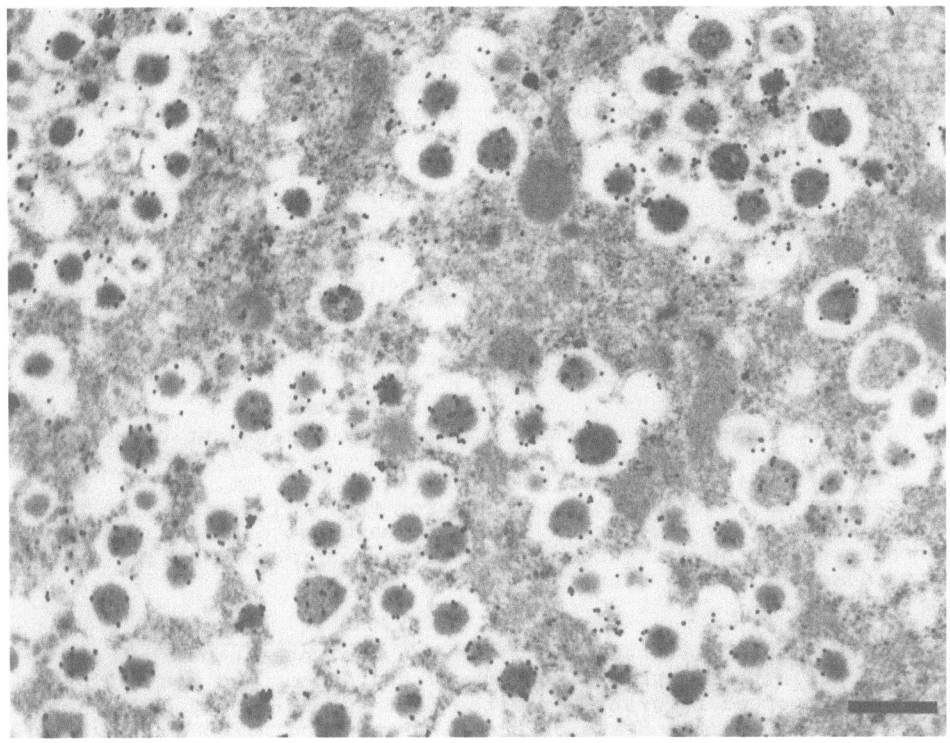

Fig. 1. Photomicrograph of an ultrathin section of a pancreatic B cell immunostained for chromogranin A by the protein A-gold technique. Bar = 0.5 um.
(Courtesy of M. Ehrhart, Abteilung Klinische Morphologie, Universität Ulm)

Functional Considerations

Low intracellular free Ca⁺⁺ concentrations are maintained in mammalian cells by Ca⁺⁺ transport systems present in the plasma membrane, endoplasmic reticulum and mitochondria. In endocrine cells, cytoplasmic Ca⁺⁺ plays a strategic role in the regulation of secretion by exocytosis and in these cells, secretory vesicles constitute an additional Ca⁺⁺ sequestration system. Secretory vesicles of practically all endocrine cells contain large amounts of Ca⁺⁺, and, in some cases, the transport system for this divalent cation across the membrane has been characterized in some detail (cf. Krieger-Brauer and Gratzl 1983; Saermark et al. 1983).

Ca⁺⁺ transported into the vesicles is largely bound to intravesicular substances (Bulenda and Gratzl 1985). Among the proteins present within chromaffin secretory vesicles, only chromogranin A binds Ca⁺⁺ (Reiffen and Gratzl 1986a,b). The presence of this protein within almost all endocrine cells, including the pancreatic islet cells may indicate that processes similar to those found in chromaffin vesicles work in many endocrine cells secreting by exocytosis. In this conjunction, the cytochemical observation that Ca⁺⁺ and chromogranin A exist together in the submembrane space of the secretory vesicle of the pancreatic islet cells is of particular interest (see previous chapters).

Markers for Endocrine Cells

A certain division of endocrine cells within the body have been described as "endocrine epithelial organs" (Feyrter 1938). The endocrine cells belonging to this division can be identified by cytochemical characteristics, by ultrastructural criteria (Pearse 1968) and by their properties related to the process of stimulus-secretion coupling (Fujita 1976).

During the investigation of the molecular events of exocytosis and its regulation, more and more constituents of these specialized endocrine cells have been identified. Besides chromogranin A, a variety of additional markers, all of which exhibit a very similar pattern of distribution in the body, have been detected in polypeptide secreting endocrine cells. Obviously, these components are linked to the function or the specific metabolic requirements of endocrine cells.

A Ca^{++} binding membrane protein with a molecular weight of 38,000 is among these cellular constituents (cf. Navone et al. 1986) as well as neuron specific enolase (Schmechel et al. 1978). Thus, a new generation of biochemical markers for the polypeptide secreting endocrine cells is available. They can be used to find out whether certain endocrine cells belong to the "endocrine epithelial organs" first described by Feyrter (1938).

Acknowledgements. I thank all my co-authors whose contributions are referenced in this paper. The work by the author and his collaborators is supported by grants from the Deutsche Forschungsgemeinschaft (Gr 681/2-2) and the State of Baden-Württemberg (Forschungsschwerpunkt No. 24). Appreciation is expressed to Miss Barbara Dell Page for wording of this manuscript and Mrs. B. Mader for typing it.

References

Banks P, Helle K (1965) The release of protein from the stimulated adrenal medulla. Biochem J 97: 40c-41c

Bulenda D, Gratzl M (1985) Matrix free Ca^{++} in isolated chromaffin vesicles. Biochemistry 24: 7760-7765

Cohn DV, Elting JJ, Frick M, Elde R (1984) Selective localization of the parathyroid secretory protein-I/adrenal medulla chromogranin A protein family in a wide variety of endocrine cells of the rat. Endocrinology 114: 1963-1974

Ehrhart M, Grube D, Bader MF, Aunis D, Gratzl M (1986) Chromogranin A in the pancreatic islet: cellular and subcellular distribution. J Histochem Cytochem 34: 1673-1682

Feyrter F (1938) Über diffuse endokrine epitheliale Organe. Barth-Verlag, Leipzig

Fujita T (1976) The gastro-enteric endocrine cell and its paraneuronic nature. In: Coupland RE, Fujita T (eds) Chromaffin, enterochromaffin and related cells. Elsevier, Amsterdam, pp 191-208

Krieger-Brauer H, Gratzl M (1983) Effects of monovalent and divalent cations on Ca^{++} fluxes across chromaffin secretory membrane vesicles. J Neurochem 41: 1269-1276

Lenzen S, Klöppel G (1984) Intracellular localization of calcium in pancreatic B-cells in relation of insulin secretion by the perfused ob/ob mous pancreas. Endocrinology 114: 1012-1020

Navone F, Jahn R, Di Gioia G, Stukenbrok H, Greengard P, De Camilli P (1986) Protein P38: an integral membrane protein specific of small clear vesicles of neurons and neuroendocrine cells. J Cell Biol (in press)

O'Connor DT, Burton D, Deftos LJ (1983) Chromogranin A: immunohisto-logy reveals its universal occurrence in normal polypeptide hormone producing endocrine glands. Life Sci 33: 1657-1663

Pearse AGE (1968) Common cytochemical and ultrastructural characte-ristics of cells producing polypeptide hormones (the APUD series) and their relevance to thyroid and ultimobranchial C cells and calcitonin. Proc Roy Soc B 170: 71-80

Reiffen FU, Gratzl M (1986a) Chromogranins, widespread in endocrine and nervous tissues, bind Ca^{++}. FEBS Lett 195: 327-330

Reiffen FU, Gratzl M (1986b) Ca^{++} binding to chromaffin vesicle matrix proteins: effects of pH, mg^{++}, and ionic strength. Biochemis-try 25: 4402-4406

Saermark T, Krieger-Brauer H, Thorn NA, Gratzl M (1983) Ca^{++} uptake to purified secretory vesicles from bovine neurohypophyses. Biochim Biophys Acta 727: 239-245

Schmechel D, Marangos PJ, Brightman M (1978) Neurone-specific enola-se is a molecular marker for peripheral and central neuroendocrine cells. Nature 276: 834-835

Varndell IM, Lloyd RV, Wilson BS, Polak JM (1985) Ultrastructural localization of chromogranin: a potential marker for the electron microscopical recognition of endocrine cell secretory granules. Histochem J 17: 981-992

Wilson BS, Lloyd RV (1984) Detection of chromogranin in neuroendo-crine cells with a monoclonal antibody. Am J Pathol 115: 458-468

Winkler H, Apps DK, Fischer-Colbrie R (1986) The molecular function of adrenal chromaffin granules: established facts and unresolved topics. Neuroscience 18: 261-290

Immunohistochemical Localization of a Nucleoside-Triphosphate-ADP-Phosphotransferase in Endocrine and Nervous Tissues

Ch. Heym[1], W. Kummer[1], G. Taugner[2], and I. Wunderlich[2]

[1] Anatomisches Institut, Universität Heidelberg, Im Neuenheimer Feld 307, 6900 Heidelberg 1, FRG
[2] Max-Planck-Institut für medizinische Forschung, 6900 Heidelberg, FRG

Introduction

Maintenance of high ATP levels in adrenal medullary chromaffin vesicles requires the transfer of energy-rich phosphate from the cytosol to intravesicular ADP. The enyzme nucleoside-ADP-phospho-transferase (NAPT) may be involved in this process (Taugner and Wunderlich 1979, 1981). The present immunohistochemical study of nervous and endocrine tissues was initiated to determine whether this enzyme is confined to secretory vesicles of chromaffin cells or whether it is also related to other amine- or peptide-storing cell types.

Materials and Methods

Ion exchange chromatography on DEAE-sephadex separated the soluble lysate of bovine adrenal medullary vesicles into 7 protein fractions (for technical details see Taugner and Wunderlich 1981). The first NAPT-enriched fraction (F_1) was re-chromatographed on Fractogel TSK WH 55. NAPT activity eluted as a single symmetric peak (FF_1), with a 100-150-fold enhanced specific activity as compared to the total soluble lysate. On SDS-PAGE the fractions with the highest specific activity (47-50) revealed aggregates of high molecular weight and a band of $\pm$60 KD. In the following protein fractions these bands waned accompanied by a decrease of NAPT specific activity and by an in-crease of the protein concentration due to the appearance of protein bands of lower molecular weight (Figs. 1, 2a). Applying antiserum to chromogranin/secretory protein I (Immuno Nuclear Corporation), F_1 was demonstrated to be devoid of chromogranin A immunoreactivity, which, however, was readily detected in the fractions five, six and seven. Antibodies against F_1 were raised in rabbits. Coupling of the antigen to bovine serum albumin via glutaraldehyde was found to be necessary to induce antibody production. Protein blots obtained from SDS-gels of I) total soluble vesicle lysate, II) F_1, III) FF_1, and IV) fractions lacking NAPT activity served to characterize the antisera. Binding of the primary antibody to the immobilized antigen was visualized using biotinylated anti-rabbit IgG, streptavidin-biotin-peroxidase complex, and 4-chloro-1-naphthol. The antiserum "Frieda" selectively recognized these two protein bands which are characteristic for FF_1 (Fig. 2b); hence, immunoreactions obtained with this antibody are referred to as NAPT-like immunoreactivity (NAPT-IR).

Immunohistochemical incubations ("Frieda" 1:2000, 24-48 h at $+4^{\circ}$C; biotin-streptavidin detection system) were performed on paraffin sections (7 um) of paraformaldehyde/picric acid-fixed guinea pig and

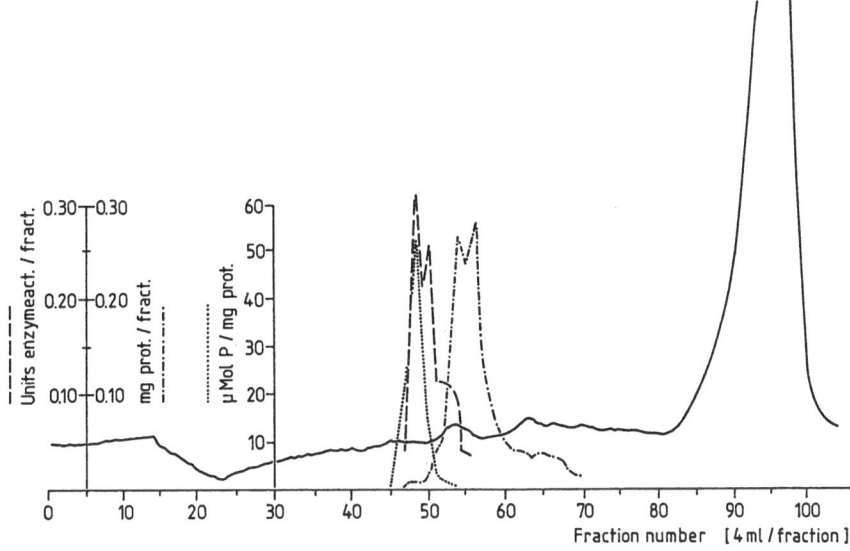

Fig. 1. Elution profile of NAPT from Fractogel TSK WH 55

human tissues. In addition, unfixed specimens were shock-frozen,
freeze-dried, exposed to hot (80°C) paraformaldehyde vapour, and
embedded in araldite. Serial semithin sections (0.5 um) cut from
these specimens were investigated fluorescence microscopically to
demonstrate catecholamines and indolamines, and were processed for
immunohistochemistry to study co-localization of NAPT-IR and other
compounds of secretory vesicles. For this purpose, antisera against
several opioid peptides (see Colombo et al., this volume), serotonin
(Immuno Nuclear Corporation; 1:5000), gastrin (code 92/2; 1:1000),
and glucagon (Amersham Buchler; 1:3000) were applied. Ultrastructural immunocytochemistry (protein A-gold technique) was performed on
guinea pig adrenal medulla, fixed in buffered glutaraldehyde
(1%)/picric acid and embedded in LR-White.

Results

Paraffin sections of guinea pig tissues displayed NAPT-IR in all
catecholamine-storing cells of the adrenal medulla, retroperitoneal
paraganglia, and carotid body. Numerous NAPT-IR cells were contained
in pancreatic islets. Single NAPT-IR cells were observed in the
adenohypophysis, thyroid gland (C-cells), tracheal epithelium (Fig.
4), basal layer of the esophageal epithelium (Fig. 5), and in the
mucosal epithelium of the stomach, duodenum, jejunum, ileum and
colon. In nervous tissue, NAPT-IR was seen only in paraganglionic
cells of autonomic ganglia (Fig. 6). Neither central (brain, spinal
cord) nor peripheral neurones (dorsal root ganglia, enteric ganglia,
pre- and paravertebral ganglia) including their terminal varicosities (e.g., in skin, spleen, gut and heart) displayed NAPT-IR. NAPT-
IR was not observed in the exocrine glands (sweat and salivary
glands) nor in the kidney. Human tissues studied included ileum,
colon and carotid body; the distribution of NAPT-IR cells was comparable to the guinea pig.

Semithin sections revealed NAPT-IR both in noradrenergic and in
adrenergic guinea pig adrenal medullary cells, most of which contai-

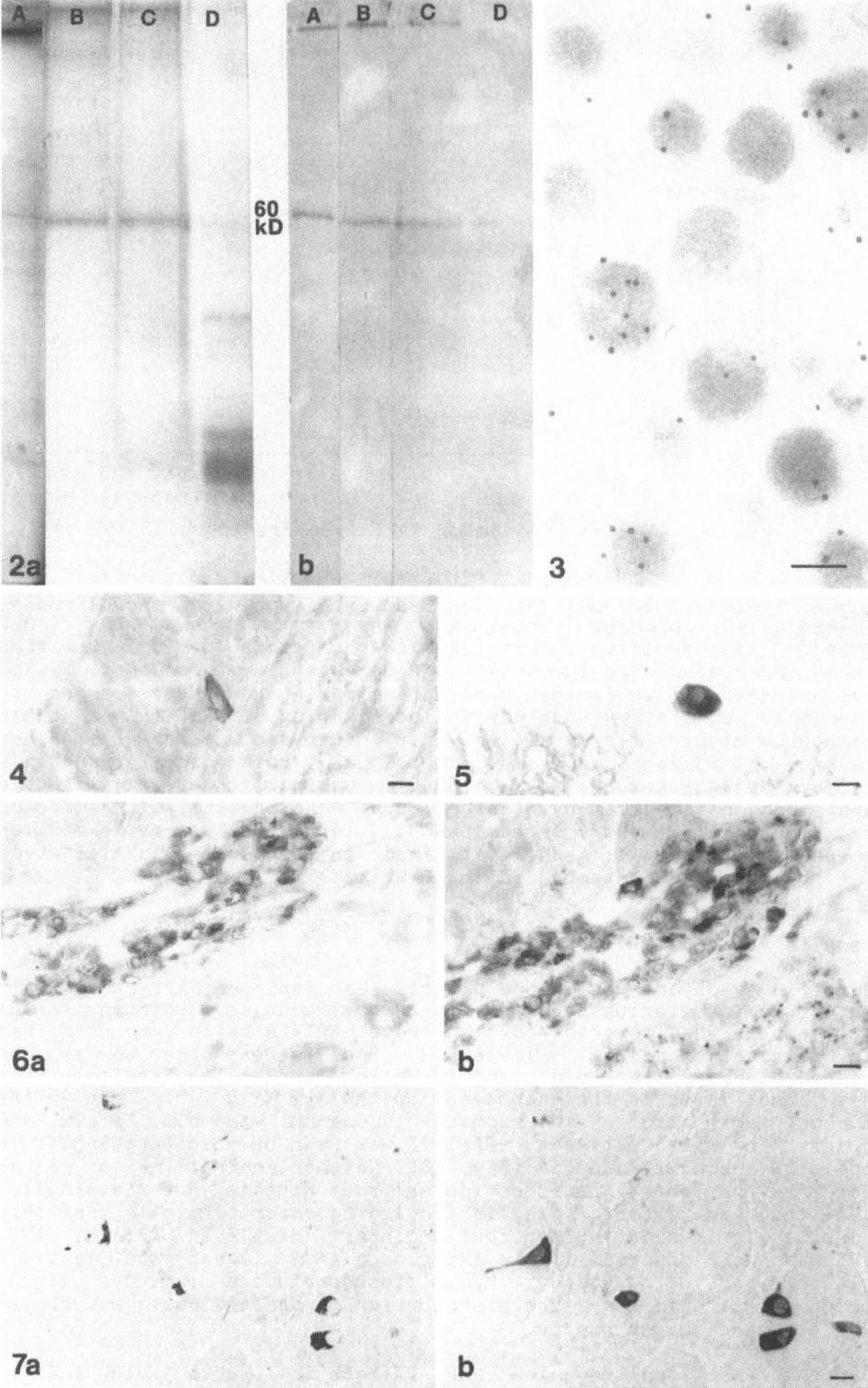

ned prodynorphin- or proenkephalin-related peptide-IR (see Colombo et al., this volume). Serotonin and NAPT-IR coexisted in all EC-cells of the human and guinea pig colon (Fig. 7) and of the guinea pig duodenum. In addition, NAPT-IR was occasionally found in duodenal gastrin-IR cells and in colonic glucagon-IR cells.

Ultrastructurally, NAPT-IR was confined to secretory vesicles in all adrenal medullary cell types, but not all granules within a single cell were labelled (Fig. 3).

Discussion

The general immunohistochemical distribution of NAPT-IR closely resembles that of chromogranins (see O'Connor et al. 1983; Rindi et al. 1986). Biochemically, however, NAPT is distinct from chromogranins by its different elution profile in ion exchange chromatography and by the lack of immunoreactive smaller molecular weight fragments, which are characteristic for proteins of the chromogranin family (Fischer-Colbrie and Frischenschlager 1985).

At present, the significance of the finding that both labelled and unlabelled dense-cored vesicles are present in single adrenal medullary cells is unclear. The negative results may be caused by reduced antigenicity due to tissue processing for electron microscopy; it is also possible that only a subset of chromaffin vesicles contains NAPT.

The widespread distribution and the variety of amines and peptides being co-localized with NAPT-IR indicate that this enzyme is not specifically linked to synthesis or processing of a particular hormone. Thus, its function may be related to more general functions, such as maintenance of intragranular environment. On the other hand, NAPT may act extracellularly following release via exocytosis. The ultrastructurally demonstrated localization of NAPT-

◀ **Fig. 2**. SDS-polyacrylamid gel-electrophoresis (a) and corresponding immunoblot (b). A: total soluble vesicle lysate. B: fraction 47. C: fraction 48. D: fraction 52. (B, C, D obtained from rechromatography of F_1 on Fractogel TSK WH 55). Significant NAPT activity was only found in fractions 47-50 (pooled as "FF₁")

Fig. 3. Immunoelectron micrograph of chromaffin vesicles in the guinea pig adrenal medulla, most of them being marked with 15 nm gold grains (bar indicates 100 nm)

Fig. 4. NAPT-immunoreactive cell body in the guinea pig tracheal respiratory epithelium (bar indicates 10 um)

Fig. 5. NAPT-immunoreactive cell body in the basal layer of the guinea pig esophageal stratified epithelium (bar indicates 10 um)

Fig. 6. Consecutive paraffin sections of a paraganglionic cell cluster in the guinea pig celiac-superior mesenteric ganglion (bar indicates 10 um) being immunoreactive to a) NAPT, b) met-enkephalin. (Note additional immunostaining of varicose nerve fibers!)

Fig. 7. Consecutive tangential semithin sections through the mucosal lamina of the human colon (bar indicates 10 um) with singly scattered immunoreactive cells to a) NAPT, b) serotonin

IR in the halo and in the core of adrenal medullary secretory vesicles is in accordance with this suggestion. Yet, direct proof for
NAPT release from endocrine cells is lacking.

References

Colombo M, Kummer W, Heym C (this volume) Immunohistochemistry of
opioid peptides in guinea pig paraganglia.

Fischer-Colbrie R, Frischenschlager I (1985) Immunological characterization of secretory proteins of chromaffin granules: chromogranins
A, chromogranins B, and enkephalin-containing peptides. J Neurochem
44: 1854-1861

O'Connor DT, Burton D, Deftos LJ (1983) Chromogranin A: immunohisto-
logy reveals its universal occurrence in normal polypeptide hormone
producing endocrine glands. Life Sci 33: 1657-1664

Rindi G, Buffa R, Sessa F, Tortora O, Solcia E (1986) Chromogranin
A, B and C immunoreactivities of mammalian endocrine cells. Histochemistry 85: 19-28

Taugner G, Wunderlich I (1979) Partial characterization of a phosphoryl group transferring enzyme in the membrane of catecholamine
storage vesicles. Naunyn-Schmiedeberg's Arch Pharmacol 309: 45-58

Taugner G, Wunderlich I (1981) Phosphoryl group transfer by a fraction of the soluble proteins of catecholamine storage vesicles. J
Neurochem 36: 1879-1892

Light and Electron Microscopic Immunocytochemical Study of the Localization of Actin and Tubulin in In-Situ Adrenal Chromaffin Cells of Normal and Restraint-Stressed Rats

H. Kondo

Department of Anatomy, School of Medicine, Kanazawa University, Kanazawa 920, Japan

Introduction

Chromaffin cells of the adrenal medulla represent modified post-ganglionic sympathetic neurons that are specialized to secrete catecholamines and neuropeptides from the chromaffin granules (Böck 1982; Kondo 1985). Release of the granule contents is considered to occur via exocytosis after a directed translocation of the chromaffin granules toward the inner surface of the plasma membrane (Pollard et al. 1978). Since the events involved in this secretory process of the chromaffin cells have many features in common with the process of the excitation—contraction coupling in muscles (Douglas 1975), it has been suggested that the secretory process of the cell is mediated by contractile proteins (Trifaro 1978). Recent immunological and immunocytochemical studies have revealed various contractile proteins, including actin, in the chromaffin cells (Aunis et al. 1980; Lee and Trifaro 1981). In addition, on the basis of findings that the microtubule—destructive drugs (such as colchicine) block the release of secretory products (Poisner 1975), numerous studies have suggested that microtubule integrity is essential for various kinds of endocrine secretion. However, detailed mechanisms of the involvement of these cytoskeletal components in the process of secretion remain to be elucidated.

Previous studies have shown that restraint—stress powerfully induces gastric ulcer and severe depletion of catecholamines from the adrenal medulla (Takagi et al. 1964; Kobayashi and Serizawa 1979). If cytoskeletal components are intimately involved in the secretory process of chromaffin cells, some changes in their intracellular organization should be visible in chromaffin cells of the stressed adrenal medulla.

The present study was undertaken to examine the precise localization of actin and tubulin in the adrenal chromaffin cells of normal and restraint—stressed rats at light and electron microscopic levels by means of the polyethylene glycol (PEG) embedding and subsequently deembedding method (Wolosewick 1980; Kondo 1984).

Materials and Methods

Restraint stress was imposed on adult albino rats weighing 200 g by tying them on a board by the limbs in a supine position for 2, 4 and 8 h. Normal and stressed rats were fixed by transcardiac perfusion for 10 min with 0.5% glutaraldehyde buffered in 0.1 M sodium cacodylate, pH 7.4. After extirpation of the adrenal medulla and additional fixation with the same fixative for 2 h, tissue was embedded in

polyethylene glycol (PEG—4000) according to a method described earlier in detail (Kondo 1984). Sections, 0.2—04 um thick, were cut,
mounted on cover slips or formvar—coated grids, and processed for
light and electronmicroscopic immunocytochemistry. Detailed proto-
cols of immunocytochemistry using the PEG method have been described
elsewhere (Wolosewick and De Mey 1982; Wolosewick et al. 1983). The
anti—actin antibody was made against native chicken gizzard actin (a
gift from Dr. I Yahara, Tokyo, Japan) and the anti—tubulin antibody
was made against SDS denatured dog brain tubulin (a gift from Dr. J.
De Mey, Brussels, Belgium). The sites of antigen—antibody reaction
were visualized with fluorescein isothiocyanate—labelled goat anti-
rabbit immunoglobulin for light microscopy and colloidal gold—conjugated IgG for electron microscopy.

Results and Discussion

In normal non-stressed adrenal medullae, immunofluorescence for
actin appeared as faint thin lines along the boundary of the chromaffin cells (Fig. 1). The intensity of the fluorescence was much
weaker than that of pericytes surrounding the sinusoidal vascular
lumen. No distinct actin fluorescence was detected in the perikaryal
cytoplasm. This circumferential distribution of actin in the in-situ
chromaffin cells is in contrast to the diffuse appearance of actin
in cultured chromaffin cells already reported by several authors
(Aunis et al.l 1980; Lee and Trifaro 1981). Tubulin immunofluorescence appeared as short strands and dots which were loosely and
randomly scattered in the entire perikarya of the chromaffin cells
except for the nuclear region (Fig. 2). The intensity of the fluorescence was weaker than that of intramedullary nerve fibers.

In stressed adrenal medullae, the actin fluorescence of almost all
the chromaffin cells was markedly intensified. The brightness was
almost equal to that of pericytes although it remained confined to
the cell margin as observed in the normal specimen (Fig. 3). Focal
prominent thickening of the fluorescent lines along cell margins was
frequently observed. The tubulin—fluorescent strands and dots appeared bright and they were densely conglomerated in the perikarya of
almost all the chromaffin cells (Fig. 4). The increase in intensity
of the actin fluorescence and in density of the tubulin-fluorescent

Figs. 1—4. Fluorescence micrographs showing the distribution of ▶
actin (Figs. 1, 3) and tubulin (Figs. 2, 4) in adrenal chromaffin
cells of normal (Figs. 1, 2) and stressed rats (Figs. 3, 4). Arrows
indicate focal prominent fluorescence for actin along the cell
margin. n = intramedullary nerve fibers. Bar indicates 25 um

Figs. 5, 6. Electron micrographs showing colloidal gold—labelings
(arrows) for actin (Fig. 5) and tubulin (Fig. 6) in chromaffin cells
of the normal rat. e = extracellular space; G = chromaffin granule;
M = mitochondria. Bar indicates 0.5 um

Fig. 7. Electron micrograph showing actin distribution in chromaffin
cells of the stressed rat. G = chromaffin granule; M = mitochondria;
P = regions of interdigitating processes. Bar indicates 0.5 um

Fig. 8. SDS—polyacrylamide gel electrophoresis of homogenates from
the adrenal medulla of normal (C) and stressed (S) rats. There is no
difference in density and width of bands for actin (A) and tubulin
(T) between the two specimens

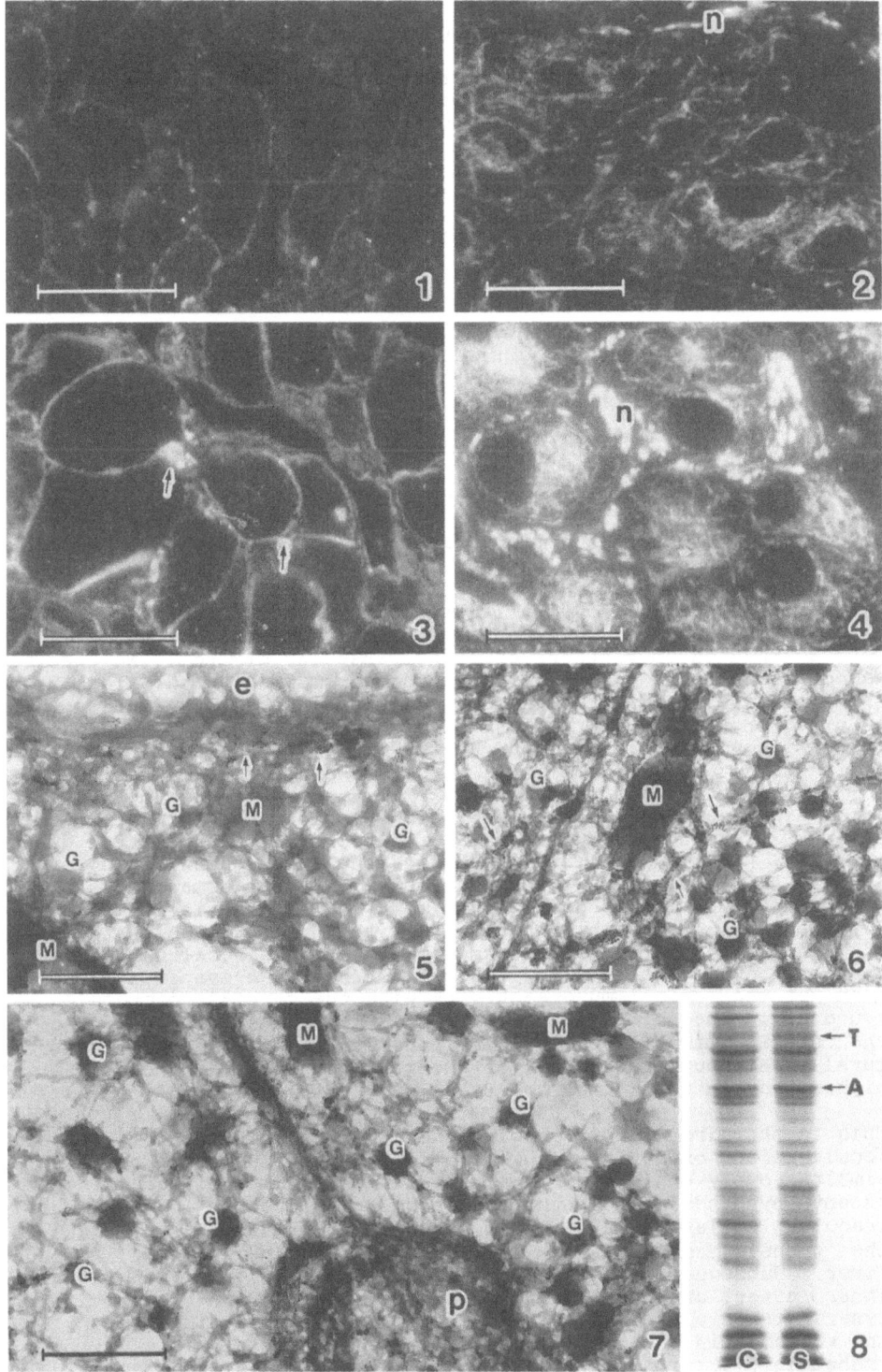

strands was recognized at 2 h after the initiation of the stress and was more pronounced with longer duration of the stress.

After electrophoreses on SDS polyacrylamide gels of the homogenates from the adrenal medulla of normal rats and those under the stress for 8 h, stained bands displaying the same electrophoretic mobility as actin and tubulin standards on the same gels showed no difference in density and width between the normal and the stressed specimens (Fig. 8). This suggests that the stress-induced increase in intensity of the actin fluorescence and in density of the tubulin fluorescent strands in the chromaffin cells represents a change in distribution, rather than a change in amount of individual cytoskeletal proteins within the cytoplasm.

The fine structure of normal chromaffin cells in PEG embedment-free sections has been fully described previously (Kondo et al. 1982). Briefly, ground cytoplasm of the chromaffin cells was fully occupied with a three-dimensional lattice of microtrabeculae that was contiguous with the surface of all intracellular organelles including chromaffin granules and the plasma membrane.

In electron microscopic immunocytochemistry using colloidal gold-IgG, the gold particles for actin were mainly deposited on the cytoplasmic cortices and microtrabeculae confluent with them underneath the plasma membrane. However, some of the gold particles were also scattered randomly on microtrabeculae far inside the cells in the normal specimen (Fig. 5). Some gold particles were deposited in a line for a short length on very thin microtrabeculae while others were aggregated on portions of microtrabecular strands. No specific association of the gold label with the chromaffin granules or mitochondria was found although some gold label were occasionally seen in close proximity to granules. The gold particles for tubulin were rather evenly distributed throughout the cytoplasm of normal chromaffin cells. They tended to be deposited on thicker microtrabeculae for a substantial length, although small aggregates of the particles were also seen randomly deposited on portions of microtrabecular strands without any topographical relation to the chromaffin granules (Fig. 6). In chromaffin cells of the stressed specimen, actin label, in cytoplasmic regions subjacent to the plasma membrane, was more dense than in the normal specimen (Fig. 7). Actin immunoreactivity was also seen within the interdigitating processes in the form of small spot aggregates, corresponding to the focal thickening of the actin fluorescence along the cell margin. No specific association of the actin immunoreactivity with the chromaffin granules was found as in the normal specimen. No marked changes in appearance of the tubulin immunoreactivity after the stress were seen ultrastructurally although the number of the gold particles in the cells was higher than in the normal specimen.

This is the first report showing the light and electron microscopic localization of actin and tubulin in in-situ chromaffin cells which simultaneously exhibit well-preserved ultrastructure. According to a recent hypothesis for the mechanism of secretion of chromaffin granules (Pollard et al. 1982), the Ca^{++}-dependent dissociation of the granule membrane from its cross-linked state with actin occurs first. Subsequently, the specific attachment of the granule to the inner aspect of the plasma membrane by means of the Ca^{++}-activated synexin takes place, resulting in the exocytosis. The finding that the chromaffin granules in the cell are held in place by the microtrabecular lattice may be compatible with this hypothesis. The marked changes in appearance of the actin and tubulin fluorescence by the stress may support the idea that these cytoskeletal proteins play some important roles in the secretion of the granule contents. However, no clear changes were found in the topographical relation of actin and tubulin with the chromaffin granules following stress. Further experimental studies remain to be performed for better understanding of the exocytotic process.

Acknowledgements. The author wishes to thank Dr. T. Aba and Mr. H. Saisu, Brain Research Institute, Niigata University, Niigata, Japan, for collaboration in SDS-PAGE. He also acknowledges Dr. I. Yahara, Tokyo Metropolitan Institute of Medical Science, Tokyo, Japan, and Dr. J. De Mey, Janssen Pharmaceutica, Brussels, Belgium, for kind gifts of anti-actin and tubulin antisera, respectively.

References

Aunis D, Guerold B, Bader M-F, Jaroslava C-T (1980) Immunocytochemical and biochemical demonstration of contractile proteins in chromaffin cells in culture. Neuroscience 5: 2261-2277

Böck P (1982) The Paraganglia. Handbuch der mikroskopischen Anatomie des Menschen, Vol 8. Springer, Berlin

Douglas WW (1975) Secretomotor control of adrenal medullary secretion: synaptic membrane and ionic events in stimulus secretion coupling. In: Blaschko H, Sayers G, Smith AD (eds) Handbook of Physiology, Endocrinology, Vol 6. American Physiological Society, pp 367-388

Kobayashi S, Serizawa Y (1979) Stress-induced degranulation accompanied by vesicle formation in the adrenal chromaffin cells of the mouse. Arch Histol Jap 42: 375-388

Kondo H (1984) Polyethylene glycol (PEG) embedding and subsequent de-embedding as a method for the structural and immunocytochemical examination of biological specimens by electron microscopy. J Electron Microsc Tech 1: 227-241

Kondo H (1985) Immunohistochemical analysis of the localization of peptides in the adrenal gland. Arch Histol Jap 48: 453-481

Kondo H, Wolosewick JJ, Pappas GD (1982) The microtrabecular lattice of the adrenal medulla revealed by polyethylene glycol embedding and stereo electron microscopy. J Neurosci 2: 57-65

Lee RW, Trifaro JM (1981) Characterization of anti-actin antibodies and their use in immunocytochemical studies on the localization of actin in adrenal chromaffin cells in culture. Neuroscience 6: 2087-2108

Poisner AM (1975) Microtubules and the adrenal medulla. Ann NY Acad Sci 253: 653-669

Pollard HB, Pazoles CJ, Cretz CE, Zinder O (1979) The chromaffin granule and possible mechanisms of exocytosis. Internat Rev Cytol 58: 159-197

Pollard HB, Cretz CE, Fowler V, Scott J, Pazoles CJ (1982) Calcium-dependent regulation of chromaffin granule movement, membrane contact, and fusion during exocytosis. Cold Spring Harbor Symp Quant Biol Vol 46: 819-834

144

Porter KR (1978) Organization in the structure of the cytoplasmic ground substance. IX Internat Congr Electr Microsc Toronto, Vol 3: 627-639

Takagi K, Kasuya Y, Watanabe K (1964) Studies on the drugs for peptic ulcer. A reliable method for producing stress ulcer in rats. Chem Pharm Bull 12: 465-472

Trifaro JM (1978) Contractile proteins in tussues originating in the neural crest. Neuroscience 3: 1-24

Wolosewick JJ (1980) The application of polyethylene glycol (PEG) to electron microscopy. J Cell Biol 86: 675-681

Wolosewick JJ, De Mey J (1982) Localization of tubulin and actin in polyethylene glycol embedded rat seminiferous epithelium. Biol Cell 44: 85-88

Wolosewick JJ, De Mey J, Meininger V (1983) Ultrastructural locali- zation of tubulin and actin in polyethylene glycol embedded rat seminiferous epithelium by immunogold staining. Biol Cell 49: 219- 226

Immunocytochemical Demonstration of GABA-Metabolizing Enzymes in the Superior Cervical and Nodose Ganglia of the Rat

O. Häppölä[1], H. Päivärinta[1], S. Soinila[1], J.-Y. Wu[2], and P. Panula[1]

[1] Department of Anatomy, University of Helsinki, Siltavuorenpenger 20A, 00170 Helsinki, Finland
[2] Department of Physiology, Pennsylvania State University, Milton S. Hershey Medical Center, Hershey, PA 170033, USA

Introduction

Increasing evidence indicates that gamma-aminobutyric acid (GABA), an inhibitory neurotransmitter in the mammalian central nervous system, is also present in different peripheral tissues. High concentrations of GABA, comparable to brain levels, have been found in the pancreatic islets, in the oviduct and in the myenteric plexus (Erdö 1985). The function and cellular location of GABA in peripheral tissues has not been well established. However, there is increasing evidence that GABA may act as a neurotransmitter in the myenteric plexus of the mammalian gut (Jessen et al. 1979). In the adrenal medulla, GABA has been localized in adrenal chromaffin cells (Kataoka et al. 1984).

In the rat superior cervical ganglion (SCG) GABA concentration is of the same order of magnitude as dopamine and noradrenaline concentrations (Bertilsson et al. 1976). The GABA-synthesizing enzyme, L-glutamic acid decarboxylase (GAD), has also been biochemically identified in the SCG (Bertilsson et al. 1976). The SCG has been demonstrated to accumulate radioactive GABA (Young et al. 1973), and to synthesize ^{3}H-GABA from ^{3}H-glutamic acid (Jessen et al. 1979). Furthermore, GABA receptors have been demonstrated in the sympathetic ganglion on neuronal cell bodies as well as on preganglionic nerve terminals. GABA has been shown to depolarize neurons of mammalian sympathetic ganglia (De Groat 1970), and to decrease neurotransmitter release from adrenergic nerve terminals in peripheral target tissues (Bowery and Hudson 1979).

GABA is also present in the rat nodose ganglion (Bertilsson et al. 1976), which contains the cell bodies of vagal afferent nerves, projecting from the thoracic and abdominal viscera to the central nervous system. In autoradiographic studies the nodose ganglion has been shown to accumulate GABA, and nodose ganglion cells have been shown to be sensitive for GABA. We now summarize our recent observations on immunohistochemical localization of GAD, and the GABA-inactivating enzyme, GABA-transaminase (GABA-T), in the rat SCG and in the rat nodose ganglion (Häppölä et al. 1987a,b).

Material and Methods

Adult Sprague-Dawley rats were perfused through the left ventricle of the heart with 4% paraformaldehyde in 0.1 M sodium phosphate buffer, pH 7.4, for 20-30 min. The tissue pieces were dissected out and immersed in the same fixative for two hours at +4°C. After fixation, the ganglia were transferred to 0.1 M sodium phosphate buffer, pH 7.4, containing 20% sucrose, for at least 24 hours at +4°C. Tissue specimens were frozen with carbon dioxide ice, and 10 um cryostat sections were cut. The sections were processed with indirect immunofluorescence technique (Coons 1958) for demonstration of GAD and GABA-T as described elsewhere in detail (Häppölä et al. 1987b). The specimens were examined with a Leitz Orthoplan or Dialux 20 fluorescence microscope equipped with epi-illumination and a specific filter block I 2 for FITC. For electron microscopical demonstration of GABA-T immunoreactivity in the SCG, the rats were perfused with 2% paraformaldehyde and 0.1% glutaraldehyde in 0.1 M sodium phosphate buffer, pH 7.4, for 20-30 min. The SCG were immediately removed and immersed in the same fixative for additional 2 hours. They were then washed in 0.1 M sodium phosphate buffer, pH 7.4, overnight, and 30 um vibratome sections were cut. The sections were incubated free floating in plastic petri dishes containing the GABA-T antiserum diluted 1:500 in PBS for 48 hours at 4°C. The peroxidase/anti-peroxidase method was applied on sections as described earlier (Häppölä et al. 1987b). Small pieces of sections were then treated with 1% OsO₄ for 1 hour, dehydrated in a graded ethanol series and embedded in an Epon-Araldite mixture. Ultrathin sections were examined without poststaining in a JEOL 100S electron microscope at 40 kV.

L-glutamic acid decarboxylase and GABA transaminase were purified and antisera produced in rabbits as described earlier (Wu 1976). These antisera have been characterized and used extensively in immunohistochemical studies in the central nervous system and they stain specifically neuronal GAD and GABA-T (Wu 1976). In the present study, no staining was observed, when the sections were incubated with normal rabbit serum or when the specific antisera were omitted.

Results and Discussion

Superior Cervical Ganglion. In the SCG, many PN cells were labelled with GAD and GABA-T antisera (Fig. 1A,B), but some batches of the GAD antiserum produced no staining in PN cells. The immunoreaction in neuronal cell bodies was moderate in intensity. In addition, small GAD- and GABA-T-immunoreactive cells were observed (Fig. 1C). The small immunoreactive cells resembled small intensely fluorescent (SIF) cells in size and morphology, and they also displayed tyrosine hydroxylase immunoreactivity in consecutive sections. GABA-T immunoreactivity was also localized in fiber-like structures around the PN cells, in fibers located in nerve trunks and around ganglionic blood vessels. Preganglionic denervation caused no change in GABA-T immunoreactivity in the SCG. In electron microscopy, GABA-T immunoreactivity was observed in the cytoplasm of PN cells, in neuronal processes and in glial satellite cells of the SCG (see Häppölä et al. 1986b). The results suggest that a subpopulation of principal neurons and SIF cells in the rat SCG contain histochemical markers for GABA synthesis and inactivation but their functional role in the ganglionic neurotransmission, as well as the location of GABA in this ganglion remains to be established.

Evidence has been presented that GABA may have a function in the peripheral target tissues of the SCG. Intravenous administration of low doses of GABA has been shown to induce hypotension and bradycardia by acting on peripheral GABA receptors in vascular and cardiac tissues. Autoradiographic studies on the target tissues of the SCG have revealed that GABA binding sites are present in the rat atria and in the rabbit pulmonary artery, as well as on sympathetic nerve terminals. Both in the pulmonary artery and in the atria of the

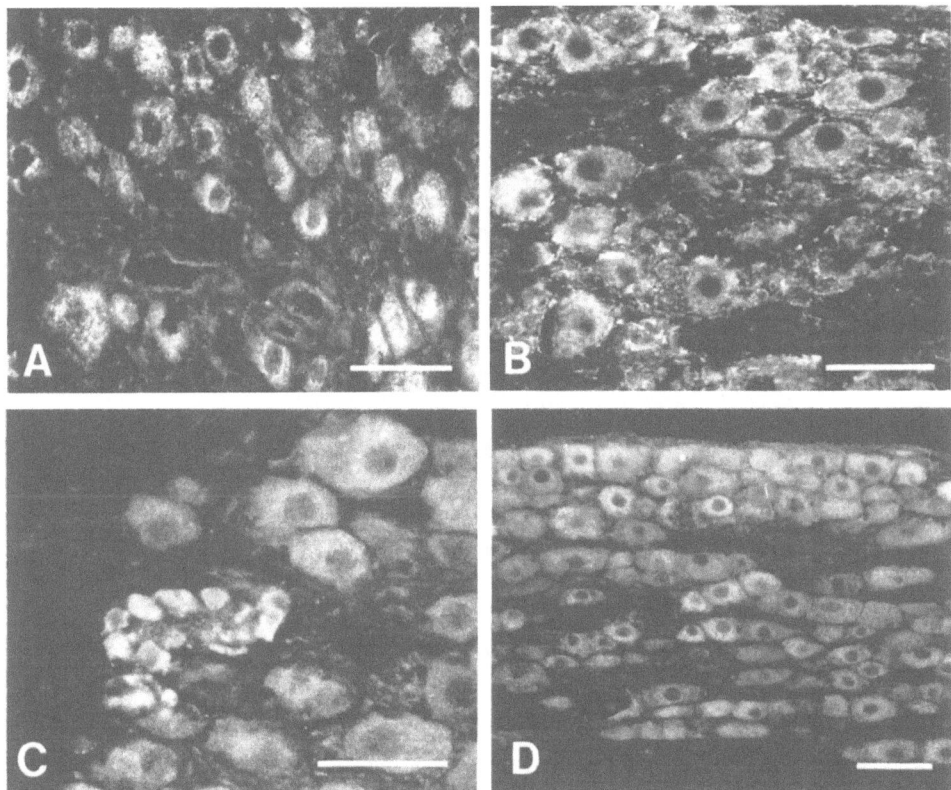

Fig. 1. A) GAD-immunoreactive principal neurons in the SCG of an adult rat. B) GABA-T-immunoreactive principal neurons surrounded by immunoreactive fiber-like processes in the SCG. C) GAD-immunoreactive SIF cells in the SCG. D) GAD-immunoreactive neurons in the rat ganglion nodosum. Bar represents 50 um

heart, GABA inhibits noradrenaline release suggesting that GABA-induced hypotension and bradycardia are in part mediated by presynaptic inhibition of postganglionic sympathetic transmission to the blood vessels and heart (for reference, see Tanaka 1985).

A synaptogenetic function for GABA in the SCG has also been suggested. Exogenous GABA treatment has been shown to result in induction and maintenance of free postsynaptic membrane thickenings of ganglionic neurons (Wolff et al. 1978), suggesting that GABA may have a role in regulation of plasticity of ganglionic neurons.

Ganglion Nodosum. The majority of neurons in the rat ganglion nodosum were immunoreactive for GAD and GABA-T (Fig. 1D). The GAD- and GABA-T-immunoreactive neurons were distributed in all parts of the ganglion. GABA-T immunoreactivity was also localized in fiber-like processes around the GABA-T-labelled neurons (Häppölä et al. 1987a).

These observations are in accordance with biochemical studies on the location of GABA in the rat ganglion nodosum and with immunohistochemical evidence for the location of GAD- and GABA-containing nerve terminals in the nucleus tractus solitarius, which is the projection zone of the neurons in the nodose ganglion (for references, see Häppölä et al. 1987a). Previously, it has also been demonstrated that the vagus nerve and the ganglion nodosum accumulate radioactive GABA (Young et al. 1973).

The present observations suggest that a subpopulation of neurons in the rat nodose ganglion contain the enzymes necessary for GABA metabolism, but the presence of GABA in the neurons of this ganglion remains to be established. However, it may be suggested that GABA may have a role in vagal afferent neurotransmission.

References

Bertilsson L, Suria A, Costa E (1976) Gamma-aminobutyric acid in rat superior cervical ganglion. Nature 260: 540-541

Bowery NG, Hudson A (1979) Gamma-aminobutyric acid reduces the evoked release of (^{3}H)-noradrenaline from sympathetic nerve terminals. Br J Pharmac 66: 108P

Coons AH (1958) Fluorescent antibody methods. In: Danielli JG (ed) General cytochemical methods. Academic Press, New York, pp 399-422

De Groat WC (1970) The actions of gamma-aminobutyric acid and related amino acids on mammalian autonomic ganglia. J Pharmacol Exp Ther 172: 384-396

Erdö SL (1985) Peripheral GABAergic mechanisms. TIPS 205-208

Häppölä O, Wu J-Y, Panula P (1987a) L-glutamic acid decarboxylase (GAD)- and GABA transaminase (GABA-T)-immunoreactive neurons in the nodose ganglion of the rat. Neurosci Lett (submitted)

Häppölä O, Päivärinta H, Soinila S, Wu J-Y, Panula P (1987b) Localization of L-glutamic acid decarboxylase (GAD) and GABA transaminase (GABA-T) immunoreactivity in the sympathetic ganglia of the rat. Neuroscience (in press)

Jessen KR, Mirsky R, Dennison ME, Burnstock G (1979) GABA may be a neurotransmitter in the vertebrate peripheral nervous system. Nature 281: 71-74

Kataoka Y, Gutman Y, Guidotti A, Panula P, Wooblewski J, Cosenza-Murphy D, Wu J-Y, Costa E (1984) Intrinsic GABAergic system of adrenal chromaffin cells. Proc Natl Acad Sci USA 81: 3218-3222

Tanaka C (1985) Gamma-aminobutyric acid in peripheral tissues. Life Sci 37: 2221-2235

Wolff JR, Joó F, Dames W (1978) Plasticity in dendrites shown by continuous GABA administration in superior cervical ganglion of adult rat. Nature 274: 72-73

Wu J-Y (1976) Purification and properties of L-glutamate decarboxylase (GAD) and GABA-transaminase (GABA-T). In: Roberts E, Chase T, Tower D (eds) GABA in nervous system function. Raven, New York, pp 7-55

Young JAC, Brown DA, Kelly JS, Schon F (1973) Autoradiographic localization of sites of (^{3}H)gamma-aminobutyric acid accumulation in peripheral autonomic ganglia. Brain Res 63: 479-486

Ultrastructure and
Synaptic Organization

Cytochemistry of Membranes in Sympathetic Ganglia

C. Andersson-Forsman and L.-G. Elfvin

Department of Anatomy, Karolinska Institutet, P.O. Box 60400, 10401 Stockholm, Sweden

Introduction

In the early electron microscopic studies of the sympathetic ganglia the structure of neurons, especially the inter- and intraganglionic connections were analyzed. In the sixties the characteristics of membranes of neurons and satellite cells have been further investigated (Elfvin 1961, 1962, 1963). These studies led to the suggestion that the plasma membranes of the two cell types are structurally different with possible differences in chemical composition.

New morphological techniques such as freeze-fracture and more sensitive and reliable histochemical as well as immunohistochemical methods have made it possible to further characterize membranes ultrastructurally. This review will present results of recent ultrastructural studies performed on the membranes of neurons and satellite cells in sympathetic ganglia. By applying the freeze-fracture technique directly or in combination with cytochemical probes further morphological evidence for structural differences between the membranes has been obtained. Furthermore, studies using cytochemical techniques applied to thin-section electron microscopy have shown a specific and distinct distribution pattern of membrane associated phosphatases (for a more comprehensive review see Andersson Forsman and Elfvin 1986b).

Freeze-Fracture

In freeze-fractured specimens of ganglia globular intramembrane particles (IMPs), 7–11 nm in diameter, were found on the fracture faces of the plasma membranes of both neurons and satellite cells. The particles were most frequent on the P face mainly as single particles with a total density of 1000–1500/um² on both membrane types (Elfvin and Forsman 1978).

In the satellite cell membrane of guinea pig and rabbit specific assemblies of particles have been found. These were small IMPs (6–7 nm) packed in orthogonal arrays of particles (OAPs) (Fig. 1) which were clearly different from the commonly found single 7–11 nm large IMPs in the membranes. Similar OAPs have been described in astrocytes of the central nervous system (Landis and Reese 1974; Gotow 1984). The exact nature and function of the OAPs is unknown. It is, however, clear that they represent protein-containing molecules. Functions such as mechanical support or ion transport have been suggested.

Experimental Brain Research Series 16
© Springer-Verlag Berlin · Heidelberg 1987

To localize cholesterol a reliable combined freeze-fracture cytochemical technique using filipin (Severs and Robenek 1983), a pylene
antibiotic, or tomatin, a saponin, was used. The distribution of
membrane deformations induced by either of the two probes has been
used qualitatively and semiqualitatively as measurement of the
amount of beta-hydroxysterols, in mammalian membranes mainly cholesterol, in the membranes of sympathetic ganglia (Andersson Forsman
1985a). The induced lesions were observed in the plasma membranes of
both neuronal and satellite cells. The membrane of the satellite
cells was found to contain a higher density of probe-induced lesions
than the neuronal membrane (Figs. 1, 2). When the relationship
between the deformations and the density of OAPs in the satellite
cell membrane was studied, no clear relation between the probe-
induced lesions and the density of OAPs was found. This is in contrast to the findings of Gotow (1984) who showed that the membrane
areas rich in OAPs in guinea pig astrocytes contain less filipin-
induced deformations.

When lesions in the neuronal membranes induced by the probes were
analyzed a non-homogeneous distribution between the perikarya and
the nerve processes was found. The perikaryal plasma membrane was
only sparsely filipin-labeled while a higher incidence of deformations was found on the processes. When the density of deformations
of various parts of the neuron was compared with the density of
IMPs, it was found that areas with a high density of IMPs contained
few probe-induced lesions and vice versa. These findings were interpreted to indicate a non-homogeneous distribution of cholesterol in
the neuronal membrane.

Taken together these results indicate possible differences in membrane composition between neurons and satellite cells in agreement
with the previous proposal based on findings on conventional thin
sectioned material of ganglia and nerve fibers that neuronal and
satellite cell membranes may structurally differ from each other
(Elfvin 1961, 1962).

Membrane-Associated Phosphatases

The transport of ions across plasma membranes is a crucial event in
the nervous system since cations are necessary for the maintenance
of the resting potential and for the propagation of the action
potential. Some of these functions are regulated by membrane-associated phosphatases. One of these, Na^+/K^+-ATPase has also been
suggested to correspond to the specific OAPs of glial plasma membranes (Landis and Reese 1974).

Morphologically the membrane-associated phosphatases can be studied
with histochemical techniques. The Wachstein-Meisel technique (1957)
based on the Gomori reaction has previously been used to localize
phosphatases. For the localization of Na^+/K^+-ATPase, in the present
material the methods according to Ernst (1972) or Mayahara et al.
(1981) have been applied. To study 5'-nucleotidase the method according to Kreutzberg et al. (1975) has been used. Finally to localize
Ca^{++}-ATPase the technique introduced by Ando et al. (1981) was used.

5'-Nucleotidase. In the sympathetic ganglia the 5'-nucleotidase was
located on the outer surface of the plasma membranes in both the
neuronal and the satellite cells (Fig. 3), although the latter
membrane appeared to exhibit more of the enzyme activity (Andersson

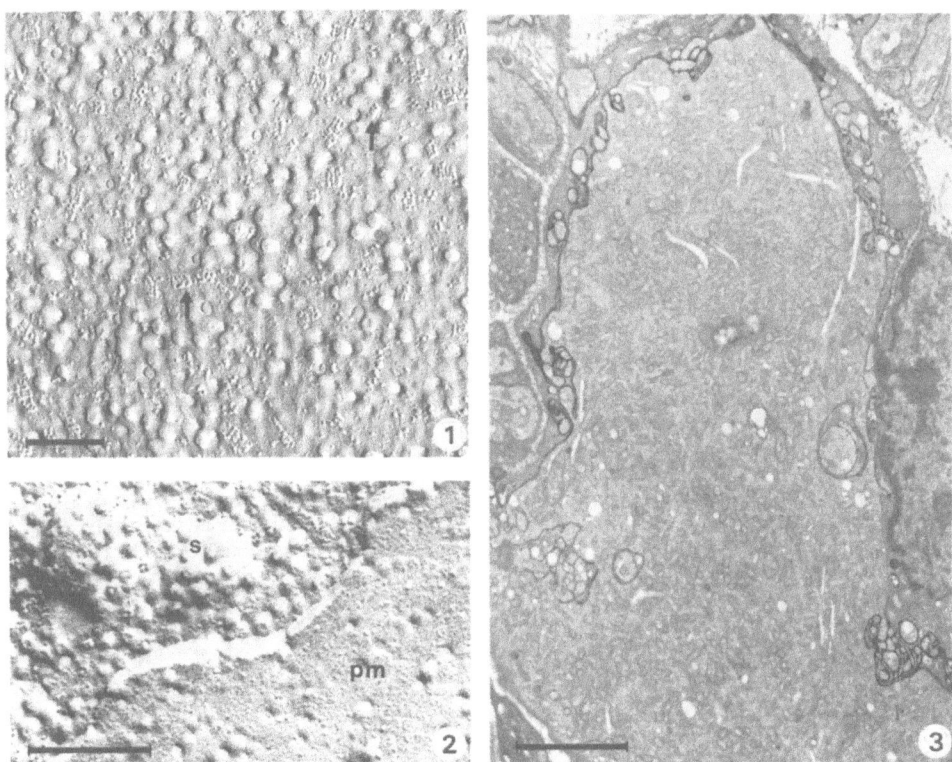

many OAPs (arrows) and showing typical filipin induced deformations.
Some of the deformations are pits while others are protuberances at
various stages. Note the very dense occurrence of deformations. The
OAPs appear unaffected by the probe. Bar indicates 0.1 um

Fig. 2. Replica showing variations in response to filipin of plasma
membranes from different cell types. The perikaryal plasma membrane
(pm) has less filipin lesions than the satellite cell plasma membrane (s). Bar indicates 0.5 um

Fig. 3. Survey electron micrograph showing boundaries between a
neuronal cell body and its surrounding elements containing activity
for 5'-nucleotidase. Bar indicates 1 um

Forsman and Elfvin 1983). Also some synaptic complexes contained
precipitates in the cleft region.

The exact function of 5'-nucleotidase in different systems is not
known but it is necessary for the general metabolism of purines in
the cells. Its function to hydrolyze nucleotide monophosphates such
as AMP to adenosine is important in many respects. Together with
other nucleosides, adenosine is important as a precursor for nucleotides, nucleic acids, various coenzymes and the second messenger
cyclic-AMP, all of which are essential for cell metabolism. In the
nervous system the enzyme has been suggested by Kreutzberg et al.
(1975) to be related to adenosine metabolism and neuromodulation.
The findings in the sympathetic ganglia of high activity located to

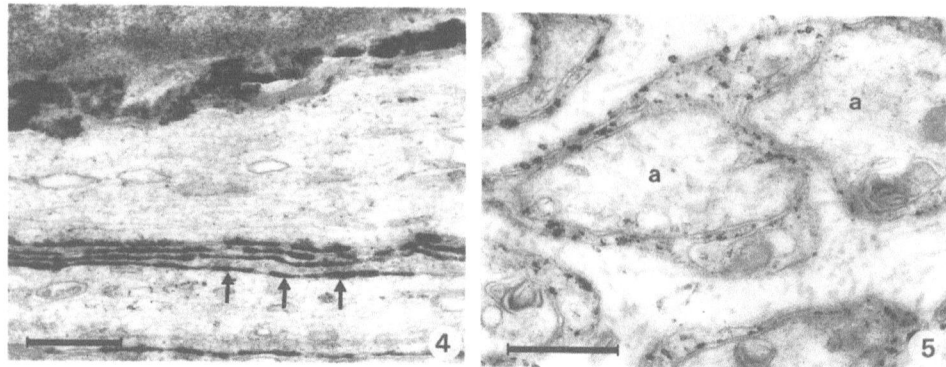

Fig. 4. Longitudinal section through unmyelinated nerve fibers. Note the staining of the cell boundaries (arrows) after incubation for Ca⁺⁺-ATPase activity. Bar indicates 0.5 um

Fig. 5. Section through a ganglion incubated for Na⁺/K⁺-ATPase activity in which satellite cells form thin layers around some nerve fibers (a). In the satellite cells electron dense material is attached to the cytoplasmic aspect of the plasma membrane. Bar indicates 0.5 um

the satellite cell plasmalemma were interpreted to indicate a high level of purine utilization by the satellite cells. The localization at synaptic sites indicates a role of purines possibly being active as neuromodulators since adenosine is known to hyperpolarize the postganglionic cells (Brown et al. 1979).

Calcium-ATPase. Ca⁺⁺-ATPase activity was localized on the outer surface of plasma membranes of both satellite cells and neurons in the sympathetic ganglia (Andersson Forsman 1985b). Most parts of the neuronal membrane contained reaction product which often was found to fill the extracellular space between closely packed cells (Fig. 4). In the synaptic cleft, electron dense deposits were often observed in the membranes.

The demonstration of Ca⁺⁺-ATPase in the plasma membranes of the sympathetic ganglia indicates an active involvement of the membrane of these cells in the regulation of intra- as well as extracellular calcium. Along with this function the enzyme is probably of significance in the relation of calcium to proteins which are dependent on this cation. The Ca⁺⁺-ATPase activity in the synaptic specializations may be related to exocytosis phenomena and transmitter release by participating in the regulation of nerve terminal calcium ion concentration. It has also been proposed that the enzyme might participate in acetylcholine uptake in the synaptic vesicles (Rothlein and Parsons 1979).

Sodium-Potassium-ATPase. The Na⁺/K⁺-ATPase activity was found on the cytoplasmic aspect of plasma membranes of both satellite cells and neurons (Andersson Forsman and Elfvin 1986a). The satellite cells clearly exhibited the enzyme activity more extensively than the neurons (Fig. 5). In the neuronal membrane, the reaction product was less frequent but specialized areas such as synapses often contained enzyme activity both in the pre- and postsynaptic membrane.

The detection of considerable amounts of reaction product in the satellite cell membrane is of particular interest because it indicates that this membrane is well equipped with enzymatic properties necessary for conducting active transport of cations. Several investigators have proposed that the glial cells play an important role in controlling the extracellular potassium ion concentration in the nervous system.

Acknowledgements. The present study was supported by grants from the Karolinska Institutet and the Swedish Medical Research Council (12X-5189).

References

Andersson Forsman C (1985a) Freeze-fracture cytochemistry of sympathetic ganglia. Distribution of filipin and tomatin induced membrane deformations in neurons and satellite cells. Histochemistry 82: 209-218

Andersson Forsman C (1985b) Ultrastructural studies on membranes in sympathetic ganglia, enteric ganglia and smooth muscle cells. Academic Thesis, Karolinska Institutet, Stockholm

Andersson Forsman C, Elfvin LG (1986a) Cytochemical localization of ouabain sensitive K^+-dependent para-nitrophenylphosphatase (transport ATPase) in the guinea pig sympathetic ganglia. J Submicrosc Cytol (in press)

Andersson Forsman C, Elfvin LG (1986b) The ultrastructure of membranes in sympathetic ganglia. Scan Electr Microsc (in press)

Ando T, Fujimoto K, Mayahara H, Miyajima H, Ogawa K (1981) A new one-step method for the histochemistry and cytochemistry of Ca^{2+}-ATPase activity. Acta Histochem Cytochem 14: 705-727

Brown DA, Caulfield MP, Kirby PJ (1979) Relation between catecholamine-induced cyclic AMP changes and hyperpolarization in isolated rat sympathetic ganglia. J Physiol 290: 441-451

Elfvin LG (1961) Electron microscopic investigations on the plasma membrane and myelin sheath of autonomic nerve fibers in the cat. J Ultrastruct Res 6: 388-407

Elfvin LG (1962) Electron microscopic studies on the effect of anisotonic solutions on the structure of unmyelinated splenic nerve fibers of the cat. J Ultrastruct Res 7: 1-38

Elfvin LG (1963) The ultrastructure of the superior cervical sympathetic ganglion of the cat. The structure of the ganglion cell processes as studied by serial sections. J Ultrastruct Res 8: 403-440

Elfvin LG, Forsman C (1978) The ultrastructure of junctions between satellite cells in mammalian sympathetic ganglia as revealed by freeze-etching. J Ultrastruct Res 63: 261-274

Ernst SA (1972) Transport adenosine-triphosphatase cytochemistry. II. Cytochemical localization of ouabain-sensitive, potassium-dependent phosphatase activity in the secretory epithelium of avian salt gland. J Histochem Cytochem 20: 23-38

Gotow T (1984) Cytochemical characteristics of astrocytic plasma membranes specialized with numerous orthogonal arrays. J Neurocytol 13: 431–448

Kreutzberg GW, Barron KD, Schubert P (1978) Cytochemical localization of 5'-nucleotidase in glial plasma-membrane. Brain Res 158: 247–257

Landis DMD, Reese TS (1974) Arrays of particles in freeze-fractured astrocytic membranes. J Cell Biol 60: 316–320

Mayahara H, Fujimoto K, Ando T, Ogawa K (1980) A new one-step method for the cytochemical localization of ouabain-sensitive potassium-dependent p-nitrophenylphosphatase activity. Histochemistry 67: 125–138

Rothlein JE, Parsons SM (1979) Specificity of association of a Ca^{++}/Mg^{++}-ATPase with cholinergic synaptic vesicles from Torpedo electric organ. Biochem Biophys Res Commun 80: 1069–1076

Severs NJ, Robenek H (1983) Detections of microdomains in biomembranes. An appraisal of recent developments in freeze-fracture cytochemistry. Biochim Biophys Acta 737: 373–408

Wachstein M, Meisel E (1957) Histochemistry of hepatic phosphatases at a physiologic pH with special reference to the demonstration of bile canaliculi. Am J Clin Pathol 27: 13–23

Synaptic Organization of the Intermediolateral Nucleus of the Thoracic Spinal Cord: Monoamine Histochemistry and Peptide Immunohistochemistry

T. Chiba and S. Masuko

Department of Anatomy, Saga Medical School, Nabeshima, Saga 840-01, Japan

Functional and morphological information concerning neurotransmitters or neuromodulators, including their candidates in the central nervous system related with autonomic functions, are accumulating. Intermediolateral nucleus (IML) of the spinal cord is one of such areas where numerous kinds of transmitters (or modulators) were found and extensive studies have been performed. A gelatinous synaptic region of about 50–75 x 100–150 um in a cross sectional plane of the IML provided a convenient locus to analyse synaptic organization of nerve terminals in relation to the preganglionic sympathetic neurons (PSN). Dendritic fields of PSN are mainly confined to this narrow longitudinal area which was confirmed by Golgi studies (Rethelyi 1972) as well as by intracellular HRP labelling in the cat (Dembowsky et al. 1985). The localization of PSN in the spinal cord was studied by retrograde transport of HRP and it was confirmed that IML pars principalis, IML pars funicularis, intercalated area and pericentral canal area were occupied by PSN in rodents (Chiba and Murata 1981; Dalsgaard and Elfvin 1979; Murata et al. 1982; Rando et al. 1981). As most PSN are located in IML pars principalis and pars funicularis in rodents (Rubin and Parves 1980), the following analyses on synaptic organization of the guinea pigs and rats will be concentrated in IML pars principalis and pars funicularis at the upper thoracic level.

PSN are small in size (10–15 x 20–30 um in diameter) and ovoid or spindle in shape. Most dendrites are thin with less than 2 um in diameter and axo-dendritic synapses with both convergent and divergent relations were commonly observed (Fig. 1). The diameter of axon varicosities ranged from 0.5–1.0 um to 1.0–2.0 um. The complex synaptic relations such as conglomerate synapses or synaptic glomeruli were sparse in the IML of guinea pigs or rats. Both symmetric and asymmetric synaptic specializations were found by an overwhelming majority of symmetric synapses (Fig. 1). Spherical, small (5 nm), clear and large (10–12 nm) dense cored vesicles were observed in axon varicosities and flat synaptic vesicles were encountered in axon varicosities of larger diameter with exclusively symmetric synaptic specializations (Fig. 1). Axon varicosities with flat synaptic vesicles were smaller in number (about 17%) and were usually seen after the fixation by an aldehyde-containing solution followed by longer rinsing in the phosphate buffer. The relative number of axons with flat synaptic vesicles increased after hemisection of the ipsilateral cervical spinal cord indicating intraspinal origin of them.

Nerve terminals containing noradrenaline (NA), dopamine (DA), adrenaline (A), as well as 5-hydroxytryptamine (5-HT) were densely distributed in the IML (see Björklund and Hökfelt 1984) and excitatory

Experimental Brain Research Series 16
© Springer-Verlag Berlin · Heidelberg 1987

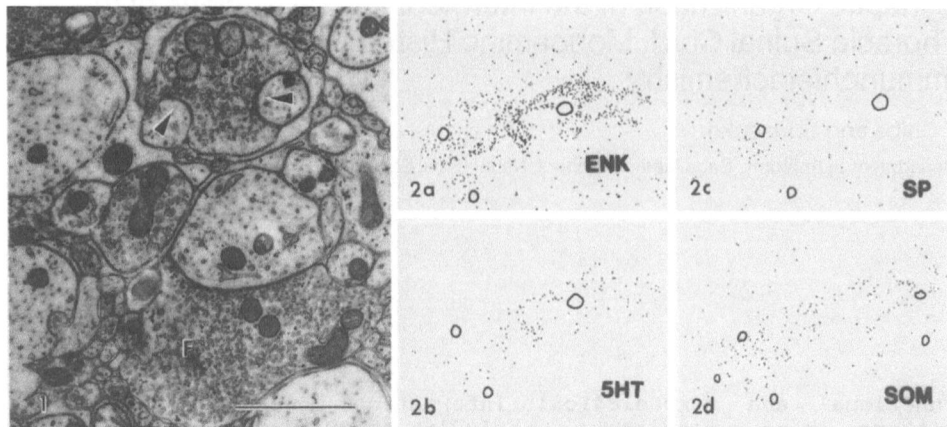

Fig. 1. Axo-dendritic synapses in guinea pig IML at Th2 segment. Divergent synapse (arrow heads) and a large axon varicosity with flat synaptic vesicles (F) forming a symmetric synapse are seen. Bar = 1 um

Fig. 2. Schematic drawing of immunoreactive axons in the 0.5 um serial sections of guinea pig IML. **a.** Enkephalin, **b.** 5-HT, **c.** Substance P, **d.** Somatostatin. Upper-right corner corresponds to dorso-lateral direction

influence on PSN was reported for NA, DA and 5-HT, while an inhibitory effect of 5-HT and A was also reported. Furthermore, modulatory influence of NA on PSN was found in the rat (Yoshimura et al. 1986). On the other hand, various peptide-containing nerve terminals were demonstrated to distribute in the IML. Substance P (SP) and enkephalin (ENK) nerves are most densely distributed, while somatostatin, oxytocin-neurophysin, vasopressin, GABA, glycine, acetylcholine, aspartate and glutamate-containing nerves are moderate to sparse in the IML (Chiba and Masuko 1986b; Holets and Elde 1982; Krukoff et al. 1985). Functional significance of peptides on PSN was suggested as either excitatory for SP, aspartate and glutamate, or inhibitory for GABA and glycine.

Distribution Pattern of Nerve Terminals Containing Peptides or Monoamines Examined by Serial 0.5 um Sections

In order to examine the relative distribution patterns of various transmitter containing nerve terminals, resin embedded thoracic spinal cords were cut serially at 0.5 um thickness and processed for immunohistochemistry after the resin was removed. Most axon varicosities were 0.5 to 2.0 um in diameter and co-localization of different transmitters in the same axon varicosity could be examined comparing adjacent 0.5 um sections with some restrictions (Figs. 2a, b, c, d). The distribution pattern of different transmitter-containing nerve terminals was considerably variable in the IML. ENK-immunoreactive (IR) axons were extremely dense and seen not only in the IML but also in the adjacent intermediate gray (Fig. 2a). 5-HT axons, on the other hand, were confined to the IML (Fig. 2b). SP axons were sparse in the lateral part of the IML and scattered in the medial part of the IML as well as in the adjacent intermediate gray (Fig. 2c). Somatostatin axons were moderately distributed and mainly in the IML (Fig. 2d). Comparing adjacent sections, coexistence of 5-HT and ENK was strongly suggested. Amon 135 5-HT axon varicosities counted in a 0.5 um section, 90 axon varicosities (67%) were confirmed to be positive for ENK-IR in the adjacent section. Co-localization of somatostatin and ENK (32%) or somatostatin and 5-HT (19%) was only partial, if any. Co-localization of 5-HT and SP (Apel et al. 1986) was unlikely at least in the IML pars principalis of the guinea pig.

Synaptic Organization of Axon Varicosities Containing Various Transmitters in the IML

To identify monoamine axons, an administration of 5-OH-DA or immunocytochemistry for NA and 5-HT were performed. 5-OH-DA was administered into the subarachnoideal space (5 mg/10 ul for 30 to 60 min.). Catecholamine (CA) axons were clearly identified by the appearance of dense core in the small spherical synaptic vesicles (Fig. 3). Labelled axons by this method were considered to be NA, DA as well as A containing axons. However, the guinea pig was devoid of A neurons in the central nervous system (Cumming et al. 1986) and only NA and DA were expected to be labelled in this particular species. Combined with the retrograde transport of HRP from the cervical sympathetic trunk, direct synaptic inputs from CA axons to the PSN were confirmed in both rats (Chiba and Masuko 1986a) and guinea pigs (Chiba and Masuko 1986b) (Fig. 3). To confirm an actual population of axon varicosities with synaptic contacts in the IML, 20 serial sections at 0.1 um thickness were cut. The diameter of CA axon varicosities ranged from 0.5 to 1.5 um and axon varicosities with more than 0.5 um in diameter observed in the middle section were followed to identify the synaptic specializations. Among 66 CA axon varicosities identified by 5-OH-DA, 54 (82%) were synaptic leaving 12 axon varicosities unidentified or nonsynaptic. Symmetric synapses were predominant and only 6 (11%) axons were with asymmetric synapses. The results strongly suggested that almost all CA axon varicosities in the IML of the guinea pig might be synaptic. NA-IR axons were confirmed to synapse on dendrites as well as soma in the IML. Possible axo-axonic contacts were observed from NA-IR as well as SP-IR axons to unidentified axon which in turn synapsed on the dendrite in the IML of rat (Figs. 4a, b, c, d). Relative density of axons containing divergent transmitters was examined by montage electron micrography of the cross sectional planes of the IML. ENK was most popular (ca. 35%), followed by 5-HT (ca. 20%), CA (ca. 20%), SP (ca. 13%) and others (Chiba and Masuko 1986b). 5-HT, ENK and SP-IR axons formed axo-dendritic as well as axo-somatic synapses in the IML (Figs. 4a, 5a,b, 6a,b). Direct synaptic contacts from SP-IR and ENK-IR axons to PSN were revealed (Chiba and Masuko 1986b) (Figs. 6a,b). Coexistence of SP and 5-HT was reported in rat IML (Appel et al., 1986) and more precise studies are necessary to elucidate other combinations of transmitters in the axons.

Thus, major inputs to the PSN seem to operate direct synaptic contacts. Taking into consideration that the PSN is devoid of axon collaterals (Dembowsky et al. 1985) and evidences for the existence of axo-axonic contacts, PSN in the IML may be controlled by feedforward rather than feedback mechanisms. Although it is far from complete to clarify the interaction of different neurotransmitters or neuromodulators in the IML, the present report added some basic morphological information concerning functional (transmitter-related) synaptic organizations in the IML.

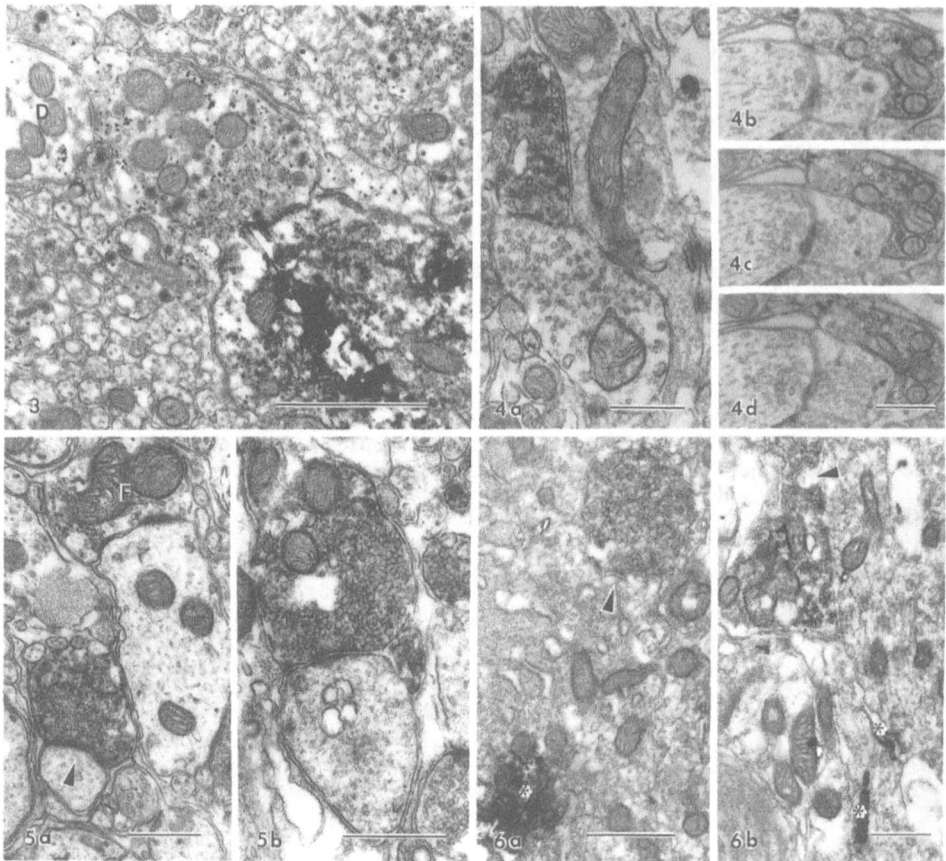

Fig. 3. Catecholamine axon varicosity labelled by 5-OH-DA forms divergent synapses on PSN soma and dendrite (D). PSN is labelled by retrogradely transported HRP. Guinea pig. Bar = 1 um

Fig. 4. Possible axo-axonic contacts from substance P axon in guinea pig (a) and NA axon in rat demonstrated in serial sections (b, c, d). Bar = 0.5 um

Fig. 5. Axo-dendritic synapse of enkephalin-IR axon (arrow head) in rat. Note that adjacent axon with flat synaptic vesicles (F) is immunonegative (a). Axo-dendritic asymmetric synaptic junction of 5-HT axon in guinea pig (b). Bar = 0.5 um

Fig. 6. Direct synaptic contacts (arrow heads) from enkephalin axon to PSN soma (a) and from substance P axon to PSN dendrite (b). PSN were labelled by retrogradely transported HRP (asterisks). Guinea pig. Bar = 0.5 um

In conclusion, axon varicosities with flat synaptic vesicles occupied 17% and had exclusively symmetric synapses. Catecholamine-(20%), 5-hydroxytryptamine- (20%), and enkephalin-containing (35%) axons were dense while substance P (13%) containing axons were moderate in population. Somatostatin, GABA, acetylcholine, oxytocin-neurophysin, vasopressin, glycine, asparatate and glutamate containing axons were sparse. Co-localization of 5-hydroxytryptamine and leu-enkephalin in the same axons was strongly suggested. Most catecholamine axon varicosities (75%) possessed synaptic specializations. Possible axo-axonic contacts were found from noradrenaline and substance P axons. Direct synaptic contacts from catecholamine, substance P and enkephalin axons to the preganglionic sympathetic neurons were demonstrated.

Acknowledgement. The authors are grateful to Miss Mariko Tsutsumi for technical assistance and for typing the manuscript.

References

Appel NM, Wessendorf MW, Elde RP (1986) Coexistence of serotonin- and substance P-like immunoreactivity in nerve fibers apposing identified sympathoadrenal preganglionic neurons in rat intermediolateral cell column. Neurosci Lett 65: 241–246

Björklund A, Hökfelt T (1984) Handbook of chemical neuroanatomy, vol 2. Elsevier, Amsterdam

Chiba T, Murata Y (1981) Architecture and synaptic relationships in the intermediolateral nucleus of the thoracic spinal cord of the rat: HRP labelling, catecholamine histochemistry and electron microscopic studies. J Neurocytol 10: 315–329

Chiba T, Masuko S (1986a) Direct synaptic contacts of catecholamine axons on the preganglionic sympathetic neurons in the rat thoracic spinal cord. Brain Res 380: 405–408

Chiba T, Masuko S (1986b) Synaptic structure of the monoamine and peptide nerve terminals in the intermediolateral nucleus of the thoracic spinal cord of the guinea pig (submitted)

Cumming P, Von Krosigk M, Reiner PB, McGeer EG, Vincent SR (1986) Absence of adrenaline neurons in the guinea pig brain: a combined immunohistochemical and high-performance liquid chromatography study. Neurosci Lett 63: 125–130

Dalsgaard CJ, Elfvin LG (1979) Spinal origin of preganglionic fibers projecting onto the superior cervical ganglion and inferior mesenteric ganglion of the guinea pig, as demonstrated by the horseradish peroxidase technique. Brain Res 172: 139–143

Dembowsky K, Czachurski J, Seller H (1985) Morphology of sympathetic preganglionic neurons in the thoracic spinal cord of the cat: an intracellular horseradish peroxidase study. J Comp Neurol 238: 453–465

Holets V, Elde R (1982) The differential distribution and relationship of serotoninergic and peptidergic fibers to sympathoadrenal neurons in the intermediolateral cell column of the rat: a combined retrograde axonal transport and immunofluorescence study. Neuroscience 7: 1155–1174

Krukoff TL, Ciriello J, Calaresu FR (1985) Segmental distribution of peptide- and 5-HT-like immunoreactivity in nerve terminals and fibers of the thoracolumbar sympathetic nuclei of the cat. J Comp Neurol 240: 103–116

Murata Y, Shibata H, Chiba T (1982) A correlative quantitative study comparing the nerve fibers in the cervical sympathetic trunk and the locus of the somata from which they originate in the rat. J Auton Nerv Syst 6: 323–333

Rando TA, Bowers CW, Zigmond RE (1981) Localization of neurons in the rat spinal cord which project to the superior cervical ganglion. J Comp Neurol 196: 73–83

Rethelyi M (1972) Cell and neuronal architecture of the intermediolateral (sympathetic) nucleus of cat spinal cord. Brain Res 46: 203–213

Rubin E, Purves D (1980) Segmental organization of sympathetic preganglionic neurons in the mammalian spinal cord. J Comp Neurol 192: 163–174

Yoshimura M, Polosa C, Nishi S (1986) Noradrenaline modifies sympathetic preganglionic neuron spike and afterpotential. Brain Res 362: 370–374

Synaptic Connections of Carotid Chemo- and Baroreceptor Primary Afferents in the Nucleus of the Tractus Solitarius in the Rat

I. Chen, R. W. Rieck, J. T. Weber, and R. D. Yates

Department of Anatomy, Tulane Medical School, 1430 Tulane Avenue, New Orleans, LA 70112, USA

Introduction

Central carotid sinus afferents (CSAs) have been shown to project primarily to the caudal half of the nucleus tractus solitarii (NTS). Although neurophysiological and light microscopic tracing studies indicate that the NTS is connected with other cardio-vascular centers within the brainstem, the morphological identity of the second order neurons of this carotid chemo- or baro-reflex circuitry has not well been established. Recently, we have developed techniques for the light and electron microscopic demonstration of both transganglionically transported horseradish peroxidase (HRP) that is localized within CSA terminals and neuromodulators or synthesizing enzymes within neurons that occupy the labeled afferent terminal fields (Chen et al. 1986). These techniques were used to investigate the synaptic structure and connections of the central CSAs to catecholaminergic neurons that are located within the caudal half of the NTS in the rat.

Materials and Methods

The proximal stump of the cut carotid sinus nerve (CSN) on one side was sealed in a polyethylene tube containing 50% HRP in eight Wistar-Kyoto rats. After a 72-hour survival period, the rats were perfused transcardially with phosphate buffered fixatives containing 4% paraformaldehyde (PFA) for 5 minutes and 2.5% glutaraldehyde + 0.5% PFA for 25 minutes. A one-in-four series of coronal vibratome sections (50 um), were collected for levels of the medulla that contain the NTS caudal to the obex. The first three series of sections were incubated with tetramethylbenzidine (TMB) at pH 4.0 followed by diaminobenzidine (DAB)-cobalt stabilization (Leman and Saper 1985). The last series of sections was incubated with the cobalt intensification of imidazole-DAB (Straus 1982). The sections from the first series were air dried on gelatin-coated slides for light microscopy. Sections from the second series were additionally processed for the immunocytochemical visualization of tyrosine hydroxylase (TH). The doubly stained sections were mounted on glass slides with buffered-glycerine, photographed, and then further processed with the 3rd and 4th series for electron microscopy.

Results

The TMB technique revealed numerous HRP-labeled fibers in the caudal half of both the ipsilateral tractus solitarius (TS) and the NTS.

Experimental Brain Research Series 16
© Springer-Verlag Berlin · Heidelberg 1987

A. Level of the Area Postrema (AP)

HRP-labeled fibers were observed primarily within three regions of the NTS.

1.) NTS ventrolateral to the TS: The number of labeled fibers in this region of the NTS was smaller than that in the other two regions. Some labeled fibers did appear to branch and terminate within this part of the NTS. TH-stained neurons were observed only occasionally in this region of the NTS.

2.) NTS dorsomedial to the TS: Many large-diameter, HRP-positive axons were located within this portion of the NTS. Some of the labeled fibers appeared to terminate within this portion of the NTS. Only a few small TH-positive neurons are located within this division of the NTS. Labeled fibers also crossed the dorsomedial border of the NTS to enter the AP which contained numerous TH-positive cells.

3.) NTS dorsal to the dorsal motor nucleus (DMN): This region of the NTS contained numerous small-diameter fibers that were HRP-positive and they often formed a mesh-like array that extended from the dorsal border of the DMN to join the group of labeled fibers within the previous region. Furthermore, many large TH-positive perikarya and processes were located within this portion of the NTS. At the caudal limit of the AP, some labeled fibers either crossed the midline or entered the DMN.

B. Caudal to the AP

The commissural subnucleus of the NTS was filled with labeled fibers which also extended to the contralateral side. In four cases, labeled fibers within the portion of the NTS ventrolateral to the TS could be traced into the A1 cell group of the ventrolateral medulla. Another group of labeled fibers were seen leaving the TS dorsolaterally to enter the spinal trigeminal nucleus (STN).

C. Ultrastructure

Afferent terminals that were HRP-positive were found in all the regions of the NTS examined. The labeled afferent terminals often occurred in linear arrays indicative of long terminals. The terminal profiles were usually irregular in shape and frequently were indented at locations where they made contact with other structures (Fig. 1). Large HRP-positive boutons (Fig. 2) were seen more frequently within the division of the NTS dorsomedial to the TS.

Fig. 1. HRP-labeled (cobalt imidazole-DAB) CSA terminal(s) in direct ▶ contact with dendrites (d) and unlabeled terminals (a)

Fig. 2. Two profiles of HRP-labeled (cobalt imidazole-DAB) terminal(s) located side by side and making synaptic contacts with two dendritic profiles (d). S, dendritic spine

Fig. 3. HRP-labeled (TMB) profiles of CSA terminal(s) intimately associated with soma and process of a small neuron. The origin of the small dendrite is indicated by an asterisk. The axo-dendritic synapse (arrow) is enlarged in the inset

Fig. 4. A very rare case of TMB-labeled (TMB) terminal in direct apposition to the TH-positive dendrite (TH). The former, in turn, synapses with the unlabeled terminal (a) and the latter, in turn, with the TH-negative dendrite (d)

165

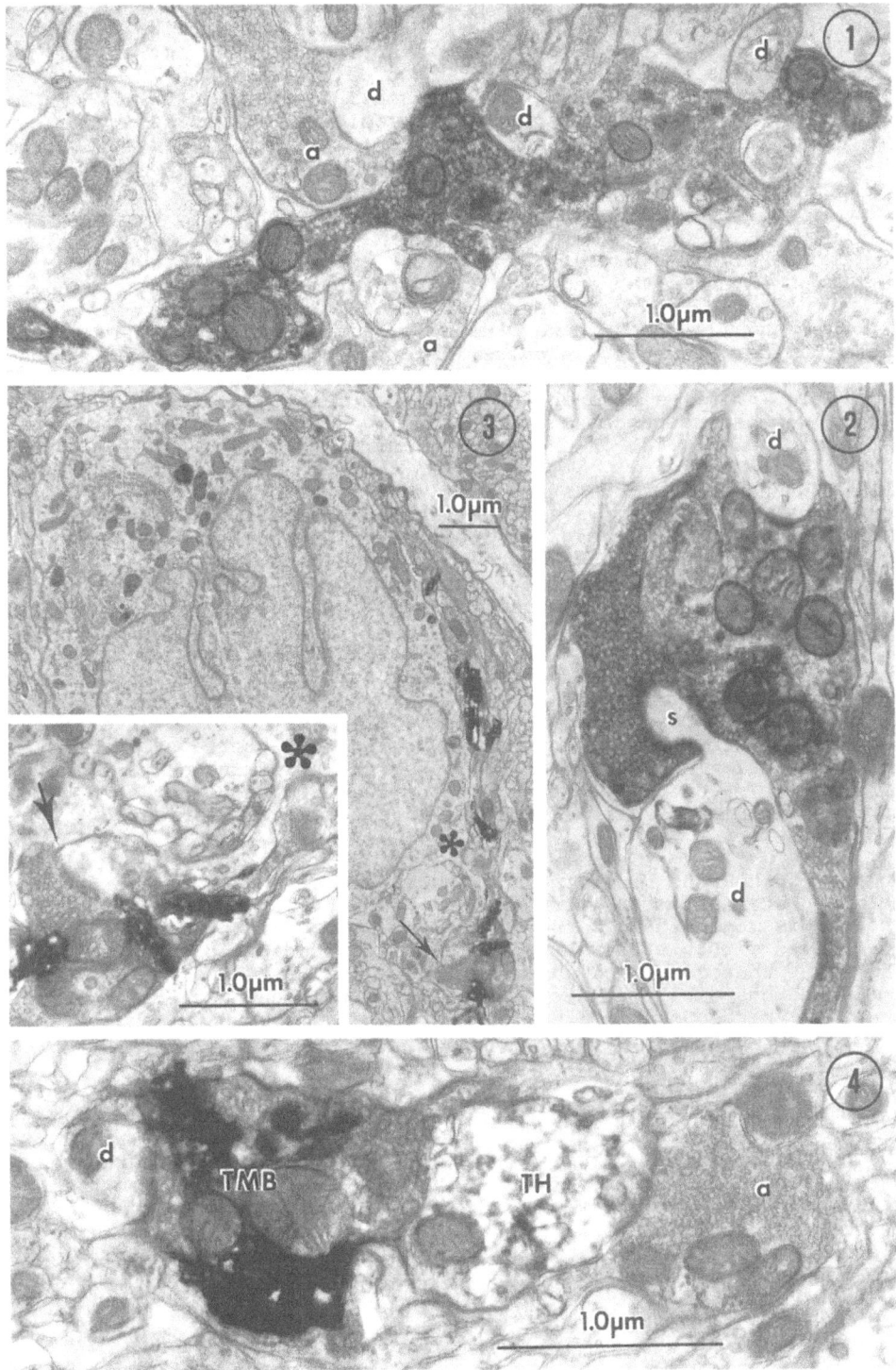

Since the large reaction product of the TMB either masked small labeled structures or distorted and displaced organelles in other profiles (Figs. 3, 4), most of the description of the morphology of labeled terminals was based on DAB reacted tissue. The HRP-positive terminal profiles contained numerous small clear vesicles (37-67 nm in diameter), occasional dense-cored vesicles, several large clear vesicles, and a few large mitochondria (Figs. 1, 2, 3, 4). Although the terminal profiles were predominantly associated with axo-dendritic synapses, they occasionally were encountered as presynaptic elements of either axo-axonal or axo-somatic synapses.

In axo-dendritic synapses, the postsynaptic dendrites exhibited a light matrix and often contained an ill-defined postsynaptic density at the active zone of the synapse. Smooth endoplasmic reticulum, vesicles and mitochondria were usually observed within small dendritic profiles, and rough endoplasmic reticulum and free polyribosomes were additionally located within large dendritic profiles. Some dendritic profiles also showed spine- or mushroom-like projections which were surrounded by a corresponding concavity within either an HRP-positive (Fig. 2) or unlabeled terminal.

Most TMB and TH-positive structures were located in separate fields. Occasionally, however, a few TMB-labeled profiles were intermingled with TH-positive dendritic profiles. The TH-positive profiles were profusely innervated by unlabeled boutons containing small clear vesicles (36 to 58 nm in diameter). Direct apposition of TMB-labeled terminals and TH-positive dendrites were only rarely seen (Fig. 4).

Direct contacts between HRP-positive and unlabeled terminals or contacts between terminal profiles that both contained HRP reaction product were identified (Figs. 1, 2). In a few cases HRP-positive terminals were presynaptic to unlabeled terminals. Although desmosome-like structures did connect apposing HRP-positive terminals, morphological indications of neurochemical transmission were never observed. In addition, one or both of the terminals that formed the apposing terminal pairs frequently did form axo-dendritic or axo-somatic synapses.

Somata of neurons receiving CSA inputs either on the soma or proximal dendritic processes were oval-, pear-, or spindle-shaped. Their short axis measured less than 8 microns. Most of these neurons were identified within either the commissural subnucleus or the region of the NTS that contained the mesh-like array of CSA fibers. These small neurons exhibited deep infoldings within the nucleus, numerous free polyribosomes, and sparse cisternae of the rough endoplasmic reticulum.

Discussion

The terminal fields of the central CSAs in the NTS of the rat have been described recently (Seiders and Stuesse 1984). The present study, however, provides the first ultrastructural demonstration of the synaptic organization of the CSAs within the NTS. In addition, the results of the present study show more extensive labeling of CSA fibers within the medulla. Furthermore, our results show that labeled fibers extend from the NTS into the AP, DMN, A1 cell groups, and the STN.

The present report shows that central CSA terminals synapse on small dendritic processes located within the NTS. Most of these dendritic processes are neither structurally different from surrounding dendritic profiles nor do they exhibit TH-immunoreactivity. Therefore, it is difficult to characterize the precise neuronal type that contributes the dendritic profiles that are postsynaptic to the CSA terminal profiles. Although a thorough statistical analysis is still required, our preliminary findings indicated that TH-positive dendritic profiles postsynaptic to HRP-labeled terminals are estimated roughly 1 in 200 TH-positive profiles. Nevertheless, we do not exclude the possibility that only a portion of the total population of CSA terminals are demonstrated with the techniques employed in the present study. The HRP-positive terminals, however, contain more large clear vesicles and large mitochondria than are evident within the unlabeled terminals that make contact with TH-positive profiles. Synaptic vesicles in central vagal afferent terminals have been shown to range from 40 to 60 nm in diameter (Sumal et al. 1983), which is similar to the range in size of the vesicle observed in the present study within unlabeled terminals synapsing on TH-positive profiles. The large clear vesicles and the large mitochondria in CSA terminals may be argued to be early signs of a transganglionic degenerative process. Even if this is the case, the present data still favor the assumption that most TH-positive dendrites within the NTS receive inputs from sources other than the CSAs, excluding the possibility that terminals on TH-positive profiles are CSAs that are selectively resistent to transganglionic degeneration or transport of HRP.

Within several sections, dendritic processes that had direct contact with HRP-positive terminals could be traced back to their parent somata. These somata were morphologically similar to the small neurons that also were shown in the present study to receive direct axo-somatic contacts from HRP-positive terminal profiles. These data corroborate a recent report showing that all of the second order baroreceptor neurons that were histologically examined within the NTS of the cat (following intracellular recording and subsequent HRP-filling) were small and spindle-shaped neurons with a short axis of 6 to 8 microns (Czachurski et al. 1981). Thus, the physiologically identified second order neurons in the latter study are very similar in shape and size to the small neurons that were identified in the present study to receive either axo-somatic or axo-dendritic input from CSA terminals. It remains to be determined whether the small second order CSA neurons within the NTS represent projection neurons or local circuit interneurons. Long latency periods between CSN stimulation and vagal efferent responses (Kunze 1971; McCloskey and Potter 1981) or between CSN activation and excitation of some ventromedullary neurons (Ciriello and Caverson 1986) may implicate the mediation of interneurons in these circuits.

In conclusion, our results indicate that small neurons within the NTS directly receive CSA inputs. The catecholaminergic neurons which are numerous in the caudal half of the NTS in the rat do not seem to be the major 2nd order neurons in the CSA pathway. The possibility that the central CSAs are linked indirectly to catecholaminergic neurons through small NTS interneurons that contain neuromodulators other than catecholamines has been taken into consideration for future experiments.

References

Chen I, Waller WP, Yates RD (1986) Simultaneous demonstration of transganglionically transported horseradish peroxidase (HRP) in central vagal afferents and neuromodulators in neurons of the nucleus tractus solitarius (NTS). Anat Rec 214: 20-21A

Ciriello J, Caverson MM (1986) Bidirectional cardiovascular connections between ventrolateral medulla and nucleus of the solitary tract. Brain Res 367: 273–281

Czachuriski J, Lackner KJ, Ockert D, Seller H (1982) Localization of neurons with baroreceptor input in the medial solitary nucleus by means of intracellular application of horseradish peroxidase in the cat. Neurosci Lett 28: 133–137

Kunze DL (1972) Reflex discharge patterns of cardiac vagal efferent fibers. J Physiol (Lond) 222: 1–5

Leman L, Saper CB, Rye DB, Wainer BH (1985) Stabilization of TMB reaction product for electron microscopic retrograde and anterograde fibers tracing. Brain Res Bull 14: 277–281

McCloskey DI, Potter EK (1981) Excitation and inhibition of cardiac vagal motorneurons by electrical stimulation of the carotid sinus nerve. J Physiol (Lond) 316: 163–175

Seiders EP, Stuesse SL (1984) A horseradish peroxidase investigation of carotid sinus nerve components in the rat. Neurosci Lett 46: 13–18

Straus W (1982) Imidazole increases the sensitivity of the cytochemical reaction for peroxidase with diaminobenzidine at a neutral pH. J Histochem Cytochem 30: 491–493

Sumal KK, Blessing WW, Joh TH, Reis DJ, Pickel VM (1983) Synaptic interaction of vagal afferents and catecholaminergic neurons in the rat nucleus tractus soitarius. Brain Res 277: 31–40

Ultrastructural Studies of the Microcircuitry of Catecholaminergic Neurons in the Nuclei of the Solitary Tract

V. M. Pickel

Department of Neurology, Laboratory of Neurobiology, Cornell University School of Medicine, New York, NY 10021, USA

The catecholaminergic (CA) neurons of the A2-group (Dahlström and Fuxe 1964) are located in portions of the medial nuclei of the solitary tracts (m–NTS) which receive baroreceptor afferents from the inferior (nodose) ganglion of the vagus nerve (Kalia and Mesulam 1980). The majority of these CA perikarya are noradrenergic as shown by the immunocytochemical localization of the synthesizing enzymes, tyrosine hydroxylase (TH) and dopamine–beta–hydroxylase (DBH) (Armstrong et al. 1981). However, a subpopulation of the CA neurons located in the dorsal m–NTS at the level of the area postrema also contain phenylethanolamine N–methyltransferase (PNMT), the enzyme used for the synthesis of adrenaline (Kalia et al. 1985). This report summarizes our recent studies of vagal and other transmitter-specific afferents to TH- and PNMT-containing neurons in cardiovascular portions of the m–NTS. The methods include the electron microscopic immunocytochemical localization of single transmitter-related antigens with the peroxidase–antiperoxidase (PAP) method of Sternberger (1979) and dual labeling using PAP and autoradiography for anterogradely transported, incorporated or immunologically bound radioactive tracers.

Vagal afferents to the m–NTS were identified autoradiographically in sections immuno–labeled for TH after collection from the brains of rats receiving unilateral injections of ^{3}H–L–proline in the nodose ganglion. By light microscopy, silver grains were seen near perikarya and processes with brown peroxidase labeling for TH. Electron microscopy showed accumulations of silver grains over selective axon terminals which formed synapses with both TH–labeled and unlabeled, presumably non–CA dendrites (Fig. 1) (Sumal et al. 1983). One of the transmitters in the vagal afferents is believed to be L–glutamate (Simon et al. 1985). The localization of glutamate in these afferents is consistent with our detection of glutamate dehydrogenase (GDH) in glia in the NTS and other areas with glutamatergic receptors (Aoki et al. 1986). The localization of GDH to certain populations of astrocytes was confirmed by dual labeling GDH with immunoperoxidase and glial fibrillary acidic protein (GFAP) with autoradiography for detection of iodinated secondary immunoglobulins (^{125}I–IgGs) (Aoki et al. 1986; Bignami et al. 1977). The GFAP-labeled glia also formed a notable band separating the solitary nuclei and the area postrema as shown in the schematic drawing (Fig. 1).

The neuropeptide substance P (SP) is contained in certain vagal afferents (Gillis et al. 1980) and in intrinsic perikarya and terminals in the m–NTS, the localization within perikarya and terminals

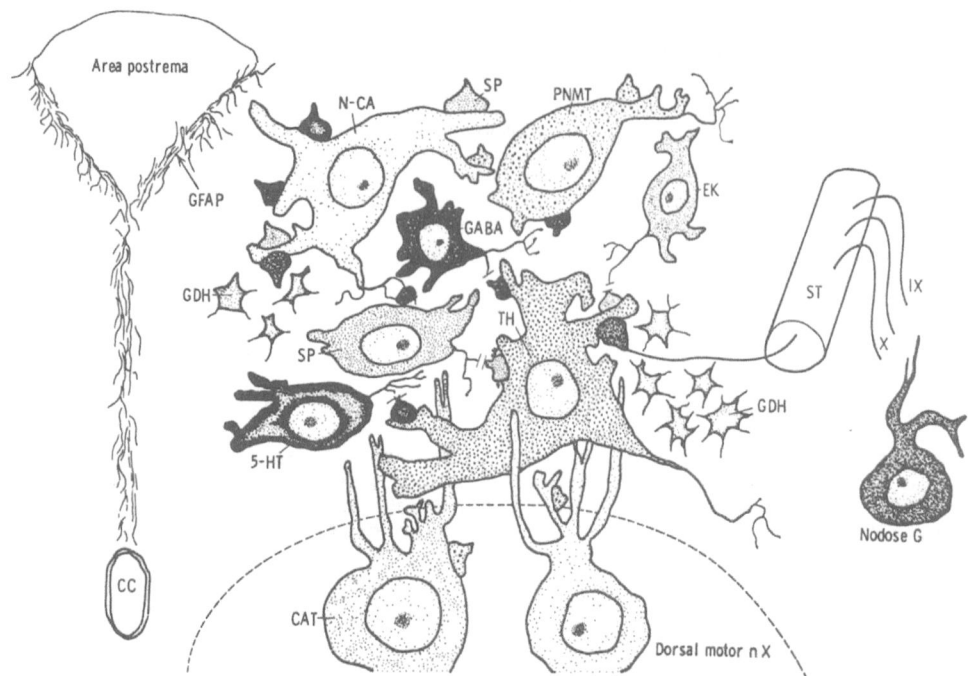

Fig. 1. Schematic drawing to illustrate the synaptic interactions of CA (TH and PNMT-containing) neurons in the m-NTS at the level of the area postrema. Afferents from IX and X cranial nerves, some of whose sensory neurons lie in the nodose ganglion (Nodose G), pass through the solitary tract (ST) and terminate upon both TH-containing and non-CA (N-CA) neurons. Glia containing glutamate dehydrogenase (GDH) and glial fibrillary acidic protein (GFAP) are also indicated. Perikarya and terminals containing neuropeptides (EK and SP), GABA and 5-HT are diagrammatically illustrated relative to TH- and PNMT-containing and to non-CA neurons. The cholinergic, CAT-labeled neurons in the dorsal motor nucleus of X and the central canal (CC) are drawn

in the NTS is shared by opioid eptides such as methionine Met[5]-enkephalin (EK) (Armstrong et al. 1981). A direct synaptic input to TH-labeled dendrites in the NTS from axon terminals containing immunoreactivity for these neuropeptides was first suggested by PAP-labeling of SP or EK in sections adjacent to those labeled for TH (Pickel et al. 1979). These synaptic relations have now been more definitively established by autoradiographic localization of ^{125}I-seondary IgGs against TH-antiserum and immunoperoxidase labeling for SP or EK on single sections prepared for electron microscopy (Fig. 1) (Pickel et al. 1986a). Furthermore, immunoperoxidase labeling for EK has been combined with anterograde transport of ^{3}H-proline from the nodose ganglia to demonstrate that vagal afferents and EK-labeled terminals converge on certain common dendrites which may or may not belong to CA neurons (Fig. 1).

Serotonin, 5-hydroxytryptamine (5-HT) is another prominent transmitter in axon terminals and a few perikarya in the m-NTS. Uptake of ^{3}H-5-HT from the ventricles of rats pretreated with a monoamine oxidase inhibitor provides selective autoradiographic labeling of axon terminals in the m-NTS (Pickel et al. 1984b). Electron microscopic autoradiographic labeling for 5-HT terminals and PAP labeling for TH established a synaptic basis for a serotonergic modulation of CA neurons in the m-NTS.

Gamma amino butyric acid (GABA) and its synthesizing enzyme glutamic acid decarboxylase (GAD) are contained within numerous perikarya and terminals in the m-NTS (Meeley et al. 1985; Simon et al. 1985). The GABAergic terminals form synapses with TH-containing neurons in the m-NTS as shown by immunogold and immunoperoxidase labeling for the respective synthesizing enzymes (Pickel et al. 1984a). Furthermore at least some of the recipient neurons have been shown to be adrenergic in dual labeling studies using PAP for a rat antiserum to GABA and immuno-autoradiographic localization of a rabbit antiserum to PNMT (Fig. 1) (Pickel et al. 1986c). The ultrastructural localization of immunoperoxidase reaction for PNMT was seen in medium sized perikarya, dendrites and axon terminals in the m-NTS. The axon terminals formed primarily symmetric synapses with unlabeled and a few PNMT-labeled dendrites in the NTS and with unlabeled perikarya and dendrites in the immediately adjacent dorsal motor nucleus of the vagus (Pickel et al. 1986b). This represents the first ultrastructural evidence that there are direct synaptic interactions between adrenergic neurons in the m-NTS and that the adrenergic neurons also innervate motor neurons of the vagal complex. The majority of the perikarya in the motor nucleus were shown to be cholinergic by PAP-labeling for choline acetyltransferase (CAT).

The synaptic relations of the A2-group of CA neurons provides morphological support for their role as common mediators of many of the cardiovascular effects of L-glutamate, SP, EK, 5-HT, GABA, and adrenaline in the m-NTS (Willette et al. 1983; Henry et al. 1985; Howe 1985). They also indicate that the adrenergic neurons may directly modulate the outflow from cholinergic neurons in the dorsal motor nucleus of the vagus.

Acknowledgements. The authors gratefully acknowledge the support provided by grants from NIH (HL18974), NIMH (MH00078 and MH40342) and NSF (BNS 8320120).

References

Aoki C, Milner TA, Sheu K-FR, Blass JP, Pickel VM (1986) Astrocytic localization of glutamate dehydrogenase in selective neuronal pathways of rat brain. J Neurosci (in press)

Armstrong DM, Pickel VM, Joh TH, Reis DJ, Miller RJ (1981) Immunocytochemical localization of catecholamine synthesizing enzymes and neuropeptides in area postrema and medial nucleus tractus solitarius of rat brain. J Comp Neurol 196: 505-517

Bignami A, Eng LF, Dahl D, Uyeda CT (1972) Localization of the glial fibrillary acidic protein in astrocytes by immunofluorescence. Brain Res 43: 429-435

Dahlström A, Fuxe K (1964) Evidence for the existence of monoamine-containing neurons in the central nervous system. Acta Physiol Scand 62: 1-55

Gillis RA, Helke CJ, Hamilton BL, Norman WP, Jacobowitz DM (1980)
Evidence that substance P is a neurotransmitter of baro- and chemoreceptor afferents in nucleus tractus solitarius. Brain Res 181:
476-481

Henry JL, Sessle BJ (1985) Effects of glutamate, substance P and
eledoisin-related peptide on solitary tract neurons involved in
respiration and respiratory reflexes. Neuroscience 14: 863-873

Howe PC (1985) Blood pressure control by neurotransmitters in the
medulla oblongata and spinal cord. J Auton Nerv Syst 12: 95-115

Kalia M, Fuxe K, Goldstein M (1985) Rat medulla oblongata. II.
Dopaminergic, noradrenergic (A1 and A2) and adrenergic neurons,
nerve fibers, and presumptive terminal processes. J Comp Neurol 233:
308-332

Kalia M, Mesulam M-M (1980) Brain stem and spinal cord projections
of vagal sensory and motor fibers in the cat: I. The cervical vagus
and nodose ganglion. J Comp Neurol 193: 435-466

Meeley MP, Ruggiero DA, Ishitsuka T, Reis DJ (1985) Intrinsic-gamma-
aminobutyric acid neurons in the nucleus of the solitary tract and
the rostral ventrolateral medulla of the rat: an immunocytochemical
and biochemical study. Neurosci Lett 58: 83-89

Pickel VM, Chan J, Joh TH, Massari VJ (1984a) Catecholaminergic
neurons in the medial nuclei of the solitary tracts receive direct
synapses from GABA-ergic terminals: combined colloidal gold and
peroxidase labeling of synthesizing enzymes. Neurosci Abs 10: 537

Pickel VM, Chan J, Milner TA (1986a) Autoradiographic detection of
(125I)-secondary antiserum: a sensitive light and electron microscopic labeling method compatible with peroxidase immunocytochemistry
for dual localization of neuronal antigens. J Histochem Cytochem 34:
707-718

Pickel VM, Chan J, Park DH, Joh TH, Milner TA (1986b) Ultrastructural localization of phenylethanolamine N-methyltransferase in sensory and motor nuclei of the vagus nerve. J Neurosci Res (in press)

Pickel VM, Joh TH, Chan J, Beaudet A (1984b) Serotoninergic terminals: ultrastructure and synaptic interaction with catecholamine-
containing neurons in the medial nuclei of the solitary tracts. J
Comp Neurol 225: 291-301

Pickel VM, Joh TH, Reis DJ, Leeman SE, Miller RJ (1979) Electron
microscopic localization of substance P and enkephalin in axon
terminals related to dendrites of catecholaminergic neurons. Brain
Res 160: 387-400

Pickel VM, Towle A, Park DH, Joh TH, Chan J, Milner TA (1986c) Pre-
and postsynaptic GABAergic terminals on adrenergic neurons in the
medial nuclei of the solitary tracts. Brain Res (submitted)

Simon JR, DiMicco SR, Aprison MH (1985) Neurochemical studies of the
nucleus of the solitary tract, dorsal motor nucleus of the vagus and
the hypoglossal nucleus in rat: topographical distribution of glutamate uptake, GABA uptake and glutamic acid decarboxylase activity.
Brain Res Bull 14: 49-53

Sternberger LA (1979) Immunocytochemistry. Wiley, Chichester

Sumal KK, Blessing WW, Joh TH, Reis DJ, Pickel VM (1983) Synaptic interaction of vagal afferents and catecholaminergic neurons in the rat nucleus tractus solitarius. Brain Res 277: 31-40

Willette RN, Krieger AJ, Barcas PP, Sapru HN (1983) Medullary gamma-aminobutyric acid (GABA) receptors and regulation of blood pressure in the rat. J Pharmacol Exp Ther 226: 893-899

Electron Microscopic Studies on the Organization of Peptide Nerve Terminals in the Inferior Mesenteric Ganglion

S. Masuko and T. Chiba

Department of Anatomy, Saga Medical School, Nabeshima, Saga 840-01, Japan

Introduction

The inferior mesenteric ganglion (IMG) receives a diversity of peptidergic fibers from different sources: substance P (SP)-immuno-reactive (IR) primary afferent nerve fibers and enkephalin (ENK)-IR preganglionic nerve fibers from the lumbar splanchnic nerves, dynor-phin (DYN)-, vasoactive intestinal polypeptide (VIP)-, cholecystoki-nin (CCK)- and bombesin-IR fibers from the colonic nerves, VIP-, CCK- and SP-IR fibers from the intermesenteric nerves and VIP-IR fibers from the hypogastric nerves (Matthews and Cuello 1982; Dals-gaard et al. 1983a,b). In the physiological study of the IMG, selec-tive functions of individual peptides, such as an excitatory action of SP (Tsunoo et al. 1982), an inhibitory action of ENK (Konishi et al. 1981) and modulatory actions of VIP and CCK (Mo and Dun 1984, 1986) have been suggested. However, their synaptic arrangement and fine structural profiles have not been studied sufficiently. The present study was undertaken to clarify the ultrastructural organi-zation of SP-, ENK-, VIP- and CCK-IR nerve fibers in the IMG with reference to the morphological features of nerve terminals derived from the lumbar splanchnic nerves or colonic nerves.

Materials and Methods

Three groups of male guinea pigs (250-300 g) were examined. In the first group the IMG was partially denervated by cutting the interme-senteric, hypogastric and colonic nerves, leaving the lumbar splanchnic nerves intact (LSN-IMG). In the second group the IMG was partially denervated by cutting the intermesenteric, hypogastric and lumbar splanchnic nerves, leaving the colonic nerves intact (CN-IMG). The third group served as unoperated normal controls. Two days after the operation, animals were perfused through the ascending aorta with Zamboni's fixative for immunohistochemistry or with 2% paraformaldehyde and 2% glutaraldehyde in 0.1 M phosphate buffer for conventional electron microscopy. The IMG was dissected out and fixed for additional 2-5 hours in the same fixative.

For the light microscopic immunohistochemistry, the IMG was embedded in Spurr's resin after dehydration. In order to compare the distri-bution of the different IR nerve fibers, serial semithin sections (0.5 um) were cut. After removal of the resin with sodium methoxide (Mayor et al. 1961) the sections were incubated for 12 hours at 4°C with one of the following antisera: rabbit anti-SP antiserum (gift of Dr. Amano, Institute of Life-Science, Mitsubishikasei) (Hatanaka and Amano 1981) at a dilution of 1:10,000, anti-leu-ENK antiserum (Immuno Nuclear Co.) at a dilution of 1:10,000, anti-VIP antiserum

(Immuno Nuclear Co.) at a dilution of 1:2,000, and anti–CCK–8 anti–serum (Immuno Nuclear Co.) at a dilution of 1:1,000. Tissue sections were processed according to the PAP method of Sternberger (1979). The specificity of antisera was examined by a preabsorption test. Leu–ENK antiserum showed partial cross–reactivity to the met–ENK and DYN, however, no cross–reactivity occurred among the four antisera employed in the present study.

For immunoelectron microscopic study, sections of 20 um in thickness were cut by a Vibratome. The sections were processed for immunoreaction using the PAP method, postfixed with 2.5% glutaraldehyde in 0.1 M phosphate buffer followed by 1% OsO_4 in the same buffer, and were embedded in Spurr's resin. Thin sections (50 nm) were examined after staining by uranyl acetate and lead citrate.

Results

Normal Control IMG. Analysis of the serial semithin sections revealed that the VIP–IR and ENK–IR fibers were most frequent (285 $\pm$ 75.7 and 273 $\pm$ 28.8 per 10,000 um^2 respectively, mean $\pm$ SD; total of 91,350 um^2 was examined), the CCK–IR fibers were moderate (168 $\pm$ 36.5) and the SP–IR fibers were sparse (59.3 $\pm$ 6.8). By electron microscopy, the following four types of varicose nerve fibers were found: (1) Small clear vesicle predominant type (SV type) which contained a cluster of numerous small clear synaptic vesicles and a few large vesicles surrounding the small vesicle cluster (Fig. 2a), (2) large vesicle predominant type (LV type) which contained a high proportion of large vesicles intermingled with small clear vesicles (Fig. 2b), (3) flat vesicle type (FV type) which was characterized by small clear flattened vesicles, and (4) granular vesicle type (GV type) containing large dense granular vesicles which resembled those of small granule–containing cells in the sympathetic ganglia. The varicosities of SV type and LV type occupied about 30% and 60% of total varicosities in the normal IMG, respectively, whereas the FV and GV types were only a few percent of the varicosities. About 25% of the total varicosities showed synaptic contacts with dendrites or dendritic spines. Axo–somatic synapse was rare. Two types of synaptic specializations, asymmetric and symmetric, were found. The asymmetric type was frequently found at the synapses of SV type varicosities, whereas the symmetric type was mainly found at varicosities of LV type (Figs. 2a, b).

LSN–IMG. The density of SP–IR fibers was essentially unaffected (45 $\pm$ 4.5/10,000 um^2) by the operation, whereas the density of ENK–IR, VIP–IR and CCK–IR fibers decreased to a level of about 50% (142 $\pm$ 3.2), 15% (41 $\pm$ 9.0) and 5% (13 $\pm$ 4.0) of the normal IMG, respectively. Thus, the ENK–IR fibers were highest in density, the SP–IR and VIP–IR fibers were moderate, and CCK–IR fibers were sparse in the LSN–IMG (Figs. 1a–d). The proportion of the SV type varicosities increased to about 70% and the LV type decreased to about 20% of the total varicosities. In the immunoelectron microscopy, the SP–IR and ENK–IR fibers showed the SV type varicosities (Figs. 3a,b) whereas the VIP–IR varicosities had the characteristics of the LV type (Fig. 3c). In a total of 45 SP–IR fibers, 17 (38%) fibers apposed directly to the neuronal profiles consisted of 12 dendrites and 9 presumable axons. Among these fibers, only 3 axo–dendritic asymmetric synapses were found (Fig. 3a). In a total of 71 ENK–IR fibers, 39 fibers (55%) were directly apposed to the neuronal profiles, 34 dendritic and 5 axonal. Eleven synaptic contacts were found and all of them were axo–dendritic synapses with asymmetrical specialization (Fig.

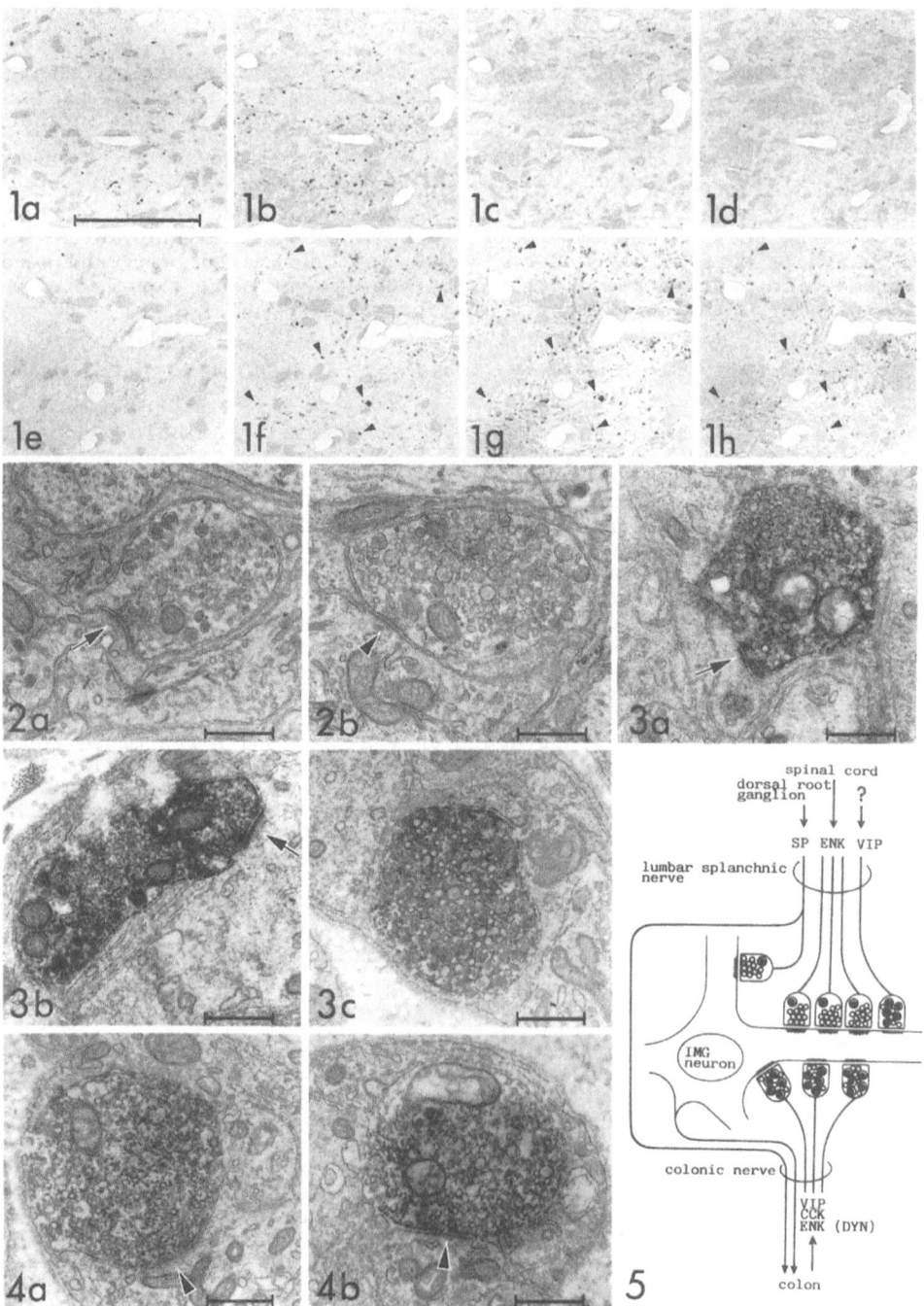

3b). In a total of 42 VIP-IR fibers, 19 fibers (45%) were directly apposed to the dendrites and 2 synpases of symmetrical type were found.

CN-IMG. The SP-IR fibers markedly decreased in number whereas the density of ENK-IR, VIP-IR and CCK-IR fibers was slightly decreased to a level of about 65% (181 $\pm$ 34.6), 70% (204 $\pm$ 38.0) and 80% (135 $\pm$ 19.8) of the normal IMG, respectively. Thus, the CN-IMG showed a relatively dense distribution of the ENK-IR, VIP-IR and CCK-IR fibers and scarce SP-IR fibers (Figs. 1e-h). Dalsgaard et al. (1983) have demonstrated a dense network of DYN-IR fibers in the guinea pig IMG which are derived predominantly from the colonic nerves. There-fore, the ENK-IR fibers in the CN-IMG (Fig. 1f) may represent the DYN-containing fibers which were cross-reacted with our anti-ENK antiserum. The ENK-IR (possibly DYN-IR), VIP-IR and CCK-IR fibers in the CN-IMG were distributed in a similar pattern, and on closer examination ENK-IR, VIP-IR and CCK-IR were frequently found in identical nerve fibers (Figs. 1f-h), suggesting the coexistence of these three peptides in the single varicosity. Recently, a similar observation has been reported in the coeliac ganglion of the guinea pig by Macrae et al. (1986). By electron microscopic observations most varicosities were LV type (about 85%) and the SV type formed approximately 10% of the total varicosities. Both the VIP-IR and CCK-IR varicosities were mainly LV type (Figs. 4a,b), and the ENK-IR (possibly DYN-IR) varicosities were also LV type (data are not shown). In a total of 75 VIP-IR fibers, 57 fibers (76%) were direct-ly apposed to the neuronal profiles including 54 dendrites and 3 axons. Twelve axo-dendritic synapses were found, 11 of which were symmetric type and only one synapse showed the asymmetric speciali-zation. In a total of 60 CCK-IR fibers 49 fibers (81%) directly apposed to the neuronal profiles consisted of 2 somata, 44 dendrites and 3 axons. In these fibers, 14 axo-dendritic synapses (12 symme-tric and 2 asymmetric type) and one axo-somatic synapse (symmetric type) were found.

◀ **Fig. 1.** Consecutive sections (0.5 um) of the LSN-IMG (a-d) and CN-IMG (e-h) immunoreacted with SP (a, e), ENK (b, f), VIP (c, g) and CCK (d, h) antisera. Arrow heads in f-h indicate colocalization of ENK-IR (possibly DYN), VIP-IR and CCK-IR in identical nerve fibers. Bar indicates 50 um

Fig. 2. Electron micrographs showing typical axon varicosities and synapses in normal IMG. (a) SV type varicosity with asymmetric synapse (arrow). (b) LV type varicosity with symmetrical synapse (arrow head). Bars indicate 0.5 um

Fig. 3. SP-IR (a), ENK-IR (b) and VIP-IR (c) varicosities in the LSN-IMG. Asymmetrical synaptic specializations (arrows) are seen in (a) and (b). Bars indicate 0.5 um

Fig. 4. VIP-IR (a) and CCK-IR (b) varicosities in the CN-IMG. Symme-trical synaptic specializations (arrow heads) are seen. Bars indi-cate 0.5 um

Fig. 5. Schematic illustration of summarized synaptic organization of the IMG

Discussion

The SP-, ENK-, VIP- and CCK-IR cell bodies have not been detected except for SIF cells in the IMG, therefore nerve terminals IR for these peptides are considered to be of extraganglionic origin (Schultzberg et al. 1983). The present study revealed distinct differences of the presynaptic profiles and synaptic specializations among various peptide-IR fibers (summarized in Fig. 5): the SP-IR and ENK-IR fibers derived from the lumbar splanchnic nerves possess the SV type varicosities with asymmetrical synapses whereas the VIP-IR fibers from the lumbar splanchnic nerves as well as VIP-, CCK- and ENK- (possibly DYN-) IR fibers derived from the colonic nerves have mainly the LV type varicosities with symmetrical synapses. These characteristic features of the SP-IR and VIP-IR varicosities are essentially similar to those reported in the coeliac ganglion of the guinea pig (Kondo and Yui 1981, 1982a). However, the observation concerning the ENK-IR varicosities by Kondo and Yui (1982b) who considered the LV type of varicosities as a characteristic feature of the ENK-IR fibers is contrary to our result of the ENK-IR fibers derived from the lumbar splanchnic nerves. This difference may be caused by the observation of normal coeliac ganglia, where the ENK-IR fibers derived from both the gastrointestinal tract and preganglionic sympathetic nerves were intermingled.

Peptide-IR fibers form a dense network surrounding the ganglion cell bodies, as seen by light microscopy. However, the present ultrastructural studies revealed that axo-somatic contacts are rare. In a physiological study, presynaptic inhibitory action of ENK on SP release has been reported (Konishi et al. 1981). No axo-axonic synaptic contacts were confirmed in the present study although the direct apposition between the ENK-IR fibers and non-IR axonal profiles were often encountered. The presence of the synaptic specialization suggests that the effects of these peptides on the ganglionic neurons are exerted through the synaptic sites, and the morphological difference of the synaptic specialization between SP-IR or ENK-IR terminals and VIP-IR or CCK-IR terminals might be related to the difference of mode of actions among these peptides. We cannot exclude the possibility that the peptides are released from varicosities without the synaptic specialization and exert their actions through the sites of direct apposition to the neuronal profiles.

Acknowledgements. The authors thank Miss M. Tsutsumi for technical and secretarial assistance. This work was supported by a grant 61770049 of the Ministry of Education, Science and Culture of Japan.

References

Dalsgaard C-J, Vincent SR, Hökfelt T, Christensson I, Terenius L (1983a) Separate origins for the dynorphin and enkephalin immunoreactive fibers in the inferior mesenteric ganglion of the guinea pig. J Comp Neurol 221: 482–489

Dalsgaard C-J, Hökfelt T, Schultzberg M, Lundberg JM, Terenius L, Dockray GJ, Goldstein M (1983b) Origin of peptide-containing fibers in the inferior mesenteric ganglion of the guinea-pig: immunohistochemical studies with antisera to substance P, enkephalin, vasoactive intestinal polypeptide, cholecystokinin and bombesin. Neuroscience 9: 191–211

Hatanaka H, Amano T (1981) A mouse neuroblastoma x rat glioma hybrid cell produces immunoreactive substance P-like material. Brain Res 215: 305-316

Konishi S, Tsunoo A, Otsuka M (1981) Enkephalin as a transmitter for presynaptic inhibition in sympathetic ganglia. Nature (Lond) 294: 80-82

Kondo H, Yui R (1981) An electron microscopic study on substance P-like immunoreactive nerve fibers in the celiac ganglion of guinea pigs. Brain Res 222: 134-137

Kondo H, Yui R (1982a) An electron microscopic study on VIP-like immunoreactive nerve fibers in the celiac ganglion of guinea pigs. Brain Res 237: 227-231

Kondo H, Yui R (1982b) An electron microscopic study on enkephalin-like immunoreactive nerve fibers in the celiac ganglion of guinea pig. Brain Res 252: 142-145

Macrae IM, Furness JB, Costa M (1986) Distribution of subgroups of noradrenaline neurons in the coeliac ganglion of the guinea-pig. Cell Tissue Res 244: 173-180

Matthews MR, Cuello AD (1982) Substance P-immunoreactive peripheral branches of sensory neurons innervate guinea pig sympathetic neurons. Proc Natl Acad Sci USA 79: 1668-1672

Mayor HD, Hampton JC, Rosario B (1961) A simple method for removing the resin from epoxy-embedded tissue. J Biophys Biochem Cytol 9: 909-910

Mo N, Dun NJ (1984) Vasoactive intesinal polypeptide facilitates muscarinic transmission in mammalian sympathetic ganglia. Neurosci Lett 52: 19-23

Mo N, Dun NJ (1986) Cholecystokinin octapeptide depolarizes guinea pig inferior mesenteric ganglion cells and facilitates nicotinic transmission. Neurosci Lett 64: 263-268

Schultzberg M, Hökfelt T, Lundberg JM, Dalsgaard C-J, Elfvin L-G (1983) Transmitter histochemistry of autonomic ganglia. In: Elfvin L-G (ed) Autonomic ganglia. Wiley, Chichester, pp 205-233

Sternberger LA (1979) Immunohistochemistry, 2nd edn. Wiley, Chichester

Tsunoo A, Konishi S, Otsuka M (1982) Substance P as an excitatory transmitter of primary afferent neurons in guinea-pig sympathetic ganglia. Neuroscience 7: 2025-2037

Anatomy and Morphology of Chromaffin Paraganglia Associated with the Inferior Mesenteric Ganglion in Cats

J. A. Mascorro and R. D. Yates

Department of Anatomy, Tulane University School of Medicine, New Orleans, LA 70112, USA

Introduction

Stilling (1890) described collections of extraadrenal "chromophil" cells which displayed a characteristic reaction with potassium di-chromate and were morphologically similar to their intraadrenal counterparts. Kohn (1902) further characterized the "paraganglia" as microscopic structures and introduced the term "chromaffin" to de-signate a particular collection of such cells in the midretroperito-neum, the paraganglion aorticum abdominale. The term paraganglia quickly gained acceptance and is used routinely to describe extrame-dullary chromaffin tissue. However, it does not completely define the vast expanse of this system as it refers mainly to minute structures that are directly apposed to sympathetic ganglia. Other studies based upon potassium dichromate mapping have shown convin-cingly that very large deposits of chromaffin tissue occur in para-aortic and non-paraganglion locations (Zuckerkandl 1902; Mascorro and Yates 1977).

Extramedullary chromaffin cells have attracted much interest and several investigators have observed paraganglion-like cells within various sympathetic ganglia (Matthews and Raisman 1969; Hervonen et al. 1979). These unique intraganglion cells are strongly fluorescent (SIF cells) and are said to function as interneurons by inhibiting or modulating ganglion transmissions via their interposition between pre- and postganglionic neurons. Other ganglion chromaffin cells exist as clusters and lack the synaptology necessary to act as interneuronal elements. Clustered chromaffin cells are vasculated, possess numerous catecholamine granules and possibly represent neu-roendocrine structures capable of releasing their product(s) into local blood vessels (Autillo-Touati and Seite 1980). This mechanism would represent a neurosecretory action by the ganglion chromaffin cell clusters; whatever the case, the exact physiological signifi-cance of extraadrenal chromaffin tissue is vague, and much of it appears to lack an innervation (Coupland et al. 1982).

It seems important now to determine the relationship between sympa-thetic ganglia and their adjacent paraganglia, and particularly between autonomic neurons and the chromaffin cells which closely border them within the ganglion. This work utilizes a gross chromaf-fin staining and fixing method (Mascorro et al. 1976) to localize paraganglia associated with the inferior mesenteric ganglion in cats and to study the cellular characteristics of these strongly chromaf-fin-positive organs.

Experimental Brain Research Series 16

Materials and Methods

Nine adult cats, irrespective of sex and weighing between 3.0 and 5.0 kg, were housed individually in spacious metal cages and allowed free access to food and water supply. The animals were anesthetized IP with 50 mg/kg of Nembutal and sacrificed by intracardiac perfusion with cold 3% glutaraldehyde in 0.1 M sodium phosphate buffer, pH 7.2-7.4. The perfusion was preceded by a phosphate-buffered saline flush which emptied the vascular tree of blood. The retroperitoneal tissue blocks (RTBs) were removed and fixed further in the perfusion fluid. The RTBs then were placed in a solution of 3% glutaraldehyde and 2.5% potassium dichromate for gross anatomical staining of chromaffin tissue for 30-60 min. Following staining, the RTBs were rinsed overnight in 0.1 M phosphate buffer containing 8% sucrose in order to wash excess dichromate ions from the RTBs so as to render the stained extraadrenal chromaffin tissues visible. The inferior mesenteric ganglion and attached paraganglion were dissected free from the RTBs, cleansed of connective and fatty tissue and photographed with Kodak Technical Pan Film 2415 or Kodachrome 40 Type A slide film. The ganglion/paraganglion tissue blocks then were bisected so that the cut surfaces each contained parts of the ganglion and attached paraganglion. The blocks were secondarily fixed in phosphate buffered osmic acid, dehydrated in absolute alcohol concentrations and routinely processed for light and electron microscopic study.

Results

Inferior Mesenteric Ganglion (IMG). The IMG was paired and firmly bound by a capsular meshwork of fibrous and fatty tissues to the inferior mesenteric artery. The capsule material was dissected away in order to better illustrate the ganglion and adjoined paraganglia (Fig. 1, upper figure).

Potassium Dichromate Mapping - Paraganglia (PG). Several paraganglia were seen adjoined to the ganglion following dichromate mapping (Fig. 1, upper figure). One particular ganglion showed 6 chromaffin bodies attached to its surface (Table 1). The PG usually measured 1 mm in length and 0.5 mm in diameter. Other chromaffin bodies measuring 13 mm in length and 1 mm in diameter regularly occurred in a paraaortic position (Fig. 1, lower figure).

Histological and Ultrastructural Characteristics. A shared connective tissue capsule enveloped and also intervened between the paraganglion/ganglion complex. The PG were comprised of small epithelial-like cells which contained pale nuclei. They exhibited a rich cytoplasmic basophilia that accounted for the darkened appearance

Table 1. Paraganglia localizations

Animal	Number of PG	Animal	Number of PG
Cat 1	5	Cat 6	4
Cat 2	4	Cat 7	3
Cat 3	3	Cat 8	3
Cat 4	3	Cat 9	3
Cat 5	6 (illustrated)		

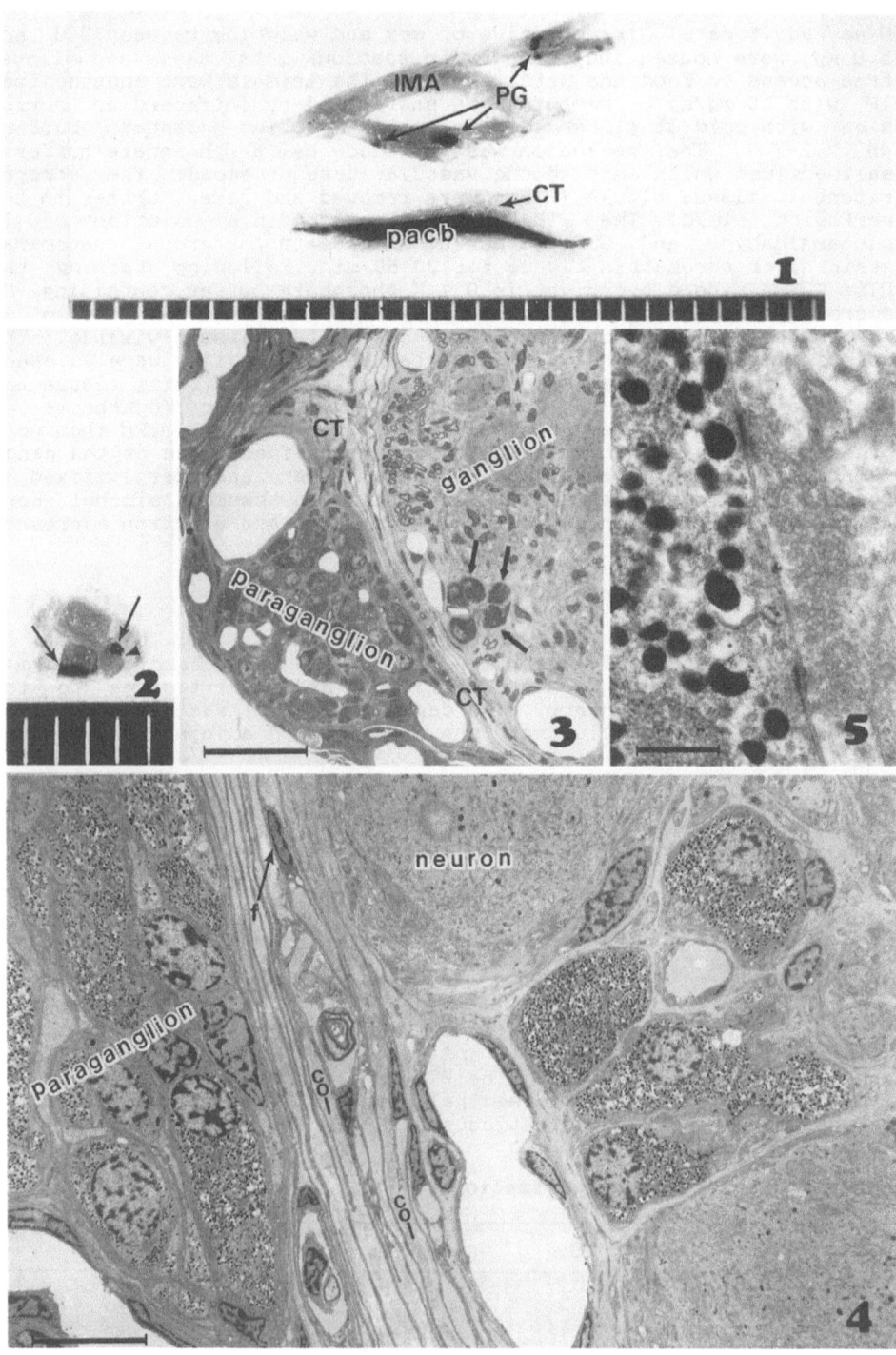

evident at the macroscopic level (PG, Fig. 3). All ganglia contained chromaffin cell clusters, some of which contained more than 60 cells (not illustrated). Most clusters, however, were smaller in terms of cell number (arrow, Fig. 3). The darkened intraganglion chromaffin cells were easily distinguished from their neural relatives as ganglion cells were larger and lacked the basophilia characteristic for chromaffin cells.

The electron microscope compared to good advantage the features of paraganglion and ganglion chromaffin cells. Favorable sections which included both cell types showed that they were ultrastructurally identical, and that both contained a voluminous amount of indivi- dually dense granules randomly dispersed throughout the cytoplasm (Fig. 4). All chromaffin cells, regardless of anatomical location, were closely related to thin-walled blood vessels. Nerve endings were observed contacting only the intraganglion chromaffin cells and those synapses contained predominantly clear vesicles (Fig. 5).

Discussion

Much has been learned about the adrenal medulla since the revelation by Henle (1865) that catecholamines reacted with chrome salts (the chromaffin reaction) to yield brown oxidation products which identi- fied catecholamine-containing cells. However, the extraadrenal com- ponent of the master chromaffin system has not been studied to the same extent and significant questions remain about this voluminous and persisting system. Paramount among these is whether extraadrenal

◄ **Fig. 1**. Ventral aspect of the inferior mesenteric ganglion (IMG) and associated paraganglia (PG) seen along the inferior mesenteric arte- ry (IMA). A large paraaortic chromaffin body (PACB) was enveloped by connective tissue (CT) and measured 13 mm in total length. (Scale = mm)

Fig. 2. Gross anatomical demonstration of 3 pieces from one IMG, two of which clearly showed dichromate stained chromaffin PG (arrows). One ganglion and adjacent paraganglion (arrowhead) were studied at the light and electron microscopic levels and appear in Figs. 3 and 4. (Scale = mm)

Fig. 3. Histological demonstration of the paraganglion and ganglion complex. The two organs were surrounded and separated by a capsule of connective tissue (CT). This opportune section contained a cluster of intraganglion cells (arrows) which appeared deeply baso- philic and similar to the adjacent chromaffin cells. Many neurons were seen throughout the ganglion proper. (Bar indicates 50 um)

Fig. 4. The appearance of the paraganglion/block was studied at the ultrastructural level as well. Collagen fibrils (col) and fibroblast cells (f) with their many elongated processes (arrows) intervened between the two organs. Intraganglion and paraganglion chromaffin cells shared fine structural similarities and were replete with dense cytoplasmic granules. (Bar indicates 5 um)

Fig. 5. A high power electron micrograph illustrating a cholinergic- like synaptic ending upon a ganglion chromaffin cell. The ending contained a majority of vesicles with clear cores and displayed two synaptic densities. (Bar indicates 0.5 um)

chromaffin cells possess the ability to release hormonal products into the circulation for general effects on distant tissues or organ systems. This may not be the case, since a recent study by Coupland et al. (1982) questions the functional significance of this non-innervated system. On the other hand, the tissue is highly differentiated and apparently persists throughout life in many species, including the human (Hervonen et al. 1976).

The present results show clearly that several chromaffin paraganglia occur around the inferior mesenteric ganglion. In addition, many more chromaffin cells in the form of clusters of varying size also are located within the ganglion. It should be emphasized, however, that extraadrenal chromaffin tissue classically known as the paraganglia represents but one constituent of a populous system of catecholamine cells distributed extensively throughout the retroperitoneum. All para- and intraganglion chromaffin cells were extensively vascularized, but synapses were noted only on the clustered chromaffin cells located within the ganglion. The question concerning the innervation of chromaffin cells that are diversely located throughout paraaortic and paraganglion environments remains unanswered. In fact, Coupland and his colleagues expressed doubt as to why the voluminous paraaortic chromaffin bodies in rabbits would persist throughout life since they are not innervated (Coupland et al. 1982). It is tempting to arbitrarily assign an endocrine role to chromaffin cells very closely related to blood vessels and replete with a hormone product. It becomes important now to produce evidence that chromaffin cells which reside in or around a sympathetic ganglion can release their hormone content and presumably exert an influence on the immediate ganglion or a distant target.

References

Autillo-Touati A, Seite R (1980) SIF cells in the cat sympathetic ganglia and associated paraganglia. In: Eränkö O et al. (eds) Histochemistry and cell biology of autonomic neurons, SIF cells and paraneurons. Raven, New York, pp 95-102

Coupland R, Kent C, Kent SE (1982) Normal function of extra-adrenal chromaffin tissues in the young rabbit and guinea pig. J Endocrinol 92: 433-442

Henle J (1865) Über das Gewebe der Nebenniere und der Hypophyse. Z Rat Med 24: 143-152

Hervonen A, Alho H, Helen P, Kanerva L (1979) Small, intensely fluorescent cells of human sympathetic ganglia. Neurosci Lett 12: 97-101

Hervonen A, Vaalasti A, Partanen M, Kanerva L, Vaalasti T (1976) The paraganglia, a persisting endocrine system in man. Am J Anat 146: 207-210

Kohn A (1902) Das chromaffine Gewebe. Z Ges Anat 3, Ergebn Anat EntwGesch 12: 253-348

Mascorro JA, Yates RD (1977) Anatomical distribution and morphology of extraadrenal chromaffin tissue (abdominal paraganglia) in the dog. Tiss Cell 9: 447-460

Mascorro JA, Yates RD, Chen I-L (1976) A glutaraldehyde/potassium dichromate tracing method for the localization and preservation of abdominal extraadrenal chromaffin tissue. Stain Technol 50: 391-396

Matthews MR, Raisman G (1969) The ultrastructure and somatic efferent synapses of granule-containing cells in the superior cervical ganglion. J Anat 105: 255-282

Stilling H (1890) A propos de quelques experiences nouvelles sur la maladie d'Addison. Rev Med 10: 808-831

Zuckerkandl E (1902) Über Nebenorgane des Sympathicus im Retroperitonealraum des Menschen. Anat Anz 15: 97-107

Phylogenesis, Ontogenesis and Ageing

Paraneurons: Immunocytochemical and Phylogenetic Aspects

T. Fujita, H. Kuramoto, R. Yui, and T. Iwanaga

Department of Anatomy, Niigata University School of Medicine, Asahi-Machi, Niigata 951, Japan

Neurons, Paraneurons, and Their Prototypes

The concept of paraneuron implies a continuity of cell types between nerve cells (neurons) and a variety of cells sharing structural and functional features with them. The latter cells, which are called paraneurons, comprise a large variety of cells that are, to a lesser or a greater extent, sensory or endocrine in nature. Figure 7 shows how the family of neurons and that of paraneurons are continuous with one another and partly overlap.

In the middle of each circle is illustrated a cell which can be regarded as the prototype for the family. One is a multipolar neuron which is found in the intramural plexuses of the intestine; this neuron possesses morphologically and electrophysiologically primitive features (Furness and Costa 1982; Wood 1984). The other is a bipolar, basal-granulated paraneuron occurring in the gut epithelium (Fujita and Kobayashi 1977). Noteworthily, cells closely resembling these prototype neuron and paraneuron can be found in coelenteral Hydra, the most primitive metazoa (Fujita et al. 1980). The bipolar paraneurons are dispersed in the ectoderm and endoderm of Hydra and have been called "sensory cells" by Lentz (1968). The multipolar neurons form a network connected by loose synapses in the mesoglia, a connective tissue separating the ectoderm and endoderm. Both of the primitive neurons and paraneurons possess large endocrine-like granules and small synaptic-like vesicles (Westfall and Kinnamon 1978). The extensive studies by Grimmelikhuijzen (1983) and Grimmelikhuijzen et al. (1982) have indicated that more than six neuropeptides cross-reactive to corresponding mammalian peptides are contained in these neurons and paraneurons.

Secretory Products of Paraneurons

Recent advances in biochemical, cytochemical and immunocytochemical studies have made it clear that the secretory products of paraneurons stored in their membrane-bounded granules mainly comprise substances of the following chemical categories (Fujita 1976, 1983):
1. **Neuropeptides.** These are products of processing of pre-pro-peptides which are synthesized in the endoplasmic reticulum-Golgi apparatus.
2. **Classical neurotransmitters.** Monoamines, i.e., catecholamines, indolamines and histamine, may be produced in some paraneurons but may lack in others. The APUD (amine precursor uptake and decarboxylation) ability proposed by Pearse (1969) as the criterion for the identification of nerve-related endocrine cells is, therefore, detectable only in some paraneurons, as it is also the case in neu-

Experimental Brain Research Series 16
© Springer-Verlag Berlin · Heidelberg 1987

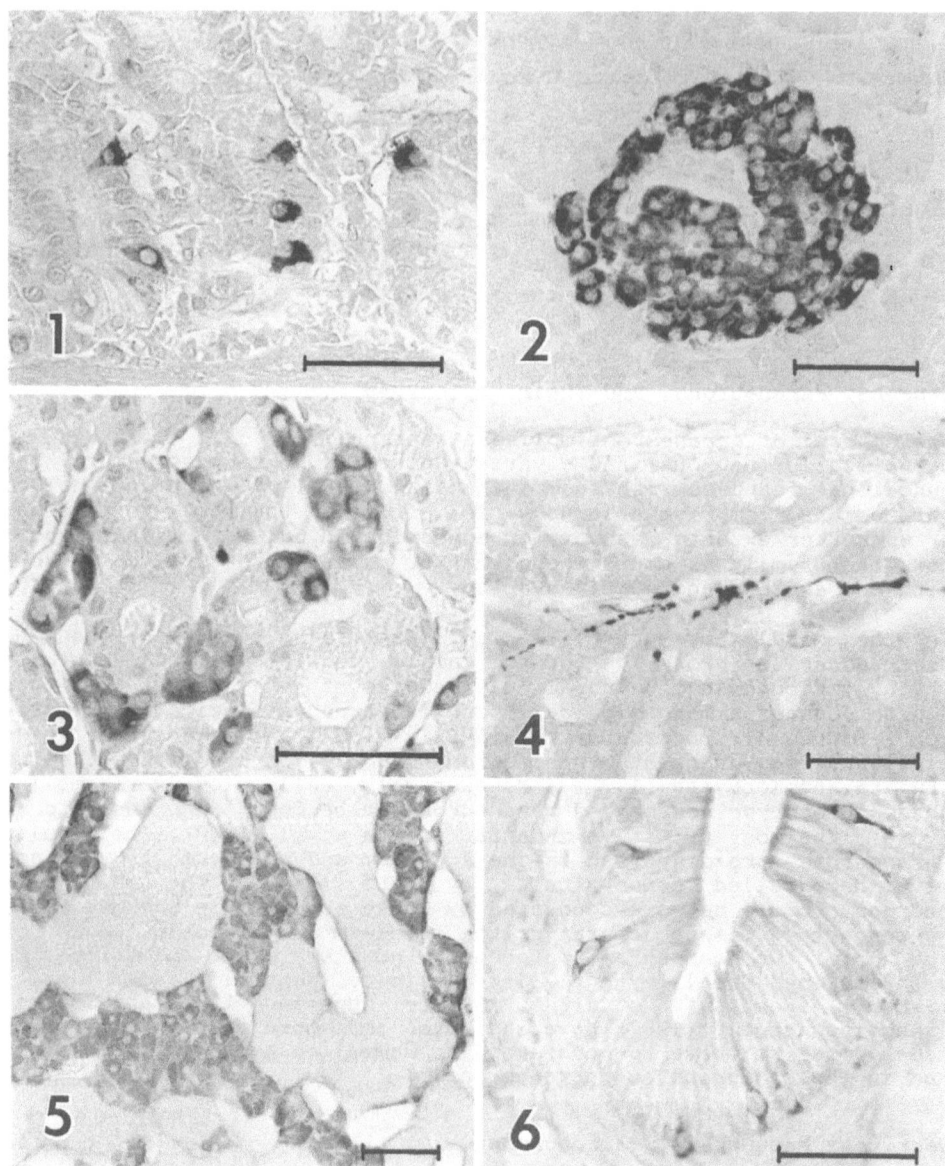

Figs. 1–6. Chromogranin A–like immunoreactivity in some vertebrate tissues demonstrated with the PAP technique. Bars indicate 50 um. Fig. 1: Human duodenum with some immunoreactive endocrine cells; Fig. 2: porcine pancreatic islet; Fig. 3: rat thyroid with immunoreactive parafollicular cells; Fig. 4: guinea pig adrenal cortex with some immunoreactive nerve fibers; Fig. 5: chick adrenal gland with immunoreactive chromaffin cells; Fig. 6: frog (<u>Rana</u> <u>catesbeiana</u>) duodenum with immunoreactive endocrine cells

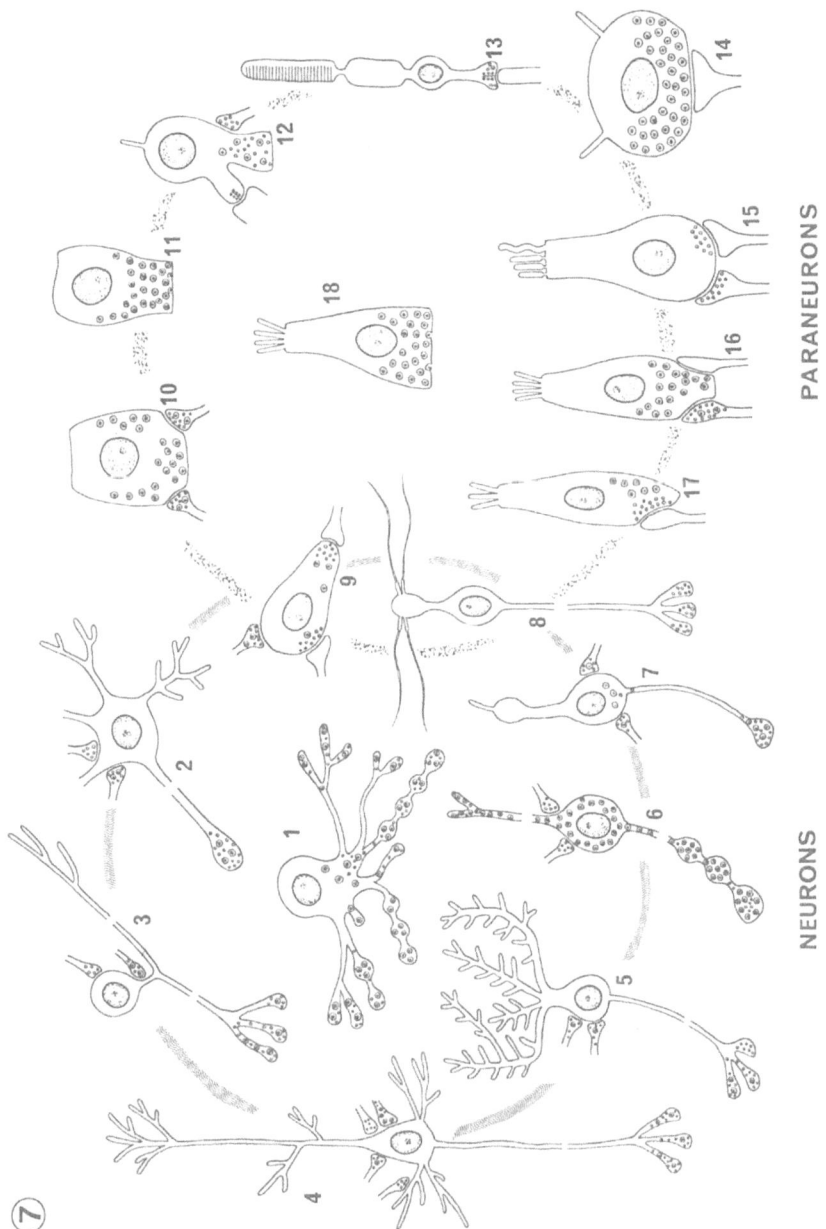

Fig. 7. Representative populations of neurons and paraneurons. 1: Intramural neuron in the intestine as the prototype of neurons; 2: sympathetic neuron; 3: sensory neuron; 4: motor neuron; 5: Purkinje cell; 6: neurosecretory or peptidergic neuron; 7: cerebrospinal fluid contacting neuron; 8: olfactory neuron; 9: SIF cell or carotid body chief cell; 10: adrenal chromaffin cell; 11: parafollicular, adenohypophyseal or pancreatic islet cell; 12: pinealocyte (in lower vertebrate); 13: visual cell; 14: Merkel cell; 15: inner ear and lateral line hair cell; 16: broncho-pulmonary endocrine cell; 17: gustatory cell; 18: gastro-enteric endocrince cell as the prototype of paraneurons

rons. Acetylcholine and amino-acids and their derivatives are messengers in some neurons; however, because of technical difficulties, it is not known whether these substances also occur in paraneurons.
3. Adenine nucleotides. ATP is the main and constant occurrence; ADP, AMP or adenosine can also be found in paraneurons.
4. Acidic carrier proteins. Chromogranins, among them, have been found to occur most extensively among different paraneurons and some neurons. Neurophysins in hypothalamic neurosecretory neurons belong to the same category of substances, though these peptides are produced as a portion of pre-pro-oxytocin and pre-provasopressin.
5. Enzymes for conversion of granule substances. Among the most important of these are the enzymes for processing of pre-prohormones and those for amine conversion, such as dopamine-ß-hydroxylase (DBH).

Most of these substances can be detected by histochemical or immunohistochemical techniques. This detection is important for identification of a given cell as a paraneuron, as the occurrence of these substances is the token of the paraneuron.

In this paper our recent findings on the occurrence of chromogranin A-like substance in a variety of paraneurons are demonstrated.

Immunohistochemistry for Chromogranin A

Using an antiserum raised in rabbit against bovine SP-1/chromogranin A (Immunonuclear Corporation, Stillwater, U.S.A.), we could confirm and extend the findings by O'Connor et al. (1983), Nolan et al. (1985) and Rindi et al. (1986) that chromogranin A-like immunoreactivity is shared by different "neuroendocrine cells" which are included in our paraneuron family. The adrenal chromaffin cells, both adrenaline (A) and noradrenaline (NA) cells, contained chromogranin A-like immunoreactivity in their secretory granules in human, pig, dog and guinea pig. In the rat, A cells were strongly immunoreactive but NA cells only weakly. A few nerve fibers supplying the adrenal cortex and medulla in the guinea pig contained chromogranin A-like immunoreactivity. In the digestive tract in these animals numerous endocrine cells dispersed in the epithelium contained chromogranin A-like immunoreactivity localized in the secretory granules. Pancreatic islet cells were partly immunoreactive in man and some animals, while in the guinea pig essentially every islet cell was immunostained. The anterior pituitary lobe in the rat and pig showed immunoreactivity in a proportion of the endocrine cells. The parafollicular cells in the dog thyroid gland were immunoreactive for chromogranin A. The majority of cells in the Zuckerkandl's organ in the guinea pig were immunopositive but the neurons of the celiac ganglion were negative. The immunoreactivity could not be demonstrated in the pulmonary endocrine cells in the human fetus, in the Merkel cells in rat hair follicles, and in the granulated atrial muscle cells in the rat. The neurosecretory neurons in the supraoptic and paraventricular nuclei and their fibers reaching the posterior pituitary were immunonegative for chromogranin A. These neurons are immunoreactive for neurophysins, acidic carrier proteins similar to chromogranins.

We examined the occurrence of chromogranin A-like immunoreactivity in some non-mammalian vertebrates. In the chicken (Aves) the adrenal chromaffin cells were mostly immunopositive; a minor population of pancreatic islet cells was immunopositive; numerous endocrine cells in the digestive tract contained the immunoreactivity. In <u>Rana catesbeiana</u> (Amphibia) a part of adrenal chromaffin cells, a part of pancreatic endocrine cells and numerous gut endocrine cells were immunoreactive. In the carp (Pisces) chromogranin A-like immunoreactivity could not be found in gut endocrine cells and other neuronal and paraneuronal cells. Neither neuronal nor paraneuronal cells in the cockroach (Insecta) showed immunoreactivity.

Although our findings are still fragmental, it is suggested that a substance cross-reactive to mammalian chromogranin A can be traced down to amphibian paraneurons. Chemical or molecular changes in the acidic carrier proteins of the neuronal and paraneuronal granules along the evolution of animals will be an interesting field for future study.

References

Fujita T (1976) The gastro-enteric endocrine cell and its paraneuronic nature. In: Coupland RE (ed) Chromaffin, enterochromaffin and related cells. Elsevier, Amsterdam, pp 191-208

Fujita T (1983) Messenger substances of neurons and paraneurons: their chemical nature and the routs and ranges of their transport to targets. Biomed Res 4: 239-256

Fujita T, Kobayashi S (1977) Structure and function of gut endocrine cells. Int Rev Cytol Suppl 6: 187-233

Fujita T, Kobayashi S, Yui R, Iwanaga T (1980) Evolution of neurons and paraneurons. In: Ishii S et al. (eds) Hormones, adaptation and evolution. Japan Sci Soc Press, Tokyo / Springer, Berlin, pp 35-43

Furness JB, Costa M (1982) Types of nerves in the enteric nervous system. Neuroscience 5: 1-20

Grimmelikhuijzen CJP (1983) Coexistence of neuropeptides in Hydra. Neuroscience 9: 837-845

Grimmelikhuijzen CJP, Dierickx K, Boer GJ (1982) Oxytocin/vasopressin-like immunoreactivity is present in the nervous system of Hydra. Neuroscience 7: 3191-3199

Lentz TL (1968) Primitive nervous system. Yale University Press, New Haven

Nolan JA, Trojanowski JQ, Hogue-Angeletti R (1985) Neurons and neuroendocrine cells contain chromogranin: detection of the molecule in normal bovine tissues by immunochemical and immunohistochemical methods. J Histochem Cytochem 33: 791-798

O'Connor DT (1983) Chromogranin: widespread immunoreactivity in polypeptide hormone producing tissues and in serum. Regul Pept 6: 263-280

Pearse AGE (1969) The cytochemistry and ultrastructure of polypeptide hormone-producing cells of the APUD series and the embryologic, physiologic and pathologic implications of the concept. J Histochem Cytochem 17: 303-313

Rindi G, Buffa R, Sessa F, Tortora O, Solcia E (1986) Chromogranin A, B and C immunoreactivities of mammalian endocrine cells: distribution, distinction from costored hormones/prohormones and relationship with the argyrophil component of secretory granules. Histochemistry 85: 19-28

Westfall JA, Kinnamon JC (1978) A second sensory-motor-interneuron with neurosecretory granules in _Hydra_. J Neurocytol 7: 365-379

Wood JD (1984) Enteric neurophysiology. Amer J Physiol 247: G585-598

Phylogenetic Aspects on Regulatory Peptides in Particular Islet Hormones in Some Lower Vertebrates

S.Falkmer[1], J.M.Conlon[2], E.Dafgård[1], and L.Thim[3]

[1] Department of Tumor Pathology, Karolinska Institutet, P.O.Box 60400, 10401 Stockholm, Sweden
[2] Klinische Forschungsgruppe für Gasteroendokrinologie, Max-Planck-Gesellschaft, 3400 Göttingen, FRG
[3] Novo Research Institute, Bagsvaerd, Denmark

Introduction

When it became possible to determine the amino-acid sequences and even the tertiary molecular structure of insulin (see Emdin et al. 1985) and some other neurohormonal peptides, it also became obvious that peptide hormones and other regulatory peptides form so-called hormone families (see Falkmer et al. 1984). This concept implies that members of such hormone families ought to have an ancestral molecule in common from which the present-day individual regulatory peptides have arisen, probably via gene duplications and mutations (Steiner et al. 1984; Thorndyke and Falkmer 1985). The insulin family, for instance, consists not only of genuine hormones but also of cell-growth factors (see Falkmer et al. 1985a, 1986). We have focused our attention on the phylogeny of insulin, somatostatin (SOM), glucagon, and pancreatic polypeptide (PP). As the islet parenchyma is an anatomical structure, unique for the vertebrates (Falkmer et al. 1984; Falkmer 1985), it was of particular interest to study the amino-acid sequences of these four islet hormones in the phylogenetically oldest vertebrates.

Islet Hormones in Invertebrates

As recently reviewed (Falkmer et al. 1985b), immunohistochemical observations of the cellular production sites of various regulatory peptides in coelenterates and in protostomian invertebrates indicate that the phylogenetical origin of most neurohormonal peptides lies in the nervous system. Even the classical gastro-enteropancreatic (GEP) hormones, such as insulin, gastrin, and secretin, appear as neurohormones in these primitive animals. However, in some of the most highly developed protostomian invertebrates, such as some arthropods (insects, crustaceans) and molluscs, immunohistochemical observations have been made, indicating that a so-called brain-gut axis has been developed for some GEP hormones (see Falkmer et al. 1984, 1985b). This is the first sign of an evolution of a disseminated endocrine cell population outside the nervous system.

A fully developed brain-gut axis for all the four islet hormones does not, however, seem to occur until at the evolutionary stage of the deuterostomian invertebrates; thus in the protochordates, such as Ciona intestinalis (the sea-squirt), insulin-, SOM-, glucagon-, and PP-immunoreactive secretion granules have been found both in neurons (and nerve fibers) in the cerebral ganglion (the "brain") and in epithelial cells of open, endocrine type in the mucosa of the gastrointestinal tract (the "gut") (Thorndyke and Falkmer 1985).

Experimental Brain Research Series 16
© Springer-Verlag Berlin · Heidelberg 1987

As regards the amino-acid sequences of insulin, SOM, glucagon, or PP, from invertebrate species, there is - to the best of our knowledge - no detailed information available, so far (Falkmer et al. 1986). It has, however, been found that at least some of the insulin immunoreactive cells found in insect brains (El-Salhy et al. 1983) seem to produce another member of the insulin family than insulin itself, viz. the pro-thoracio-tropic hormone (PTTH) (Nagasawa et al. 1984a,b). Such an observation makes it imperative to look for other members of the insulin family, as well, such as the insulin-like growth factors (IGF 1 and 2) (see Falkmer et al. 1985a), in the nervous system of invertebrates.

Islet Hormones in Vertebrates

1. Agnatha (Cyclostomes)

The first islet parenchyma ("endocrine pancreas") in evolution appears already in the phylogenetically oldest vertebrates, viz. the jaw-less fisch (Agnatha; Cyclostomes) (Falkmer 1985). Of the two extant main subphyla of this class of vertebrates, viz. the Petromyzontia (the lampreys) and the Myxinoids (the hagfishes), the latter are best known with regard to the histophysiology of their neuroendocrine system (see Falkmer et al. 1984). Thus, in Myxine glutinosa (the Atlantic hagfish), a grossly visible separate islet organ occurs but no exocrine acinar pancreas. It is a 2-hormone islet parenchyma, consisting of about 99% insulin cells and 1% SOM cells (see Falkmer et al. 1984). The SOM cells - but not the insulin cells - also appear as cells of closed and open type in the gut mucosa, together with glucagon and PP immunoreactive cells (see Falkmer 1985).

Hagfish insulin has previously been comprehensively investigated (see Emdin et al. 1985). Now, the hormonal peptides of the SOM cells have also been analyzed more closely. In our preliminary observations (Conlon JM, Thim L and Falkmer S, unpublished) both the SOM cells in the gut and those in the islet organ were found to produce predominantly the high-molecular form (approx. 28-32 amino-acid residues) of SOM and only to a less extent the classical SOM-14 (Figs. 1 and 2). Thus, the islet parenchyma and the gut of the hagfish have the same SOM profile. A similar lack of tissue specificity in processing pro-SOM has recently been observed also in another lower vertebrate, viz. the Elasmobranchian cartilaginous fish, Torpedo marmorata (the electric ray); here, however, SOM-14 was the only component detected in extracts of pancreas, stomach, gut, and brain (Conlon et al. 1985).

As to the molecular composition of the hormonal peptides produced by the glucagon and PP immunoreactive cells in the hagfish gut, there is no new information available, so far.

2. Holocephalan Cartilaginous Fish

It is not until at the evolutionary stage of the earliest Gnathostomian vertebrates, viz. the Holocephali, that a genuine exocrine pancreatic gland occurs, equipped with typical islets of Langerhans (Falkmer 1985). In all the three extant Holocephalan species available, these islets form essentially a 3-hormone-parenchyma, producing, in addition to insulin and SOM, also members of the glucagon family; most of the PP cells still "lag behind" in the gut mucosa and in the epithelium of the long pancreatic duct (Falkmer et al. 1984).

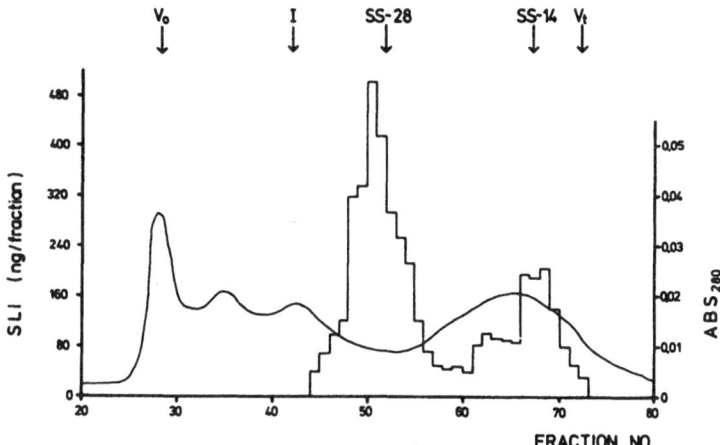

Fig. 1. Elution pattern of an acid-ethanol extract of hagfish islet organs, subjected to gel filtration on Sephadex G-50. Using a radio-immunoassay with an antiserum directed against the central residues of mammalian SOM-14, crossreacting fully with SOM-28, the islet parenchyma and the gut were found to contain 2 nmol/g wet weight (ww) and 2 pmol/g ww respectively, of SOM-like immunoreactivity. In both extracts the SOM-28-sized component represented the predominant molecular form (about 70% of the total immunoreactivity)

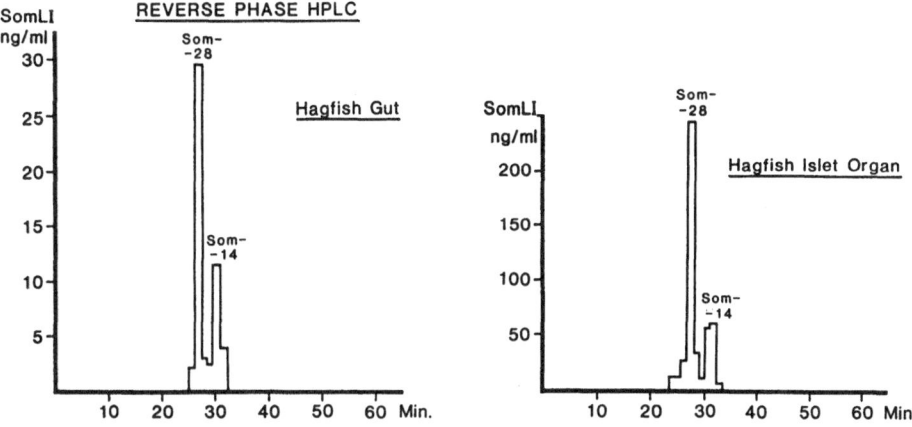

Fig. 2. Results of efforts to purify to homogeneity by reversed phase high-performance liquid chromatography (HPLC) (using an octa-decyl-silyl-silica column) of both the islet and gut SOM components. The predominant form of SOM-like immunoreactivity in both organs was found to be in a peak with a retention time slightly less than that of synthetic SOM-28, whereas the classical SOM-14 formed only a minor fraction

In one of these three Holocephalan species, <u>Hydrolagus colliei</u> (the Pacific ratfish), the amino-acid sequence of its insulin has now been clarified (Conlon JM, Thim L, Dafgard E, Falkmer S, unpublished).

The ratfish insulin amino-acid sequence was found to be:

```
     5    10    15    20    25    30
```

A-chain: GIVEQCCHNTCSLANLEGYCN

B-chain: VPTQRLCGSHLVDALYFVCGERGFFYSPKPIREL

Rather surprisingly, it was found that a tetrapeptide COOH-terminal extension of the B-chain existed. This sequence (Ile-Arg-Glu-Leu) was found to be homologous to the N-terminal region of the connecting peptide of proinsulin from the guinea-pig (Arg-Arg-Glu-Leu) and man (Arg-Arg-Glu-Ala). This finding was consistent with the occurrence of a single base mutation in the region of the gene, encoding one of the dibasic residue processing sites, i.e. the B 31 Arg (AGA) to the B 31 Ile (AUA). As a result, in the ratfish proinsulin an alternative cleavage site, situated within the C-peptide region of proinsulin, has been utilized to the conventional one where proinsulin is converted to insulin by proteolytic cleavages at sites of pairs of basic amino-acid residues (see Steiner et al. 1984).

Otherwise, it was found that the ratfish insulin had A-chain with the conventional number of 21 amino-acid residues, present already in hagfish insulin (see Emdin et al. 1985). In contrast to Myxine insulin, however, ratfish insulin was found to have a histidine residue in the B 10 position. This implies that ratfish insulin might be the first in evolution to be able to form zinc-containing hexamers; the presence of an alanin residue at the A 14 position also gives additional strength to the probability that ratfish insulin can hexamerize (Emdin et al. 1980). Whether or not this actually is the case, remains, however, to be settled.

As to the amino-acid composition of SOM, glucagon and PP in the Holocephali, no information is available, so far.

3. Elasmobranchian Cartilaginous Fish and Bony Fish

The islet parenchyma of all other cartilaginous fish and also that of bony fish (as well as that of higher vertebrates) is a 4-hormone organ (Falkmer 1985), where from our laboratories some reports have already been given of the amino-acid sequences of insulin (Bajaj et al. 1983; Cutfield et al. 1986), SOM (Conlon et al. 1985), glucagon and its biosynthetically related peptides (Conlon and Thim 1985; Conlon et al. 1987a) and PP (Conlon et al. 1987b). As some of these investigations are still in progress, their results will be reviewed in a subsequent report.

Acknowledgements. This work was supported by grants from the Stiftung Volkswagenwerk, the Swedish Medical Research Council (Project No. 12x-718), the Swedish Diabetes Association, the Cancer Society of Stockholm and the King Gustaf V Jubilee Fond. The authors also thank the staff of the Kristineberg Marine Biology Station at Fiskebäckskil, Sweden, for their fine cooperation.

References

Bajaj M, Blundell TL, Pitts JE, Wood SP, Tatnell MA, Falkmer S, Emdin SO, Gowan LK, Crow H, Schwabe C, Wollmer A, Strassburger W (1983) Dogfish insulin. Primary structure, conformation and biological properties of an elasmobranchial insulin. Eur J Biochem 135: 535-542

Conlon JM, Thim L (1985) Primary structrue of glucagon from an elasmobranchian fish, Torpedo marmorata. Gen Comp Endocr 60: 398-405

Conlon JM, Agoston DV, Thim L (1985) An elasmobranchian somatostatin: primary structure and tissue distribution in Torpedo marmorata. Gen Comp Endocr 60: 406-413

Conlon JM, Falkmer S, Thim L (1987a) Posttranslational processing of proglucagon in the pancreatic islets of the daddy sculpin (Cottus scorpius). Eur J Biochem (submitted)

Conlon JM, Schmidt WE, Gallwitz B, Falkmer S, Thim L (1987b) Characterization of an amidated pancreatic polypeptide from the daddy sculpin (Cottus scorpius). Regul Peptd (submitted)

Cutfield JF, Cutfield SM, Carne A, Emdin SO, Falkmer S (1986) The isolation, purification and amino-acid sequence of insulin from the teleost fish, Cottus scorpius (daddy sculpin). Eur J Biochem 158: 117-123

El-Salhy M, Falkmer S, Kramer JK, Speirs RD (1983) Immunohistochemical investigations of neuropeptides in the brain, corpora cardiaca, and corpora allata of an adult Lepidopteran insect, Manduca sexta (L). Cell Tiss Res 232: 295-317

Emdin SO, Dodson GG, Cutfield JM, Cutfield SM (1980) Role of zinc in insulin biosynthesis. Some possible zinc-insulin interactions in the pancreatic B-cell. Diabetologia 19: 174-182

Emdin SO, Steiner DF, Chan SJ, Falkmer S (1985) Hagfish insulin: evolution of insulin. In: Foreman RE, Gorbman A, Dodd JM, Olsson R (eds) The evolutionary biology of primitive fishes. Plenum, New York, pp 363-378

Falkmer S (1985) Comparative morphology of pancreatic islets in animals. In: Volk BW, Arquilla ER (eds) The diabetic pancreas, ed 2. Plenum, New York, pp 17-52

Falkmer S, El-Salhy M, Titlbach M (1984) Evolution of the neuroendocrine system in vertebrates. A review with particular reference to the phylogeny and postnatal maturation of the islet parenchyma. In: Falkmer S, Hakanson R, Sundler F (eds) Evolution and Tumor Pathology of the Neuroendocrine System. Elsevier, Amsterdam, pp 59-87

Falkmer S, Dafgard E, El-Salhy M, Engström W, Grimelius L, Zetterberg A (1985a) Phylogenetical aspects on islet hormone families: a minireview with particular reference to insulin as a growth factor and to the phylogeny of PYY and NPY immunoreactive cells and nerves in the endocrine and exocrine pancreas. Peptides 6, Suppl 3: 315-320

Falkmer S, Gustafsson MKS, Sundler F (1985b) Phylogenetic aspects on the neuroendocrine system. A minireview with particular reference to cells storing neurohormonal peptides in some primitive protostomian invertebrates (flatworms, annelids). Nord Psykiatr Tidsskr 39, Suppl 11: 21-30

Falkmer S, Dafgard E, Engström W (1986) Phylogeny of insulin. "Primitive" insulins and the cell cycle. Chem Scripta 26B: 209-212

Nagasawa H, Kataoka H, Hori Y, Isogai A, Tamura S, Suzuki A, Guo F, Zhong X, Mizoguchi A, Fujishita M, Takahashi SY, Ohnishi E, Ishizaki H (1984a) Isolation and some characterization of the prothoracico-tropic hormone from Bombyx mori. Gen Comp Endocr 53: 143-152

Nagasawa H, Kamito T, Fugo H, Suzuki A, Ishizaki H (1984b) Amino-terminal amino-acid sequence of the silkworm prothoraciocotropic hormone: homology with insulin. Science 22: 1344-1345

Steiner DF, Chan SJ, Docherty K, Emdin SO, Dodson GG, Falkmer S (1984) Evolution of polypeptide hormones and their precursor processing mechanisms. In: Falkmer S, Hakanson R, Sundler F (eds) Evolution and Tumor Pathology of the Neuroendocrine System. Elsevier, Amsterdam, pp 203-223

Thorndyke MC, Falkmer S (1985) Evolution of gastro-entero-pancreatic endocrine system in lower vertebrates. In: Foreman RE, Gorbman A, Dodd JM, Olsson R (eds) The evolutionary biology of primitive fishes. Plenum, New York, pp 379-400

Endocrine-Autonomic Relationships: Neural Crest and Placodal Contributions to the Enteric Nervous System

A. G. E. Pearse

Royal Postgraduate Medical School, Hammersmith Hospital, Du Cane Road, London W12 0HS, Great Britain

Introduction

In mammalia, more than 40 types of neuroendocrine (formerly endocrine) cells have been described. Diffusely distributed throughout the body, they have as their common function the production of physiologically active peptides and amines. They share with neurons, moreover, an increasing number of common cytochemical (funtional) characteristics and have therefore been referred to as a single system, the so-called diffuse neuroendocrine system (DNES).

Some 13 of the cells are of proven neuroectodermal origin but for the remainder origins from endoderm and even mesoderm are not only proposed, but generally accepted. In particular, it is proposed that the intestinal and pancreatic neuroendocrine cells arise from an endodermal stem cell whose common progeny includes all the differentiated cells of the region. In unmodified form this hypothesis is incompatible with the status of the cells as a group of independent neuroregulators.

If the DNES is truly a system, it must be made up of cells derived from precursors which are themselves irrevocably determined as neuroendocrine. This is the paradox which is exemplified by the endocrine-autonomic relationships of the lamina propria mucosae.

In any discussion of the autonomic innervation of the gastrointestinal tract and of its relationship to the endocrine (neuroendocrine) cells of the gastrointestinal mucosa there are clearly three anatomical zones of interest. These are 1) the submucosa and its neurons, 2) the mucosa and its neuroendocrine cells, and 3) the lamina propria and its neuroendocrine cells and nerve plexuses.

For the first zone, the facts are virtually beyond dispute and can be set forth briefly and succinctly. The second zone is a region where active controversy has prevailed for at least two decades, and controversial views on the matter have been expressed for far longer than that. The most cogent problem, however, concerns the third zone. What are the origins of the neuroendocrine cells and nerve tracts in this buffer zone between mucosa and submucosa, what is their nature and what are their relationships? Here, a mass of undigested data leaves only a problem. It is to be hoped that this problem will prove to be amenable to resolution by the application of some of the newer techniques of immuno- and hybridocytochemistry. Until such illumination is achieved, only speculative views can be advanced.

Experimental Brain Research Series 16
© Springer-Verlag Berlin · Heidelberg 1987

The Submucosa

The Autonomic System: Origins, Development and Constitution

"The origin of enteric ganglion cells is not yet clearly established"; so wrote Le Douarin and Teillet (1973). Only ten years later,
substantially due to the embryological studies of these authors (Le
Douarin and Teillet 1974) Gershon et al. (1983) could write "there
is widespread agreement that the ganglion cells of the enteric
nervous system are neuroectodermal derivatives, as are other peripheral ganglion cells". In short, the enteric nervous system has
been shown to be derived from neural crest precursors arising between the levels of somites 1-7 and caudal to somite 26. These
precursors are committed to neuronal differentiation at the time of
colonisation of the gut wall but still retain some diversity with
regard to their final phenotype (Black 1982).

Cytochemical studies have shown as many as fifteen types of neurotransmitter substances in the enteric nerve supply (Furness and
Costa 1980; Smith and Madson 1981). These transmitters, two or more
of which may be present in one and the same nerve fibre, today
include acetylcholine, noradrenaline, 5-hydroxytryptamine, vasoactive intestinal peptide (VIP), cholecystokinin (CCK), somatostatin,
substance P, gamma-aminobutyric acid (GABA), dynorphin, met-enkephalin, galanin, peptide histidine isoleucine (PHI), neuropeptide tyrosine (NPY), gastrin releasing peptide (GRP) and calcitonin gene-
related peptide (CGRP). Thus the enteric nervous system demonstrates
a degree of phenotypic diversity which greatly exceeds that shown by
any other region subject to autonomic nervous control of its
functions. This degree of diversity with regard to final expression
and synthesis of hormonal or transmitter products it shares, however, with the endocrine (neuroendocrine) cells of the enteric
mucosa, whose tally of peptides (but not of amines) greatly exceeds
the score registered above for autonomic neurons.

The Mucosa

Endocrine Cells: Origins and Constitution

The endocrine nature of the cells of Nicolas-Kultschitsky (yellow
cells of Schmidt, enterochromaffin cells of Ciaccio, argentaffin
cells of Masson, clear cells of Feyrter) was first established by
Masson (1914). Since then, the most popular theory for their origin,
and by extrapolation for all the other types of gut endocrine cells,
has held that their precursor is an endodermal stem cell (Masson
1928; Cheng and Leblond 1974).

The only alternative theory, deriving them from neuroectodermal
elements originating outside the gut epithelium, has received little
suport from experimental studies, with particular reference to those
of Andrew (1976) and Le Douarin (1978). These two authors, or rather
groups, demonstrated conclusively that cells derived from the neural
crest played no part in the development of gut endocrine cells.
There remained, therefore, the sole possibility that these could
arise from neuroendocrine-determined pre-crest precursors derived
from neuroectoderm. That such a group of potential precursors exists
has been shown by the work of Kessler et al. (1981) who treated rat
embryos at day E11.5 with nerve growth factor and demonstrated
catecholaminergic cells "ectopically scattered throughout the embryo
from head to toe". This work was extended by Teitelman et al. (1981)
who identified the earliest glucagon and insulin synthesising cells

in the embryonic dorsal pancreas (at days E12 and E15 respectively) as dopaminergic. They failed, however, to link them conclusively to the ectopic dopaminergic cells described by Kessler et al. (1981)

If the endodermal stem cell theory is accepted, as the one which most closely fits the observed facts, credit can again be afforded to the French-Canadian pathologist, Pierre Masson. In his 1928 paper he described the argentaffin cells as constituting a "neuroectoderm, an entodermic placode".

This hypothesis can bring the gastrointestinal and pancreatic endocrine cells into line with the other principal group of neuroendocrine cells belonging to the APUD (amine precursor uptake and decarboxylation) series which includes hypothalamic and olfactory neurons, pinealocytes, pituitary, thyroid C and parathyroid chief cells, Merkel, carotid body types 1 and 2 and adrenomedullary adrenalin and noradrenalin cells (Pearse 1986). These are all demonstrably of either placodal or neural crest origin.

It is necessary here to draw attention to an essential property exhibited by stem cells. This is their "independence" (Till 1982) and hence their ability to give rise to new stem cells (self-renewal) having an identical irreversibly determined status. There is ample evidence to show that the endocrine cells of the gut and pancreas are capable of self renewal independently of their presumptive archetypal precursor. One must therefore postulate that this primary archetype, the crypt base cell of Cheng and Leblond (1974), must give rise, after an unknown number of doublings, to independent races of separately determined and therefore secondary stem cells (for enterocyte, goblet, Paneth and endocrine cells). A model for this is seen in the haemopoietic system (Burton et al. 1982) where, after some 17-19 doublings from the original clonogenic stem cell, substantial clones of determined erythroid stem cells successively maintain erythropoiesis. On this model the crypt base stem cell would thus give rise to four independent clonogenic stem cell lines, precursors of each of the four groups of cells in gut epithelium.

The Lamina Propria

Endocrine Cells and Nerve Plexuses: Origins and Relationships

Extra-epithelial argentaffin cells were first described in the appendix by Masson (1921), in a paper dealing with obliterative appendicites. His findings, specifically in this disorder and in neurogenic appendicopathy were readily confirmed by fellow pathologists. Simard (1934) recognised that argentaffin cells could be found in the lamina propria of the whole gastrointestinal tract, as did Feyrter (1934). Van Campenhout (1983) studying human embryos of crown-rump length 20-30 cm, confirmed Masson's pathological observations and extended them to the developing vermiform appendix. Endocrine cells other than argentaffin or argyrophil were not recognised by any of the above quoted authors, for obvious reasons.

Much later, EM studies by Matsuo and Seki (1976, 1978) revealed numerous cells in the lamina propria of the gastric fundus in rats and dogs which, on account of their content of endocrine-type secretory granules, were described by the authors as neuroendocrine. They were noted to be surrounded by numerous non-myelinated nerve fibres and supporting cells resembling Schwann cells, and were considered to constitute an endocrine (APUD) autonomic complex. A typical neuroendocrine cell in the human gastric lamina propria is shown in Fig. 1.

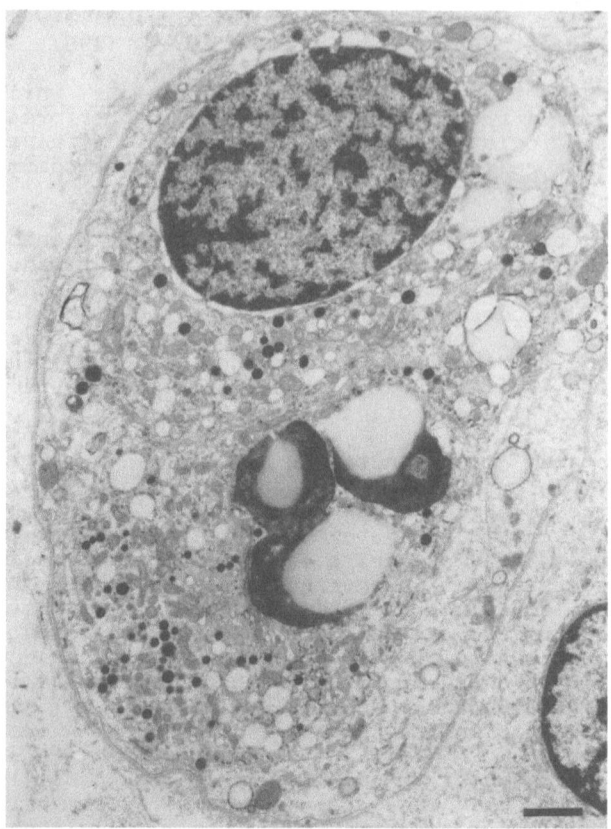

Fig. 1. Neuroendocrine cells in human gastric mucosa (pylorus). Situated in the lamina propria, these two cells contain secretory type granules 100–200 nm in diameter. They share a common basement membrane, which often includes elements of the nerve plexus. Lipofuscin-type pigment (and lipid inclusions) are commonly observed. Bar indicates 1 um

Until the publication of these experimental studies the consensus of opinion with regard to the origin of the lamina propria endocrine cells was that they were budded off from the basal epithelium of the intestine by a process known as bourgeonnement or endophytie, depending on linguistic attachments. Matsuo and Seki (1976, 1978), however, considered that their neuroendocrine cells were derived, together with the associated nerve plexuses, from invading cells belonging to the neural crest. No direct evidence for this view was forthcoming, however.

While discussing their nature, many later authors (e.g. Stachura et al. 1981) expressed no opinions on the origin of the extraepithelial neuroendocrine cells. Others, however, (Aubock and Ratzenhofer 1982; Rode et al. 1982, 1983) considered them to be an integral part of the subepithelial nerve plexus and derived, together with that complex, from the neural crest. Since cytochemical neuroendocrine markers would be expected to be, and indeed are, essentially the same in the case of either origin, and since the necessary experimental embryology has not been carried out, the matter remains in equipoise. It may be observed, however, that several illustrations to their studies provided by Le Douarin and coworkers clearly show marked (quail) cells from the donor neural crest in developing chick gut subepithelial tissues. These cells would be obvious precursors for the lamina propria neuroendocrine cells. The intimate relationship between the nerves of the plexus and the neuroendocrine cells (shared basement membrane) supports this view but, of course does not prove it.

A Placodal Hypothesis

The germ layer concept has been an exceedingly useful one in embryology but today it needs to be used with less rigidity than in former times and as loosely descriptive rather than factually binding.

Evidence for an origin of the extrinsic and intrinsic nerve supply to the gut from neural crest is complete. Evidence for the lamina propria neuroendocrine cells as derivatives of either crest or placodes is inconclusive. Evidence for an intraepithelial stem cell origin for the neuroendocrine cells of the mucosa is strong and compelling. As such, even today, this stem cell has to be described as endodermal. But since the definitive endoderm originates from ectoderm it may now be acceptable to stretch the term placode beyond its formerly rigidly ectodermal status, and to adopt the proposal of Masson (1928). As already noted, this concept regards the mucosal endocrine cells as "placodal neuroentoderm". In this way it would be possible to unify the entire enteric nervous (and neuroendocrine) system, which would become a joint responsibility of placodes on the one hand and of neural crest on the other.

Phylogenetically, the relationship between these two entities is close. Both are homologous with the epidermal and perivisceral neural plexuses of our early chordate ancestors (Northcutt and Gans 1983). Both exhibit the property of migration and both give rise to sensory neurons and to special sense organs (in this context it should be noted that visceromotor neurons are the sole provenance of the neural crest). In this way they combine to produce an effective controlling system not only for the motility of the gastrointestinal tract but also, by virtue of an extensive list of amine and peptide neurotransmitters and hormones, of finer functions such as the transfer of ions and molecules through the luminal epithelium.

Acknowledgement. I am indebted to my former student, Professor J. Stachura, for the electron micrograph reproduced as Figure 1.

References

Andrew A (1976) An experimental investigation into the possible neural crest origin of pancreatic APUD (islet) cells. J Embryol exp Morphol 35: 577–593

Aubock L, Ratzenhofer M (1982) "Extraepithelial enterochromaffin cell-nerve fibre complexes" in the normal human appendix and in neurogenic appendicopathy. J Pathol 136: 217–226

Black IB (1982) Stages of neurotransmitter development in autonomic neurons. Science 215: 1198–1204

Burton DI, Ansell JD, Gray RA, Micklem HS (1982) A stem cell for stem cells in murine haematopoiesis. Nature 298: 562–563

Cheng H, Leblond CP (1974) Origin differentiation and renewal of the four main epithalial cells in the mouse small intestine: V. Unitarian theory of the origin of the four epithelial cell types. Amer J Anat 141: 537–562

Feyrter F (1934) Carcinoid and Carcinom. Ergebn allg Path path Anat 29: 305–489

Furness JB, Costa M (1980) Types of nerves in the enteric nervous system. Neuroscience 4: 305–310

Gershon MD, Payette RF, Rothman TP (1983) Development of the enteric nervous system. Fed Proc 42: 1620–1625

Le Douarin NM (1978) The embryological origin of the endocrine cells associated with the digestive tract: experimental analysis based on the use of a stable cell marking technique. In: Bloom SR (ed) Gut hormones, 1st edn. Churchill–Livingstone, Edinburgh, pp 49–56

Le Douarin NM, Teillet M-A (1973) The migration of neural crest cells to the wall of the digestive tract in avian embryo. J Embryol exp Morphol 30: 31–48

Le Douarin NM, Teillet M-A (1974) Experimental analysis of the migration and differentiation of neuroblasts of the autonomic nervous system and of neuroectodermal mesenchymal derivatives using a biological cell marker technique. Dev Biol 41: 162–184

Masson P (1914) La glande de l'intestin chez l'homme. C R Acad Sci Paris 158: 59–60

Masson P (1921) Nevromes sympathiques et l'appendicule obliterante. Lyon chir 18: 281–299

Masson P (1928) Carcinoids (argentaffin tumors) and nerve hyperplasia of the appendicular mucosa. Amer J Pathol 4: 181–211

Matsuo Y, Seki A (1976) Integration of hormonal and neural control of gastrointestinal secretion. Asian Med J 19: 589–630

Matsuo Y, Seki A (1978) The coordination of gastrointestinal hormones and the autonomic nerves. Amer J Gastroenterol 69: 21–50

Northcott RG, Gans C (1983) The genesis of neural crest and epidermal placodes: a reinterpretation of vertebrate origins. Quart Rev Biol 58: 1–28

Pearse AGE (1986) The diffuse neuroendocrine system: peptides, amines, placodes and the APUD theory. In: Hökfelt T, Fuxe K, Pernow P (eds) Progress in Brain Research, Vol 68. Elsevier, Amsterdam, pp 25–31

Rode J, Dhillon AP, Papadaki L (1983) Serotonin–immunoreactive cells in the lamina propria plexus of the appendix. Hum Pathol 14: 464–469

Rode J, Dhillon AP, Papadaki L, Griffiths D (1982) Neurosecretory cells of the lamina propria of the appendix and their possible relationship to carcinoids. Histopathology 6: 69–79

Simard L-C (1934) Sur les relations des cellules argentaffines de l'intestin avec les nerfs chez l'embryon de veau. Arch d'Anat micr 30: 235-248

Smith PH, Madson KLL (1981) Interactions between autonomic nerves and endocrine cells of the gastroenteropancreatic system. Diabetologia 20: 314-324

Stachura J, Krause WJ, Ivey KJJ (1981) Ultrastructure of endocrine-like cells in lamina propria of human gastric mucosa. Gut 22: 534-541

Teitelman G, Joh TH, Reis DJ (1981) Linkage of the brain-skin-gut-axis: islet cells originate from dopaminergic precursors. Peptides 2, Suppl 2: 157-168

Till JE (1982) Stem cells in differentiation and neoplasia. J Cell Physiol Suppl 1: 3-11

Van Campenhout ME (1953) Les cellules argentaffines de l'appendice ileo-caecal de l'embryon. Bull Acad Med Belg 18: 160-183

Developmental Aspects of Carotid Body Glomus Cells

J. T. Hansen

Department of Neurobiology and Anatomy, School of Medicine and Dentistry, University of Rochester, 601 Elmwood Avenue, Rochester, NY 14642, USA

Introduction

Glomus (chief, type I) cells of the mammalian carotid body are similar in morphology to the small intensely fluorescent (SIF) cells found in various autonomic ganglia. Glomus cells appear to be of neural crest origin (LeDouarin and Teillet 1974) and, in the adult, exhibit morphological characteristics of both endocrine and neuronal cells. Evidence obtained from human fetuses suggests that presumptive glomus cells migrate to their definitive site at the carotid bifurcation in concert with the developing sympathetic trunk (Korkala and Hervonen 1973). However, any connection between the developing sympathetic trunk and the carotid body primordia is lost early in development, although the two components continue to develop in close proximity to one another.

In this study, the morphology of the glomus cells was examined at a time when these cells have already reached the carotid bifurcation, but still possess a number of features of undifferentiated neural crest cells. This study is limited to the glomus cell population; for a more comprehensive overview of mammalian carotid body development, the reader is referred to a study by Kondo (1975).

Materials and Methods

Fetuses at E18 and E20 (term: E22) were obtained from the uteri of decapitated time-pregnant Sprague-Dawley rats, were themselves decapitated, and immersion fixed in 3% glutaraldehyde, 2% paraformaldehyde, 1% acrolein, and 2.5% DMSO in 0.1 M cacodylate buffer (Kalt and Tandler 1971). Rat pups at 1, 3 and 7 days postpartum were perfusion fixed with 3% glutaraldehyde, 1% paraformaldehyde in 0.1 M cacodylate buffer, following ether anesthesia. The carotid bodies were removed and routinely processed for electron microscopy including postfixation in osmium tetroxide, en bloc staining in uranyl acetate, and embedding in Spurr's resin. Thin sections were stained on the grid with lead citrate.

For catecholamine histofluorescence, carotid bodies were removed from decapitated fetuses and pups, frozen on dry ice, and stored in liquid nitrogen. Cryostat sections (20 um) were reacted according to the SPG (sucrose-potassium phosphate-glyoxylic acid) method of de la Torre (1980), and viewed in a Nikon microscope equipped with epifluorescence.

Experimental Brain Research Series 16
© Springer-Verlag Berlin · Heidelberg 1987

Results

Glomus cells at day E18 were loosely consolidated adjacent to the internal carotid artery, but at this stage were penetrated by only a few small capillaries (Fig. 1). By day E20, the glomus cells formed a tighter mass of cells, surrounded by a thin capsule and penetrated by more capillaries (Fig. 2). Glomus cells from 1 day old rats were organized into discrete lobules typical of the adult carotid body (Fig. 3).

Ultrastructurally, glomus cells at day E18 often appeared as individual cells lying next to one another (Fig. 4), or separated from one another by intercellular spaces which contained connective tissue elements and numerous nerve fibers. Supporting (sustentacular, type II) cells were not abundant; generally, both glomus cells and nerve fibers wer not ensheathed with processes from supporting cells until day E20. The glomus cell cytoplasm contained abundant dilated rough endoplasmic reticulum (RER), numerous polysomes, and very few dense-cored vesicles (Fig. 4). Mitotic glomus cells were observed at day E18 (Fig. 5), but were not evident at later times. Occasionally, presumptive ganglion cells were observed interspersed among glomus cells, and their morphology was similar except that they were larger in diameter.

At day E20, glomus cells had dilated RER, polysomes, and a greater number of dense-cored vesicles (Fig. 6). However, the glomus cells were now adjacent to one another or separated only by intervening nerve endings and processes of the supporting cells (Fig. 6). Day E20 glomus cells appeared to represent intermediates between the relatively undifferentiated cells of E18 carotid bodies and the cells of neonates which were almost morphologically identical to adult glomus cells (Fig. 7). By day 7 postpartum, glomus cells were similar to adult cells in that the dilated RER and abundant polysomes were absent, and the cytoplasm contained numerous dense-cored vesicles and mitochondria.

Despite the relative paucity of dense-cored vesicles in E18 and E20 glomus cells, these cells still exhibited an intense catecholamine histofluorescence which was indistinguishable from neonate or adult glomus cells.

Discussion

Glomus cells from day E18 fetuses are surprisingly undifferentiated considering that these cells have already reached the definitive site of the carotid body by this stage. The cells are not unlike presumptive ganglion cells in their morphology, except that they are smaller. The characteristic dense-cored vesicles of neonatal and mature glomus cells are relatively rare in cells at day E18, even though the cells are intensely fluorescent when reacted with the SPG method. The presence of dilated RER and numerous polysomes suggests that protein synthesis is actively occurring at this time, but the packaging of secretory product (the catecholamines dopamine and norepinephrine in the rat glomus cells) into dense-cored vesicles is not evident to a significant degree until day E20. However, from day 1 onward, the glomus cells rapidly "mature" in their morphology, until by day 7 they appear similar to adult cells.

Adrenergic neural crest derivatives are of three types: sympathetic ganglion cells, adrenal medullary (chromaffin) cells, and SIF cells. Glomus cells appear to be derived from the neural crest (LeDouarin

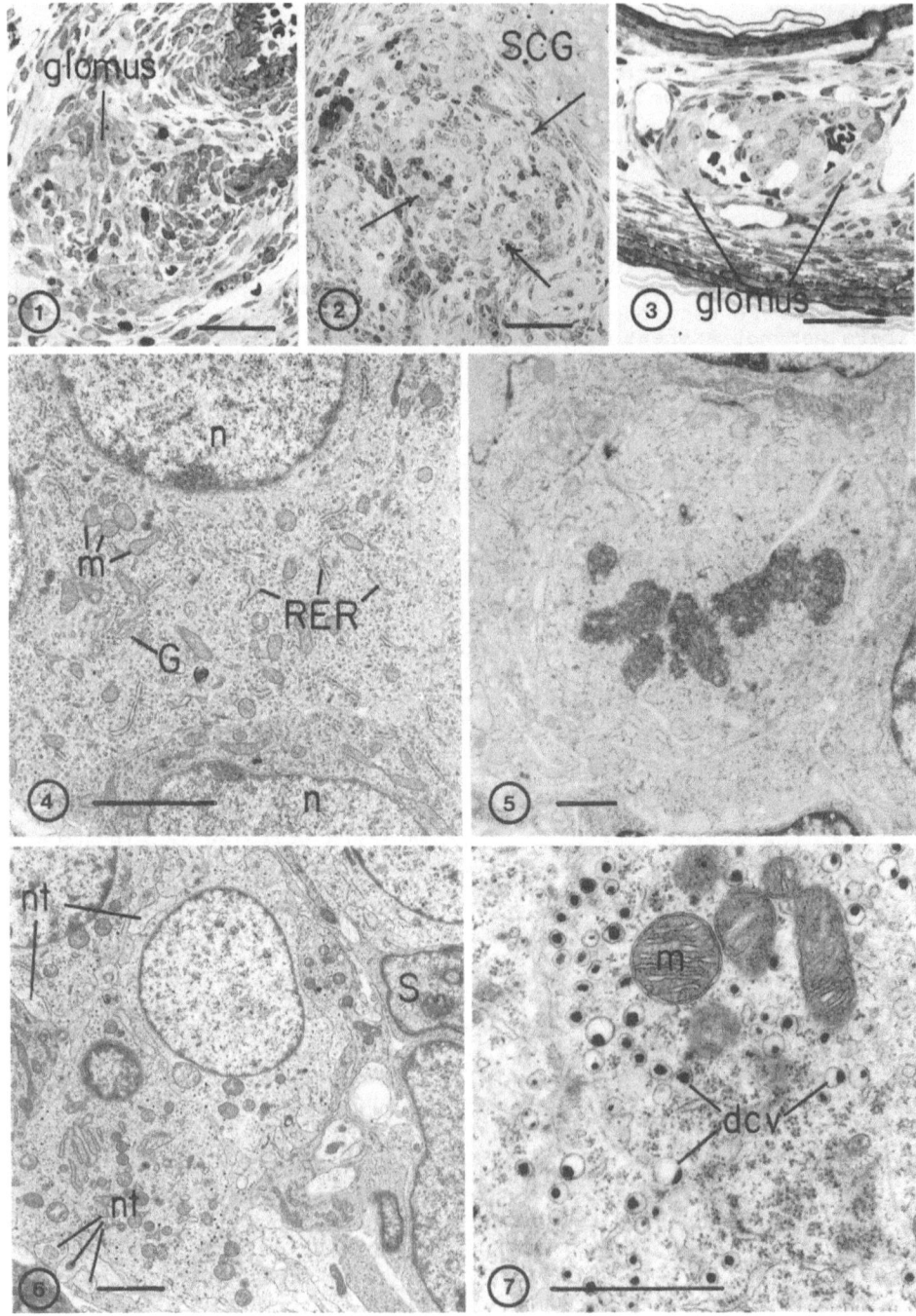

and Teillet 1974), and exhibit neuron-specific enolase-like immuno-reactivity (Kondo et al. 1982). SIF, and by analogy glomus, cells are hypothesized to represent intermediates between adrenergic neurons and adrenal chromaffin cells (Landis and Patterson 1981). Based upon the local environment, these cells may develop into either principal ganglion cells or endocrine adrenal cells. The glomus cells at day E18 appear relatively undifferentiated, resembling a sympathoadrenal precursor (Hervonen 1975), although they do fluoresce at this stage. By day E20, glomus cells begin to differentiate into the interneuronal phenotype which is characteristic of these cells in the adult carotid body. Nevertheless, glomus cells, not unlike SIF cells, retain many morphological features which are intermediate between endocrine and neuronal phenotypes. How the adjacent vasculature, ingrowth of axons, and humoral factors influence this phenotypic transformation remains unknown, but recent studies into SIF cell development point to an important role for environmental "signals" (Doupe et al. 1985). Glomus cells provide a unique population of paraneurons which may be further studied to elucidate how adrenergic derivatives develop from SIF cell intermediates.

Acknowledgement. This work was supported by grant 83 733 from the AHA.

References

de la Torre JC (1980) An improved approach to histofluorescence using the SPG method for tissue monoamines. J Neurosci Meth 3: 1-5

Doupe AJ, Patterson PH, Landis SC (1985) Small intensely fluorescent cells in culture: role of glucocorticoids and growth factors in their development and interconversions with other neural crest derivatives. J Neurosci 5: 2143-2160

◄ **Fig. 1.** Loosely consolidated glomus cells at day E18 adjacent to the internal carotid artery. Scale bar - 50 um

Fig. 2. Glomus cell mass at day E20 (demarcated by arrows) adjacent to ganglion cells of the superior cervical ganglion (SCG). Scale bar - 50 um

Fig. 3. Glomus cells of the carotid body from a 1 day old rat. Note organization of lobules with intervening capillaries. Scale bar - 50 um

Fig. 4. Portions of several glomus cells at day E18. Note the dilated rough endoplasmic reticulum (RER), polysomes, Golgi complex (G), mitochondrion (m), nucleus (n) of the cells, and paucity of cytoplasmic dense-cored vesicles. Scale bar - 2 um

Fig. 5. Mitotic glomus cell at day E18. Scale bar - 2 um

Fig. 6. Portions of glomus cells and a supporting cell (S) at day E20. Note the dilated RER and presence of cytoplasmic dense-cored vesicles. Supporting cell processes now partially surround the glomus cells and nerve terminals (nt). Scale bar - 2 um

Fig. 7. Glomus cell cytoplasm from a 1 day old rat. Note the relative decrease in polysomes, and the increase in dense-cored vesicles (dcv) of various sizes. Mitochondrion (m). Scale bar - 1 um

Hervonen H (1975) Differentiation of sympathicoblasts in cultures of chick ganglia. Anat Embryol 146: 225-243

Kalt MR, Tandler B (1971) A study of fixation of early amphibian embryos for electron microscopy. J Ultrastruct Res 36: 633-645

Kondo H (1975) A light and electron microscopic study on the embryonic development of the rat carotid body. Am J Anat 144: 275-294

Kondo H, Iwanaga T, Nakajima T (1982) Immunocytochemical study on the localization of neuron-specific enolase and S-100 protein in the carotid body of rats. Cell Tissue Res 227: 291-295

Korkala O, Hervonen A (1973) Origin and development of the catecholamine-storing cells of the human fetal carotid body. Histochemie 37: 287-297

Landis SC, Patterson PH (1981) Neural crest cell lineages. Trends Neurosci 4: 172-175

LeDouarin N, Teillet M-A (1974) Experimental analysis of the migration and differentiation of neuroblasts of the autonomic nervous system and of neuroectodermal mesenchymal derivatives using a biological cell marking technique. Dev Biol 41: 162-184

Developmental Signals Controlling Sympathoadrenal Cell Differentiation in the Rat:
The Role of Glucocorticoid Hormones and Neuronotrophic Proteins

H.-D. Hofmann, K. Seidl, C. Grothe, and K. Unsicker

Abteilung für Anatomie und Zellbiologie, Universität Marburg, Robert-Koch-Strasse 6, 3550 Marburg, FRG

Introduction

Among neurons and endocrine cells originating from the neural crest sympathetic neurons, small granule-containing (SGC = SIF) cells and adrenal chromaffin cells, collectively named sympathoadrenal cells, constitute a special entity. It has been suggested that the hypothetical precursor cell of sympathoadrenal cells resembles the SGC (SIF) cell of sympathetic ganglia. This proposition was mainly based on in vitro studies using early postnatal rat adrenal chromaffin (Unsicker et al. 1978) or SIF cells from rat superior cervical ganglia (Doupe et al. 1985), i.e. cells which had to be considered as models rather than true precursor cells. Very recently pure preparations of these presumptive precursor cells isolated from the embryonic rat adrenal gland (Seidl and Unsicker 1985; Seidl 1986) have enabled us to re-investigate several issues related to signals that govern sympathoadrenal cell development. In this article we review data mainly from our laboratory concerning the role of glucocorticoid hormones and neuronotrophic factors in determining phenotype expressions of early postnatal chromaffin cells and their embryonic precursors.

Glucocorticoid Hormones (GC)

GC have been shown to profoundly influence the morphological and transmitter phenotype of cultured chromaffin cells from early postnatal rats and their embryonic precursor cells. Most importantly, GC crucially control the initial acquisition of the adrenergic phenotype of adrenal chromaffin precursor cells in vitro. Cells isolated and plated at embryonic day (E) 16.3 fail to express epinephrine and its synthesizing enzyme phenylethanolamine N-methyltransferase (PNMT) during 4 days in culture unless GC (dexamethasone) is supplied at a concentration of at least 0.1 nM. These studies using highly purified (>95%) chromaffin precursor cells clearly document that the postulate of a novel key factor or an intrinsic genetic signal that induces initial adrenergic characteristics and is different from GC (cf. Bohn et al. 1981) can no longer be maintained. Subsequently to the initial expression of the adrenergic phenotype GC trigger a variety of biochemical and morphological responses including (i) induction of tyrosine hydroxylase (TOH) and PNMT activities, (ii) concomitant increase in the cellular epinephrine content and (iii) increases in cellular volume, numbers and size of storage granule suggesting GC effects on granular storage proteins. GC also support in vitro survival of embryonic and postnatal chromaffin cells (Grothe et al. 1985; Seidl 1986). All data obtained from cell culture studies point at a decisive role of GC in determining

the fate of immature sympathoadrenal cells, triggering further differentiation of chromaffin cells and maintaining their specific
phenotype. It will have to be determined next, what activates the GC
receptor at E 16.3 and causes the subsequent increase in GC binding
capacity in adrenal chromaffin cells. It is tempting to speculate
that GC themselves might be important in this process.

Evidence from in vivo studies also strongly supports the notion of a
crucial role of GC in chromaffin cell development. The appearance of
PNMT and epinephrine at E 17.3 is preceded by a dramatic increase in
adrenal weight as well as adrenal and plasma corticosterone between
E 16.3 and E 17.3. The subsequent increase in the adrenal epinephrine content and activities of both TOH and PNMT is paralleled by
increasing levels of adrenal corticosterone and GC receptors (Seidl
and Unsicker 1985; Seidl 1986).

Neuronotrophic Factors (NTFs)

Rat adrenal chromaffin cells are target cells for two purified NTFs,
nerve growth factor (NGF) and ciliary neuronotrophic factor (CNTF)
(Unsicker et al. 1985b,c). In addition, they respond to NTF activities contained in conditioned media and cell extracts from fibroblasts, glial and chromaffin cells (Unsicker et al. 1984, 1985b;
Unsicker and Lietzke 1986).

Outgrowth of neuritic processes with typical growth cones and expression of tetanus toxin binding sites is the most remarkable
morphological response of chromaffin cells to NTFs.

Laminin, a neurite promoting factor enhances neurite length and
branching as compared to polyornithine as a substratrum and during
prolonged culture periods (up to 4 days) also increases the proportion of neurite-bearing cells grown in the presence of an NTF (Unsicker et al. 1985b,c). NTFs also enhance chromaffin cell survival
in vitro during a transient period of postnatal development (8 day
postnatally; Unsicker et al. 1985b), but in contrast to GC have only
little effect on newborn and embryonic chromaffin cell survival
(Seidl 1986).

An effect consistently seen with NGF on sympathetic neurons is an
induction of TOH activity. This may also be observed when cultured
rat chromaffin cells of certain ages are grown in the presence of
NGF.

The TOH response requires that the cells have previously been exposed to GC since it cannot be elicited from cells isolated before the
physiological rise in adrenocortical GC secretion occurs (E 17).
Chromaffin cells isolated from 3 month-old rats do not longer respond to NGF with an increase in TOH activity. This developmental
decrease in susceptibility to NGF also applies to the neurite outgrowth behaviors of rat chromaffin cells at the adult age. Since
administration of both NGF and CNTF may elicit neuritic growth from
adult cells it will be interesting to determine whether both NTFs
have also synergistic effects on the induction of TOH.

NGF affects catecholaminergic traits of cultured early postnatal rat
chromaffin cells not only with respect to TOH, but also shifts the
epinephrine/norepinephrine ratio towards norepinephrine (Grothe et
al. 1985; Müller and Unsicker 1986). Both phenomena, NTF-induced
neuritic growth and predominance of norepinephrine, the genuine
sympathetic transmitter foster the notion of NTFs being potent

inductors of <u>neuronal</u> phenotypic expressions in sympathoadrenal cells.

The responsiveness of chromaffin cells to NTFs raises the question as to whether this reflects a physiological rather than a pharmacological behavior that may be explained by the close ontogenetic relationship of chromaffin cells with sympathetic neurons. Administration of antibodies to NGF to embryonic and early postnatal rats has very little effect on the development of the adrenal medulla (Bode et al. 1986).

NTFs other than NGF, however, may be present in the adrenal glands, among them activities stored and released by chromaffin (cf. Unsicker et al., this volume) and possibly also by Schwann cells. Their putative relevance for chromaffin cell development still remains to be demonstrated.

Synergistic Versus Antagonistic Effects of GC and NTFs

As described in the previous sections both GC and NTFs may regulate phenotypic expressions of chromaffin cells in vitro. Inhibition of NTF-induced neuritic growth by GC (Unsicker et al. 1978) and their pronounced effect on PNMT activity and epinephrine levels (Müller and Unsicker 1986; Seidl 1986) suggest an indispensable role for GC in the determination and differentiation of the endocrine chromaffin as opposed to the neuronal phenotype in the sympathoadrenal cell lineage. Moreover, GC apparently have a permissive effect for NGF in the induction of TOH. Contributions to chromaffin cell development provided exclusively by NTFs have been sorted out by exposing cultured chromaffin precursor cells to NGF and C6-conditioned medium before they have been primed by endogenous GC (Seidl 1986). Such cells extend neurites without altered TOH activity, as PC12 cells do in the absence of GC, i.e. they fail to display a typical feature of sympathetic neurons. The extent to which chromaffin cells and their precursor in vivo are influenced by NTFs that have so spectacular effects on them in vitro is difficult to be evaluated as long as GC influences in vivo cannot be dissected from overall chromaffin cell performances.

In conclusion, it appears that the development of neuronal and endocrine phenotypes in the sympathoadrenal cell lineages depends on a labile balance between GC and NTF influences requiring the simultaneous presence of both categories of factors. However, either GC or NTFs have to become dominant in order to channel the stem cells in one direction or the other.

Acknowledgement. Work described in this article was supported by the German Research Foundation (Un 34/10-1).

References

Bode K, Hofmann HD, Müller TH, Otten U, Schmidt R, Unsicker K (1986) Effect of pre- and postnatal administration of antibodies to nerve growth factor on the morphological and biochemical development of the rat adrenal medulla: a reinvestigation. Dev Brain Res 27: 139-150

Bohn MC, Goldstein M, Black IB (1981) Role of glucocorticoids in expression of the adrenergic phenotype in rat embryonic adrenal gland. Dev Biol 82: 1-10

Doupe AJ, Patterson PH, Landis SC (1985) Small intensely fluorescent cells in culture: role of glucocorticoids and growth factors in their development and interconversions with other neural crest derivatives. J Neurosci 5: 2143-2160

Grothe C, Hofmann HD, Verhofstad AAJ, Unsicker K (1985) Nerve growth factor and dexamethasone specify the catecholaminergic phenotype of cultured rat chromaffin cells: dependence on developmental stage. Dev Brain Res 21: 125-132

Müller TH, Unsicker K (1986) Nerve growth factor and dexamethasone modulate synthesis and storage of catecholamines in cultured rat adrenal medullary cells: dependence on postnatal age. J Neurochem 46: 516-524

Seidl K (1986) Purified embryonic chromaffin precursor cells in dissociated cell culture: morphological and transmitter phenotype expression. PH D Thesis, Marburg

Seidl K, Unsicker K (1985) Glucocorticoids trigger initial expression of phenylethanolamine N-methyl-transferase (PNMT) and adrenaline (A) in chromaffin precursor cells of embryonic rats. Eur J Cell Biol 39, Suppl 12: 33

Unsicker K, Krisch B, Otten U, Thoenen H (1978) Nerve growth factor-induced fiber outgrowth from isolated rat adrenal chromaffin cells: impairment by glucocorticoids. Proc Natl Acad Sci USA 75: 3498-3502

Unsicker K, Vey J, Hofmann H, Müller T, Wilson AJ (1984) C6 glioma cell-conditioned induces neurite outgrowth and survival of rat chromaffin cells in vitro: comparison with the effects of nerve growth factor. Proc Natl Acad Sci USA 81: 2242-2246

Unsicker K, Skaper SD, Varon S (1985b) Neuronotrophic and neurite-promoting factors: effects on early postnatal chromaffin cells from rat adrenal medulla. Dev Brain Res 17: 117-129

Unsicker K, Skaper SD, Varon S (1985c) Developmental changes in the responses of rat chromaffin cells to neuronotrophic and neurite-promoting factors. Dev Biol 111: 425-433

Unsicker K, Lietzke R (1986) Chromaffin cells: modified neurons that are both targets and storage sites of neuronotrophic and neurite promoting factors. Proceedings of "Glial-Neuronal Communication in Development and Regeneration. Castle Ringberg, Bavaria, FRG, June 9-15, 1985. Springer, Berlin (in press)

5-Hydroxytryptamine in Developing Sympathetic Tissue

S. Soinila[1], M. Ahonen[2], O. Häppölä[2], H. Päivärinta[2], L. Eränkö[1], H. W. M. Steinbusch[3], and T. H. Joh[4]

[1] Neurobiological Research Unit, Department of Anatomy, University of Helsinki, Siltavuorenpenger 20A, 00170 Helsinki, Finland
[2] Department of Anatomy, University of Helsinki, Siltavuorenpenger 20A, 00170 Helsinki, Finland
[3] Department of Pharmacology, Faculty of Medicine, Free University, Van der Boechorststraat 7, 1081 BT Amsterdam, The Netherlands
[4] Laboratory of Molecular Neurobiology, Cornell University, Medical College, 1300 York Avenue, New York, NY 10021, USA

Introduction

The sympathetic system is derived from neural crest cells that migrate to the dorsolateral aspects of the aorta (LeDouarin 1973). Some of these cells aggregate to form the sympathetic chains while others continue migration more ventrally. Some of the latter cells penetrate into the developing adrenal gland to form the medulla, while others move toward the ventral aspect of the aorta and form the prevertebral ganglia, the retroperitoneal paraganglia and eventually the enteric ganglia (De Champlain et al. 1970; Fernholm 1972). Catecholamine synthesis is initiated first when the migrating cells have formed the sympathetic chains. This occurs in the rat on day 11 of gestation (E11) in the cervical region (Cochard et al. 1978; Soinila 1984) and on day E12 in the thoracolumbar region (Teitelman et al. 1979; Ahonen et al. 1986). The first catecholamine-containing cells in the preaortic region are seen on day E13 (Ahonen et al. 1986) and in the adrenal gland on day E14 (Verhofstad et al. 1979).

During the subsequent development the catecholamine pattern changes in some sympathetic organs. In addition to tyrosine hydroxylase (TH) and dopamine-beta-hydroxylase (DBH), phenylethanolamine N-methyl-transferase (PNMT) appears in a high percentage of the medullary cells on day E17 (Teitelman et al. 1979; Verhofstad et al. 1979), and in some retroperitoneal paraganglionic cells on day E21 (Ahonen et al. 1986). These events indicate that the synthesis of dopamine by TH and noradrenaline by DBH are initiated simultaneously, while adrenaline synthesis by PNMT is regulated by a different mechanism.

In addition to the catecholamine, each sympathetic cell type contains one or more neuroactive substances the functions of which are not completely understood. In the adult rat, 5-hydroxytryptamine (5-HT) is present in some principal nerve (PN) cells (Häppölä 1986; Häppölä et al. 1986), in the ganglionic small intensely fluorescent (SIF) cells (Eränkö and Eränkö 1971; Verhofstad et al. 1981) and in the adrenaline cells of the adrenal medulla (Verhofstad and Jonsson 1983).

The present article summarizes our recent studies concerning (1) the stages of prenatal development at which the sympathetic cells express 5-HT immunoreactivity; and (2) correlation of the time-course of 5-HT appearance with that of the catecholamine-synthesizing enzymes. These studies were performed on thoracolumbar transections of prenatal rats fixed with paraformaldehyde and processed according to the indirect immunofluorescence method of Coons (1958). In this manner series of sections were obtained in which the developing sympathetic chain ganglia, adrenal gland and retroperitoneal paraganglia of the same animal could be examined.

Initial Appearance of 5-HT (E12-E14)

The first indication of catecholamine synthesis was observed on day E12 as intense TH immunoreactivity in cells that were exclusively located in the chain ganglia (Fig. 1). Examination of consecutive sections revealed that all ganglionic TH-reactive cells also were immunoreactive to 5-HT (Fig. 1).

On day E13 TH-immunoreactive cells were seen, except for the chain ganglia, on the lateral sides of the aorta where they formed two perirenal ganglia. Furthermore, some solitary cells were located on the ventral side of the aorta (Fig. 1). These cells are the precursors of the prevertebral ganglia and the retroperitoneal paraganglia. Comparison of consecutive sections showed that all TH-reactive cells also showed 5-HT immunoreactivity.

On day E14 the number of TH-immunoreactive cells was greatly increased in each above-mentioned location (Fig. 1). Moreover, there was wide variation between the cells in terms of fluorescence intensity. TH and 5-HT immunoreactivities showed essentially similar distribution. Due to the continuous range of intensities it was not possible to distinguish developing PN cells from the SIF cells or chromaffin cells. At this age, some solitary, intensely TH-reactive cells were seen for the first time in the developing adrenal gland (Fig. 1). These cells, which are the precursors of the adrenal medulla, were also immunoreactive to 5-HT.

The observations indicate that 5-HT immunoreactivity is expressed initially in each sympathetic tissue in conjunction with the appearance of TH immunoreactivity. Furthermore, initially all catecholamine-containing cells seem to contain 5-HT.

Subsequent Developmental Changes (E15-E21)

Sympathetic Ganglia. The distribution of TH immunoreactivity changed gradually with age so that from the continuous range of fluorescence intensities on E14 chain ganglia two distinct cell types became distinguishable in E15 ganglia: some cells showed intense TH immunoreactivity, while the majority were only weakly TH-immunoreactive (Fig. 1). Only the former cells were immunoreactive to 5-HT (Fig. 1). During the subsequent development, the relative number of 5-HT-immunoreactive cells decreased gradually. In terms of cell size, wide variation was obvious among the 5-HT cells of prenatal ganglia,

Fig. 1. Schematic reconstruction of serial sections of E12, E13 and E14 rat embryos, and E16 adrenal gland. In the three first figures, the left and right sides illustrate TH and 5-HT immunoreactivity, respectively. Open circles stand for nonreactive cells, half-filled circles for moderate and filled circles for intense immunoreactivity. GGL = chain ganglion, PRG = perirenal ganglion, ADR = adrenal gland. Asterisk marks the aorta. N = notochord.
The lower panel photomicrographs are illustrating TH (A), 5-HT (B) and PNMT (C) immunoreactivity in E21 paraganglionic tissue. Bar = 100 um

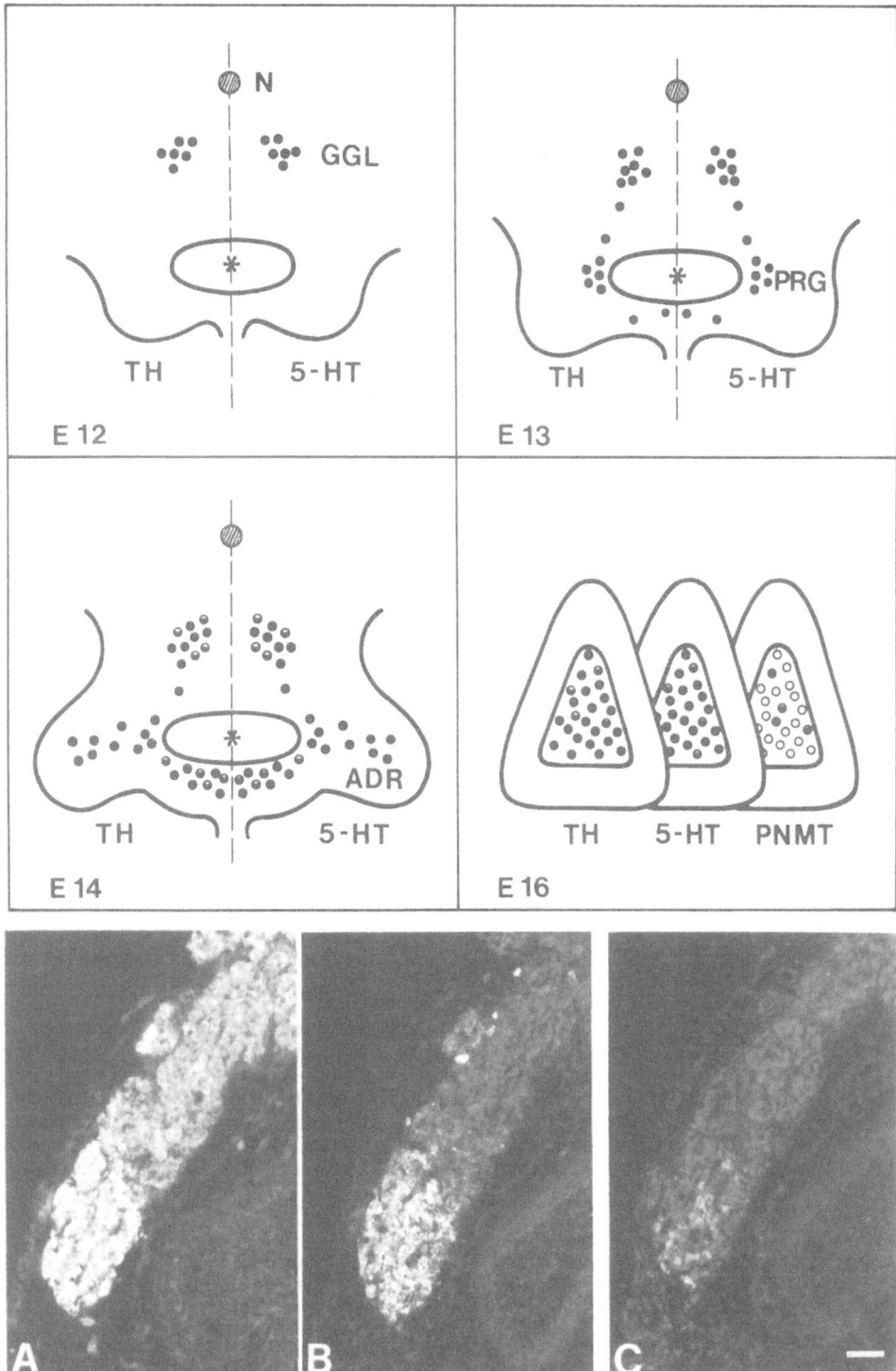

while in newborn ganglia both small and large 5-HT cells were present. The former correspond to the SIF cells, all of which become 5-HT-immunoreactive by the adult age (Päivärinta et al. 1987). The large 5-HT cells are observed until postnatal day 35 and they presumably represent a subpopulation of PN cells (Häppölä et al. 1986). Notably, the great majority of the PN cells remained nonreactive to 5-HT after day E15 (Soinila et al. 1986).

Retroperitoneal Paraganglia. On day E15, the retroperitoneal paraganglionic tissue comprised of a large population of cells in which 5-HT immunoreactivity was confined to those cells showing intense TH reactivity (Fig. 1). This tissue reached its maximum size around the time of birth and consisted then of three catecholamine-containing cell types. The majority of cells were weakly or moderately immunoreactive to TH and DBH and nonreactive to 5-HT. A dense group of cells showed intense immunoreactivity to both TH, DBH and 5-HT. A subpopulation of this cell group was, in addition, immunoreactive to PNMT (Fig. 1).

Adrenal Medulla. After the first TH- and 5-HT-immunoreactive cells had penetrated the developing gland on day E14, the number of medullary cells increased considerably and, as in the ganglia and paraganglia, a stage was reached when all degrees of fluorescence intensities were represented. On day E16, a few PNMT-immunoreactive cells were seen for the first time (Fig. 1). The number of PNMT cells increased rapidly and all PNMT cells also were 5-HT-reactive. On day E21 most cells were intensely immunoreactive to both TH, PNMT and 5-HT, while a few cell islets were weakly TH-reactive but nonreactive to either PNMT or 5-HT.

Conclusions

5-HT immunoreactivity described in the above cited studies may represent synthesis and/or uptake of the amine (Soinila et al. 1986). The observations indicate that in the rat the initial expression of 5-HT immunoreactivity occurs in each sympathetic tissue simultaneously with TH immunoreactivity. Later, however, 5-HT immunoreactivity is retained (1) in the SIF cells which contain either dopamine, noradrenaline or adrenaline (Koslow 1976; Gerold et al. 1982); (2) transiently in some PN cells (Häppölä 1986; Häppölä et al. 1986); (3) in some noradrenaline cells and in all adrenaline cells of the retroperitoneal paraganglia (Ahonen et al. 1986; Soinila et al. 1986); (4) in the medullary adrenaline cells (Verhofstad and Jonsson 1983). Based on these findings, it is suggested that the expression of 5-HT immunoreactivity is regulated independently rather than coupled to the synthesis of a particular catecholamine. Glucocorticoids may have a regulatory function since they induce an increase in the number of 5-HT-immunoreactive cells in vitro (Eränkö and Eränkö 1981), as well as in vivo (Päivärinta et al., this volume).

Acknowledgements. We thank Mrs. M.-L. Piironen and Mrs. H.-L. Wennäkoski for technical assistance, Mrs. T. Närvänen for illustration and Orion Corporation Research Foundation for financial support.

References

Ahonen M, Soinila S, Joh TH (1986) Pre- and postnatal development of rat retroperitoneal paraganglia. J Auton Nerv Syst (submitted)

Cochard P, Goldstein M, Black IB (1978) Ontogenetic appearance and disappearance of tyrosine hydroxylase and catecholamines in the rat embryo. Proc Natl Acad Sci USA 75: 2986-2990

Coons AH (1958) Fluorescent antibody methods. In: Danielli JF (ed) General cytochemical methods. Academic Press, New York, pp 399-422

De Champlain J, Malmfors T, Olson L, Sachs C (1970) Ontogenesis of peripheral adrenergic neurons in the rat: pre- and postnatal obser-vations. Acta Physiol Scand 80: 276-288

Eränkö O, Eränkö L (1971) Small, intensely fluorescent granule-containing cells in the sympathetic ganglion of the rat. In: Eränkö O (ed) Histochemistry of nervous transmission. Elsevier, Amsterdam (Progr Brain Res Vol 34), pp 39-51

Eränkö O, Eränkö L (1981) Catecholamine storage and synthesis sites in the sympathetic system: histochemical aspects. In: Stoward PJ, Polak JM (eds) Histochemistry: the widening horizons. Wiley, Chiche-ster, pp 31-46

Fernholm M (1972) On the appearance of monoamines in the sympathetic systems and the chromaffin tissue in the mouse embryo. Z Anat Ent-wickl-Gesch 135: 350-361

Gerold N, Enz A, Schröder H, Heym Ch (1982) Biochemical analysis of catecholamines in small intensely fluorescent (SIF) cell clusters of the rat superior cervical ganglion. J Neurosci Meth 6: 287-292

Häppölä O (1986) 5-hydroxytryptamine-immunoreactive neurons in the superior cervical ganglion of the rat. Brain Res (submitted)

Häppölä O, Päivärinta H, Soinila S, Steinbusch HWM (1986) Pre- and postnatal development of 5-hydroxytryptamine-immunoreactive cells in the superior cervical ganglion of the rat. J Auton Nerv Syst 15: 21-31

Koslow S (1976) Mass fragmentographic analysis of SIF cell catechol-amines of normal and experimental rat sympathetic ganglia. In: Eränkö O (ed) SIF cells. Structure and function of the small inten-sely fluorescent sympathetic cells. DHEW Publ No (NIH) 76-942. US Government Printing Office, Washington DC, pp 82-88

LeDouarin N (1973) A biological cell labelling technique and its use in experimental embryology. Devl Biol 30: 217-222

Päivärinta H, Häppölä O, Joh TH, Panula P, Steinbusch HWM, Watanabe T (1987) Coexistence of histamine, histidine decarboxylase, 5-hydro-xytryptamine and tyrosine hydroxylase in the sympathetic cells of the rat. Histochm J (submitted)

Soinila S (1984) Pre- and postnatal development of the small inten-sely fluorescent cells in the rat superior cervical ganglion. Int J Devl Neurosci 2: 65-76

Soinila S, Ahonen M, Joh TH, Steinbusch HWM (1986) 5-hydroxytrypt-amine and catecholamines in developing sympathetic cells of the rat. J Auton Nerv Syst (submitted)

Teitelman G, Baker H, Joh TH, Reis D (1979) Appearance of catechol-amine-synthesizing enzymes during development of rat sympathetic nervous system: possible role of tissue environment. Proc Natl Acad Sci USA 76: 509-513

Verhofstad AAJ, Jonsson G (1983) Immunohistochemical and neurochemi-cal evidence for the presence of serotonin in the adrenal medulla of the rat. Neuroscience 10: 1443-1453

Verhofstad AAJ, Hökfelt T, Goldstein M, Steinbusch HWM, Joosten HWJ (1979) Appearance of tyrosine hydroxylase, aromatic amino-acid de-carboxylase, dopamine-beta-hydroxylase and phenylethanolamine N-methyltransferase during the ontogenesis of the adrenal medulla. Cell Tissue Res 200: 1-13

Verhofstad AAJ, Steinbusch HWM, Penke B, Varga J, Joosten HWJ (1981) Serotonin-immunoreactive cells in the superior cervical ganglion of the rat. Evidence for the existence of separate serotonin- and catecholamine-containing small ganglionic cells. Brain Res 212: 39-49

Coexistence of Noradrenaline and Neuropeptides in Human Fetal Paraganglia

A. Hervonen[1] and I. Linnoila[2]

[1] Section of Gertontology, Department of Public Health, University of Tampere Medical School, P.O. Box 607, 33101 Tampere, Finland
[2] National Cancer Institute, Navy Medical Oncology Branch, Naval Hospital, Bethesda, MD 20814, USA

Introduction

The coexistence of enkephalin and catecholamines in the adrenomedullary cells was first described by Schultzberg et al. (1978). Since then, the chromaffin cells have become a useful biological model for studies on synthesis, storage and secretion of enkephalins and other peptides originating from the same precursor molecule (for review see Viveros and Wilson 1983).

Many neuropeptides have been localized also in the extra-adrenal catecholamine storing cells. Enkephalin-like immunoreactivity was found in the small, intensely fluorescent (SIF) cells of rat, guinea-pig and cat (Schultzberg et al. 1978) as well as in man (Hervonen et al. 1980, 1981a,b). Recently, Heym et al. (1984) demonstrated the coexistence of neurotensin- and enkephalin-like immunoreactivities in the paraganglia (PG) of cat, supporting the expression of several peptide phenotypes in the same PG. Human fetal (Hervonen et al. 1980) and adult (Vaalasti et al. 1985) PG are known to contain enkephalin-like immunoreactivity. There is also evidence presented for the occurrence of more than one regulatory peptide in the same cell (Hervonen et al. 1980, 1981a). In the current study we have immunohistochemically localized Met(5)-enkephalin, vasoactive intestinal polypeptide (VIP), substance P, somatostatin and ACTH 1-24 in the human fetal abdominal paraganglia.

Material and Methods

The human fetuses were obtained from legal interruptions of pregnancy performed by laparotomy for special indications in Finland. Gestational ages of the five fetuses used for the study ranged from 15-23 weeks.

The thoracic sympathetic trunk and a retroperitoneal tissue block from the level of the diaphragm to the bifurcation of the aorta were dissected, frozen in liquid nitrogen and freeze dried. The tissues were then exposed to paraformaldehyde vapour for 60 min at 60-80°C. A detailed description of the antibodies used has been published elsewhere (Linnoila et al. 1980; Hervonen et al. 1981b).

Following the demonstration of formaldehyde-induced fluroescence (FIF) for catecholamines, immunohistochemistry was carried out on the same or consecutive deparaffinized sections. The improved immunoglobulin-enzyme bridge method for light microscopy (Petrutz et al. 1975) was used.

Experimental Brain Research Series 16
© Springer-Verlag Berlin · Heidelberg 1987

To assess the specificity of the staining, one serial section was incubated with non-immune rabbit serum and another section was treated with preabsorbed antiserum (10-50 ug of the corresponding peptide in 1 ml of antiserum diluted 1:1000).

Results

The fluorescence micrographs are used to demonstrate the size and shape of the PG (Figs. 1a, 2b, 3, 4a, 5a). The PG consist of cords of fluorescent cells separated by capillaries (Fig. 3). All cells exhibit strong FIF indicating the presence of high concentrations of noradrenaline. Enkephalin-like immunoreactivity (ELI) was demonstrated in all PG. The number of labelled cells varied between 10-20% of all cells (Figs. 1b and 2a). Met^5-ELI and Leu^5-ELI was localized in the same areas of the PG, possibly in the same cells (Fig. 1b,c). VIP-like immunoreactivity was demonstrated in several PG also showing ELI (Fig. 1d). Somatostatin-like immunoreactivity was demonstrated in several small PG (Fig. 4) and serial sections showed positive

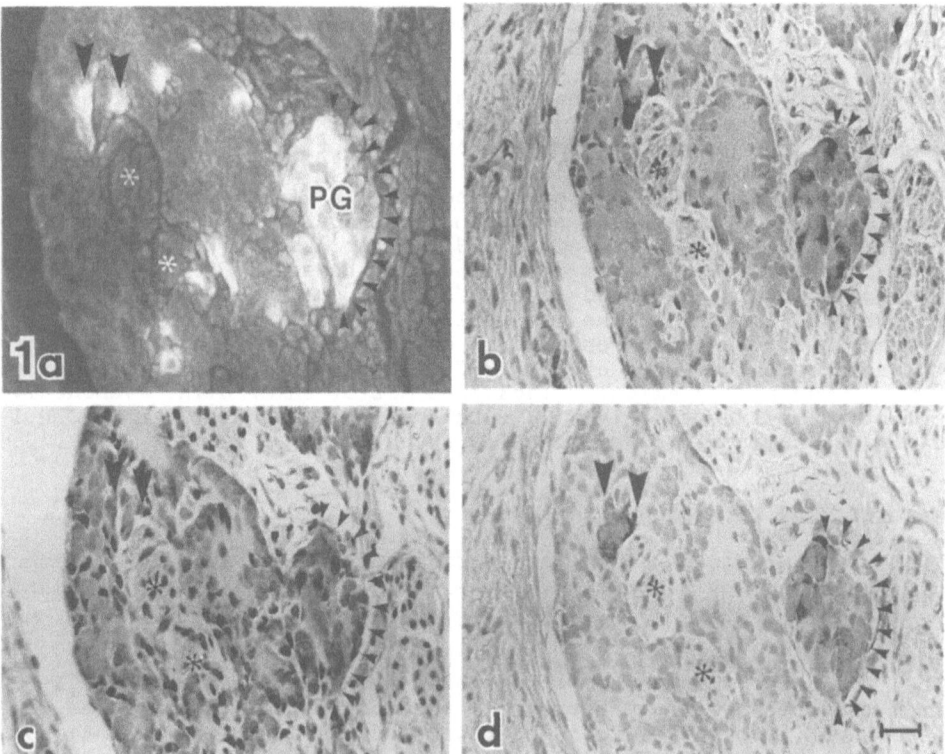

Fig. 1. Serial section of a small paraganglion and several small intensity fluorescent (SIF) cells in the coeliac mesenteric complex. The paraganglion (PG) is surrounded by small arrowheads and the SIF-cells are pointed out by larger arrowheads. Asterisk marks a cross section of a nerve bundle. The catecholamine histofluorescence (a) corresponds to the distribution of met-enkephalin- (b), leu-enkephalin- (c) and VIP-like immunoreactivities (d). Bar indicates 20 um

immunoreaction with both ACTH 1-24 (Fig. 5) and substance P antibodies.

Discussion

Preliminary work suggested the coexistence of derivatives of pro-enkephalin and pro-opiomelanocortin in the same cells of human fetal extra-adrenal chromaffin tissue (Hervonen et al. 1980).

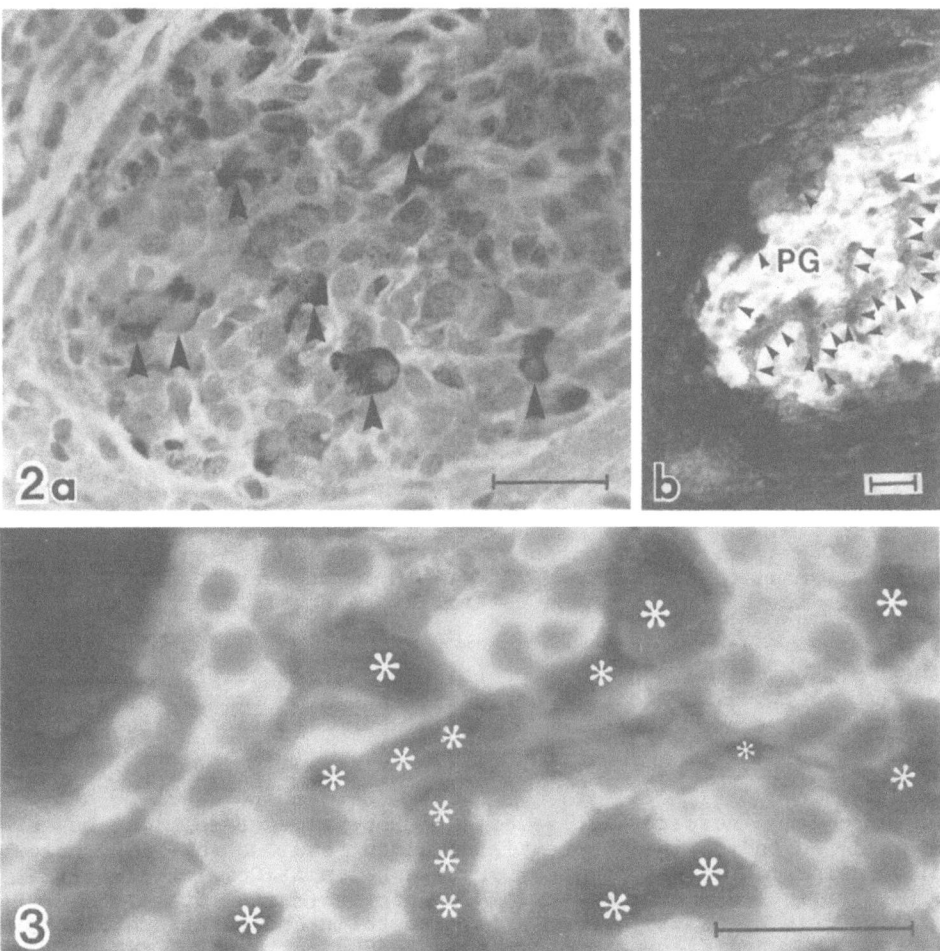

Fig. 2. A higher magnification of immunoreaction with met(5)-enkephalin antibody (a) in a large, fluorescent solitary abdominal paraganglion (b). Arrowheads in (a) point at enkephalin-like immunoreactivity; arrowheads in (b) point out the abundant vascular supply of the organ. Bars indicate 30 um

Fig. 3. Large magnification of an abdominal paraganglion. Asterisks mark sinusoidal capillaries of varying diameter. Bar indicates 40 um

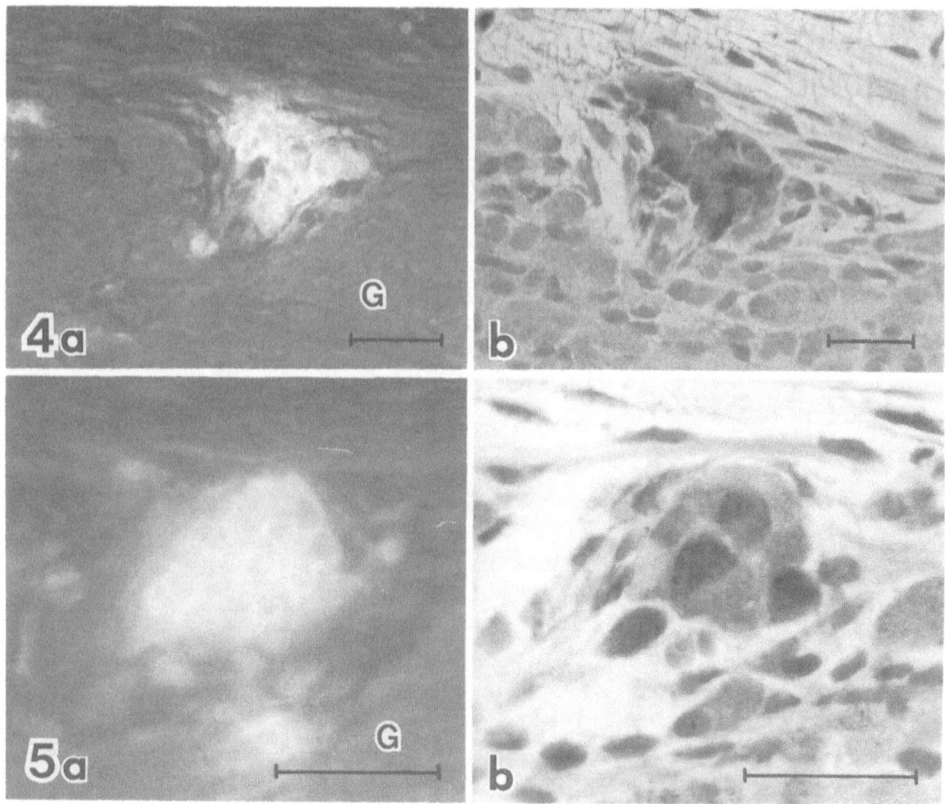

Fig. 4. Consecutive sections of a small paraganglion embedded in the coeliac ganglion (G). The somatostatin immunoreaction (b) corresponds to the distribution of catecholamine histofluorescence. Bar indicates 30 um

Fig. 5. Consecutive section of the same paraganglion shown in Fig. 4. The catecholamine histofluorescence and ACTH immunoreaction are in the same cells. Note the granular ACTH reaction product. Bar indicates 30 um

Our results indicate that several neuropeptide genes might be expressed in the newly differentiated human fetal chromaffin cells. Enkephalins, originating from proenkephalin A (Comb et al. 1982), ACTH originating from pro-opiomelanocortin (Mains et al. 1977), somatostatin originating from prosomatostatin and substance P originating from a different prohormone (see Steiner et al. 1984) were demonstrated in the same cell groups. Furthermore, serial sectioning demonstrated the presence of at least enkephalin and VIP in some cells and enkephalin, ACTH, somatostatin and substance P in small PG as well as the presence of enkephalin and ACTH in the solitary large PG.

The staining properties of the chromaffin cells varied depending on their location. The small clusters of cells closely related to sympathetic ganglia exhibited four peptides simultaneously. It is possible that microenvironmental influences are responsible for these differences. Chromaffin cells derive from the neural crest and assume the phenotypic characteristics during ventrally directed migration (Hervonen 1971). The phenotype adopted by the primitive sympathetic cells appears to be influenced by the environment through which these cells migrate as well as by their final location (LeDouarin 1980; Patterson 1982).

Pheochromocytomas, the proliferative lesions of chromaffin tissue, provide further data on multiple phenotypes. Enkephalin has been demonstrated in the human pheochromocytoma (see DeLellis et al. 1983, 1984) In addition to enkephalin, ACTH, MSH and beta-endorphin (Giraud et al. 1981; Suda et al. 1984), dynorphin (Yoshimasa et al. 1981), VIP and neurotensin (Tischler et al. 1982), somatostatin (see Sano et al. 1983) as well as neuropeptide Y (DeLellis et al. 1984) have been demonstrated in different cell lineages of mammalian pheochromocytomas. Three known precursor systems for opioid peptides (enkephalins, dynorphin and beta-endorphin) were found to be expressed in the same tumors (Casselin et al. 1984). These findings indicate that at least after neoplastic transformation, an activation of neuroendocrine genes takes place.

The fetal PG mature early (Coupland 1965; Hervonen 1971) while the adrenal medulla histologically and biochemically remains immature during fetal life (Phillippe 1983). The chromaffin cells studied represent fully differentiated neuroendocrine cells which persist for life (Hervonen et al. 1978). Pelto-Huikko et al. (unpublished) observed somatostatin in human adrenomedullary cells known to contain enkephalins (Linnoila et al. 1980), and VIP was found in human SIF-cells (Helen et al., unpublished) containing enkephalins (Hervonen et al. 1980, 1981b). Furthermore, denervation was found to increase the number of enkephalin-containing cells in the rat adrenal medulla (Schultzberg et al. 1978) and to induce the appearance of ACTH- and neurotensin-positive cells, which normally are absent (Pelto-Huikko et al. 1987). DeLellis (1984) found that normal human adrenomedullary cells are capable of producing VIP. These data suggest that the expression of multiple neuroendocrine phenotypes may occur also in non-neoplastic chromaffin tissue. A molecular-biological approach is needed to further characterize gene expression in the human fetal paraganglion model.

References

Casselin F, Pique L, Bertagna C, Benlot C, Antreassinan J, Proeschel MF, Girard F, Zogbi F, Legrand JC, Luton JP, Dauchy J, Thibault J (1984) Simultaneous evaluation of the catecholamine pathway and three opioid peptide-producing system in human pheochromocytomas. Neuropeptides 4: 175–182

Comb M, Seeburg PH, Adelman J, Eiden L, Herbert E (1982) Primary structure of the human met- and leu-enkephalin precursor and mRNA. Nature 295: 663–666

Coupland RE (1965) The natural history of chromaffin cell. Longmans, London

DeLellis RA, Tischler A, Lee AK, Blount M, Wolfe HJ (1983) Leu-enkephalin like immunoreactivity in proliferative lesions of the human adrenal medulla and extra-adrenal paraganglia. Am J Surg Path 7: 29–37

DeLellis RA, Tischler AS, Wolfe HJ (1984) Multidirectional differentiation in neuroendocrine neoplasms. J Histochem Cytochem 32: 899-904

Giraud P, Eiden LE, Audigier Y, Gillioz P, Colte-Devolx B, Boureque F, Eskay R, Olivier C (1981) Enkephalins, ACTH, MSH and beta-endorphin in human pheochromocytomas. Neuropeptides 1: 237-252

Hervonen A (1971) Development of catecholamine-storing cells in the human fetal paraganglia and adrenal medulla. Acta Physiol Scand Suppl 368: 1-94

Hervonen A, Partanen S, Vaalasti A, Partanen M, Kanerva L, Alho H (1978) The distribution and endocrine nature of the abdominal paraganglia of adult man. Am J Anat 153: 563-572

Hervonen A, Pickel VM, Joh TH, Reis DJ, Linnoila I, Kanerva L, Miller RJ (1980) Immunocytochemical demonstration of catecholamine synthesizing enzymes and neuropeptides in the catecholamine storing cells of human fetal sympathetic nervous system. In: Eränkö O, Päivärinta H, Soinila S (eds) Histochemistry and cell biology of autonomic neurons, SIF-cells and paraneurons. Adv Biochem Psychopharm 25. Raven, New York, pp 373-378

Hervonen A, Linnoila I, Pickel VM, Helen P, Pelto-Huikko M, Alho H, Miller RJ (1981a) Localization of (Met5)- and (Leu5)-enkephalin-like immunoreactivity in nerve terminals in human paravertebral sympathetic ganglia. Neuroscience 6: 323-330

Hervonen A, Pickel VM, Joh TH, Reis DJ,Linnoila I, Miller RJ (1981b) Immunohistochemical localization of the catecholamine synthesizing enzymes, substance P and enkephalin in the human fetal sympathetic ganglion. Cell Tissue Res 214: 33-42

Heym Ch, Reinecke M, Weihe E, Forssmann WG (1984) Dopamin-beta-hydroxylase-, neurotensin-, substance P-, vasoactive intestinal polypeptide- and enkephalin-immunohistochemistry of paravertebral and prevertebral ganglia in the cat. Cell Tissue Res 235: 411-418

LeDouarin NM (1980) The ontogeny of the neural crest in avian embryo chimeras. Nature 286: 663-669

Linnoila I, DiAugustini RP, Hervonen A, Miller RJ (1980) Distribution of (Met-5)- and (Leu-5)-enkephalin, vasoactive intestinal polypeptide and substance-P-like immunoreactivities in human adrenal glands. Neuroscience 55: 2247-2259

Mains RE, Eipper BA, Ling N (1977) Common precursor to corticotropins and endorphins. Proc Natl Acad Sci USA 74: 3014-3018

Patterson PH (1982) Cellular and hormonal interactions in the development of sympathetic neurons. In: Schmitt FO, Bird SJ, Bloom FE (eds) Molecular genetic neurosciences. Raven, New York, pp 437-443

Pelto-Huikko M, Salminen T, Partanen M (1987) Neuropeptides in the denervated rat adrenal medulla. Cell Tissue Res (in press)

Petrutz P, Dimeo P, Ordonneau P, Weaver C, Keefer DA (1975) Improved immunoglobulin-enzyme bridge method for light microscopic demonstration of hormone-containing cells of the rat adenohypophysis. Histochemie 46: 9-26

Philippe M (1983) Fetal catecholamines. Am J Obstet Gynec 146: 840–855

Sano T, Saito H, Inoba H, Hizawa K, Saito S, Yamanoi A, Mizunuma Y, Matsumura M, Yuasa M, Hiraishi K (1983) Immunoreactive somatostatin and vasoactive intestinal polypeptide in adrenal pheochromocytoma. An immunohistochemical and ultrastructural study. Cancer 52: 282–289

Schultzberg M, Lundberg J, Hökfelt T, Terenius L, Brandt J, Elde RP, Goldstein M (1978) Enkephalin like immunoreactive gland cells and nerve terminals of the adrenal medulla. Neuroscience 3: 1169–1186

Steiner DF, Docherty K, Carrol R (1984) Golgi/granule processing of peptide hormone and neuropeptide precursors: a minireview. J Cell Biochem 24: 121–130

Suda T, Tomori H, Demura H, Shizume K, Mouri T, Miura Y, Sasano N (1984) Immunoreactive corticotropin and corticotropin releasing factor in human hypothalamus, adrenal, lung cancer and pheochromocytoma. J Clin Endocrinol Metab 58: 919–924

Tischler A, Lee YC, Slayton VW, Bloom SR (1982) Content and release of neurotensin in PC12 pheochromocytoma cell cultures: modulation by dexamethasone and nerve growth factor. Reg Peptides 3: 415–450

Vaalasti A, Pelto-Huikko M, Tainio H, Hervonen A (1985) Light and electron microscopic demonstration of enkephalin-like immunoreactivity in the human paraganglia. Cell Tissue Res 239: 683–687

Viveros HO, Wilson S (1983) The adrenal chromaffin cell as a model to study the co-secretion of enkephalins and catecholamines. J Auton Nerv Syst 7: 41–58

Yosimasa T, Nakao K, Oki S, Tanakoa I, Nakai Y, Imura H (1981) Presence of dynorphin-like immunoreactivity in pheochromocytomas. J Clin Endocrinol Metab 53: 213–214

Histochemical and Cytochemical Localizations of Catecholamine and Peptide-Like Reactivity in Aged Primate Stellate Ganglia

J. G. Wood, C. A. Westmoreland, and K. Best

Department of Neurobiology and Anatomy, The University of Texas Medical School, P.O. Box 20708, Houston, TX 77025, USA

Introduction

The aging process reportedly has an adverse effect on the functional activity of the autonomic nervous system (ANS). Reis et al. (1977) and Partanen et al. (1985) have shown very precise, not all diminished, alterations in ANS metabolic activity with increasing age, and Schmidt and co-workers (1983) acknowledge that aging changes are not uniform throughout the ANS. Helen (1983) and others (Hervonen et al. 1978; Partanen et al. 1980) have found that, along with a lipofuscin increase, there is a general decrease in catecholamine (CA) histofluorescence. Case and Matthews (1985) and Govoni et al. (1983) present evidence for established ANS alterations during the latter life developmental processes. Chiba and Williams (1975) have provided a background for anatomical study of autonomic ganglia, while Dail and Barton (1983) and Matthews (1983) present more recent information and insights into the structural organization of autonomic ganglia.

The immunolocalization of a number of peptides and the relationship of the peptide-like immunoreactivity to particular segments of the ANS has contributed significantly to the recent ANS literature. Enkephalin-like (MetE and LeuE) immunoreactivity has been localized in nerve fibers and terminals (Hervonen et al. 1980) and cell bodies (Schultzberg et al. 1979; Pelto-Huikko et al. 1980). Substance P (SP) is expressed in sympathetic neurons that may be noradrenergic (Bohn et al. 1984). Neuropeptide Y (NPY)-like activity has also been shown to be associated with adrenergic ganglion cell bodies (Lundberg et al. 1983; Jarvis et al. 1986), while vasoactive intestinal polypeptide (VIP)-like immunoreactivity may be associated with cholinergic neurons (Lundberg et al. 1979).

Catecholamines (CA) were determined to be present in certain ganglion cells quite early (Eränkö and Härkönen 1963) and it was postulated by Eränkö (1972) that the intracellular CA had two pools, a granular one and a non-granular one. Other attempts were made to localize CA (Richards and Tranzer 1975) in autonomic ganglia, and the granule types of autonomic ganglion cells have been classified (Taxi et al. 1983) following osmium tetroxide fixation. It seems important to establish a relationship of the implicated peptide-like reactivity with the CA neurons.

Methods

All the animals utilized in these experiments were adult male primates 20–24 years of age. Each animal, under deep sodium pentobarbital anesthesia, was perfused via the ascending aorta with Ham's F-12 buffer followed by the appropriate fixative and buffer. The fixative for the immunocytochemistries was 4% paraformaldehyde (CHO) and 0.05% glutaraldehyde (Glut) in o.1 M phosphate (PO₄) buffer for 1 h followed by 4% CHO in PO₄ buffer at pH 10, while that for CA was either 4% or 6% Glut in 0.2 M sodium cacodylate buffer (pH 7.4). Antibodies for the peptides MetE, LeuE, VIP, SP and NPY were purchased commercially. Incubations were carried out on thinly sliced pieces of tissue at a dilution of 1:1000 for 48 h. Both immunofluorescent and horseradish peroxidase (HRP) tags were used for light and electron microscopy, respectively. CA incubation was conducted with a 0.2 M potassium dichromate (DC) solution in the cacodylate buffer at pH 4.1. In all cases, appropriate control sections or blocks of tissue were processed. X-ray energy dispersive spectroscopic analysis (EDS) was conducted on a JEOL-100CX microscope with a Kevex X-ray detector and multichannel analysis system.

Results

MetE and LeuE have the same general intercellular distribution, with prominent fiber networks coursing through the ganglion and surrounding the ganglionic cell bodies. LeuE-like reactivity is shown in Fig. 1 with immunofluorescence around and in between the ganglion cells. The electron micrograph indicates the details of the reactive fibers in proximity to the processes of the ganglion neurons. Neuron cell bodies were enkephalin-negative. LeuE showed a much greater degree of reactivity than did MetE. In a similar manner, SP immuno-like reactivity (Fig. 2) is confined to the intercellular and peri-cellular areas and the nerve trunks. There is no neuronal cell body reactivity for SP. HRP electron microscopy reveals that SP-like immunoreactive elements establish synaptic contact with nonreactive neuronal end organs in the vicinity of nonreactive cell bodies. VIP-positive fibers with varicosities could be seen forming punctate spots on or near ganglion cells (Fig. 3). In addition, VIP-like immunoreactivity was present in some of the smaller ganglion cells on the pleomorphic dense bodies (PDB) normally described as lipo-fuscin granules. This VIP-like reactivity is evident both at the light and the electron microscopic levels (Fig. 3). NPY-like immuno-reactivity was confined to clusters of principal neurons (PNs) but not to all PNs (as indicated by "x"). Fig. 4 shows the diffuse immunostaining in some PNs but not in others. This intracellular staining (arrows) which is located away from the PDBs follows the pattern of intracellular CA distribution and can also be seen electron microscopically (Fig. 4). There is also NPY-like reactivity in a number of varicose processes throughout the ganglion.

Mast cells (Fig. 4 - lower left) illustrated bright immunofluorescence with each of the peptides tested. Light microscopy with the GDC method revealed bright red mast cell granules. Mast cells are known to exist in the capsular and perivascular areas in autonomic ganglia (Weinreich 1985).

CA, with the method employed, can be visualized at both the light and the electron microscopic levels. It was found that the CA-positive PNs occurred throughout the stellate ganglion, but that not all PNs were CA-positive. CA reactivity was determined by the EDS localization of chromium (Cr) which is a result of the glutaraldehyde-potassium dichromate (GDC) reaction. The method selectively reacts with norepinephrine (NE) and positive Cr-NE sites are located in lacunae or patches in the CA neurons (Fig. 5). The positive Cr reaction product follows the granular distribution pattern.

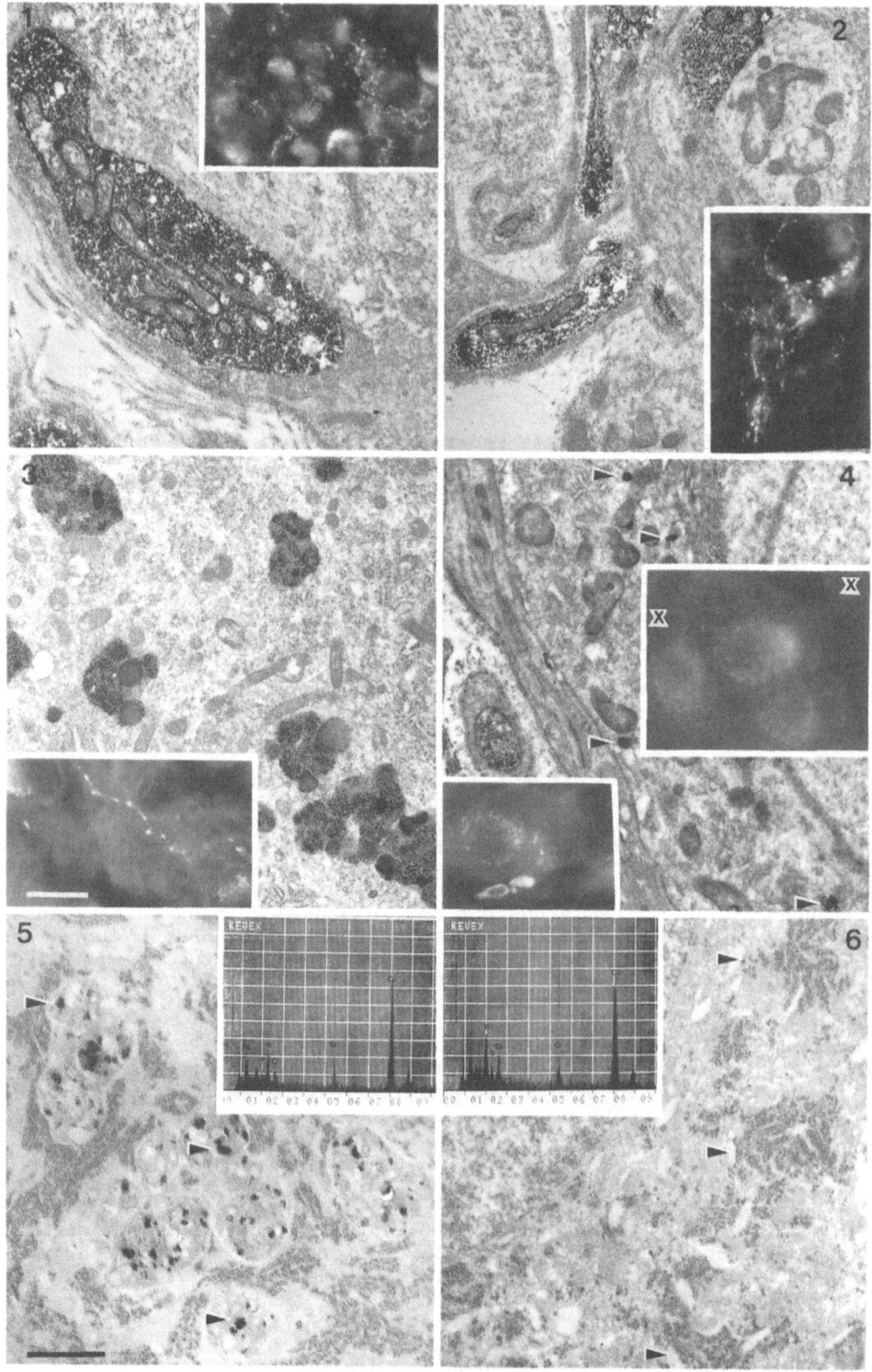

A striking feature elicited by this histochemical technique and confirmed by EDS is the visualization of marked densities in the rough endoplasmic reticulum (RER) or Nissl substance (Fig. 6) not readily seen in conventional methods. These dense areas of RER are Cr-positive indicating the presence of NE. Thus, NE is present both in the granules and the RER of the PNs. Such a distribution and the frequent location of the RER at the periphery of the neuron indicate that this RER locus is no doubt Eränkö's (1972) non-granular pool of NE. This NE pool was not previously visualized because the conventional fixatives presumably dispersed the RER. This fact has been mentioned by Taxi (1975). Other evidence favoring two different NE pools is the strongly sulfur (S)-positive granular pool (Fig. 5) possibly indicating sulfonated NE, disulfide bridges or sulfhydryl groups, and the RER pool which is strong for phosphor (P) (Fig. 6) indicating the presence of phosphonucleotides.

Summarily, this particular set of studies has illustrated that MetE- and LeuE-like reactivity is present intercellularly in the aged primate, with LeuE being more reactive. VIP-like and NPY-like reactivity are intracellular. VIP reactivity is associated with PDBs, while NPY-like reactivity may be associated with PNs containing NE. SP-positive fibers demonstrate synaptic connections in the neuropil areas. NE is found in two areas of the PN cytoplasm, the granular area and the RER or non-granular areas. Further comparisons should be made in younger primates and in addition, in other species.

Acknowledgements. This work was supported in part by Grants NS10326 and BRSRR-05745. The authors wish to express appreciation to Dr. David Marshak for his advice and assistance with the immunohistochemistry.

References

Bohn MC, Kessler JA, Adler JE, Markey K, Goldstein M, Black IB (1984) Simultaneous expression of SP-peptidergic and noradrenergic phenotypes in rat sympathetic neurons. Brain Res 298: 378-381

Fig. 1. Electron micrograph of a LeuE-like immunoreactive nerve fiber. Inset: LeuE-immunofluorescent fiber network

Fig. 2. Electron micrograph of SP-like immunoreactive nerve fibers. Inset: SP-immunofluorescent fiber network

Fig. 3. Electron micrograph of VIP-like immunoreactive dense bodies in a ganglion cell. Inset: VIP-immunofluorescent fibers and intracellular granules

Fig. 4. Electron micrograph of intracellular NPY-like immunoreactive granules (arrows). Insets: NPY-immunofluorescent ganglion cells. Lower left: 2 mast cells

Figs. 5 and 6. Electron micrographs of chromium reaction product in CA neurons by aid of the glutaraldehyde-potassium-dichromate reaction. Insets: EDS localization of elements

Scale bars indicate 1 um for all electron micrographs and 50 um for all light micrographs

Case CP, Matthews MR (1985) A quantitative study of structural features, synapses and nearest-neighbour relationships of small, granule-containing cells in the rat superior cervical sympathetic ganglion at various adult stages. Neuroscience 15: 237-282

Chiba T, Williams TH (1975) Histofluorescence characteristics and quantification of small intensely fluorescent (SIF) cells in sympathetic ganglia of several species. Cell Tissue Res 162: 331-341

Dail WG, Barton S (1983) Structure and organization of mammalian sympathetic ganglia. In: Elfvin L-G (ed) Autonomic ganglia. Wiley, Chichester, pp 3-25

Eränkö O (1972) Light and electron microscopic histochemical evidence of granular and non-granular storage of catecholamines in the sympathetic ganglion of the rat. Histochem J 4: 213-224

Eränkö O, Härkönen M (1963) Histochemical demonstration of fluorogenic amines in the cytoplasm of sympathetic ganglion cells of the rat. Acta Physiol Scand 58: 285-286

Govoni S, Missale C, Castaletti L, Spano PF, Trabucchi M (1983) Decreased content of met enkephalin-like peptides in the superior cervical and celiac ganglia in aged rats. Neurobiol Aging 4: 147-150

Helen P (1983) Fine-structural and degenerative features in adult and aged human sympathetic ganglion cells. Mech Aging Dev 23: 161-175

Hervonen A, Pelto-Huikko M, Helen P, Alho H (1980) Electron-microscopic localization of enkephalin-like immunoreactivity in axon terminals of human sympathetic ganglia. Histochemistry 70: 1-6

Hervonen A, Vaalasti A, Partanen M, Kanerva L, Hervonen H (1978) Effects of aging on the histochemically demonstrable catecholamines and acetylcholinesterase of human sympathetic ganglia. J Neurol 7: 11-23

Järvi R, Helen P, Pelto-Huikko M, Hervonen A (1986) Neuropeptide Y (NPY)-like immunoreactivity in rat sympathetic neurons and small granule-containing cells. Neurosci Lett 67: 223-227

Lundberg JM, Hökfelt T, Schultzberg M, Uvnäs-Wallenstein K, Kohler C, Said SI (1979) Occurrence of vasoactive intestinal polypeptide (VIP)-like immunoreactivity in certain cholinergic neurons of the cat: evidence from combined immunohistochemistry and acetylcholinesterase staining. Neuroscience 4: 1539-1559

Lundberg JM, Terenius L, Hökfelt T, Goldstein M (1983) High levels of neuropeptide Y in peripheral nor-adrenergic neurons in various mammals including man. Neurosci Lett 42: 167-172

Matthews MR (1983) The ultrastructure of junctions in sympathetic ganglia of mammals. In: Elfvin L-G (ed) Autonomic ganglia. Wiley, Chichester, pp 27-66

Partanen M, Santer RM, Hervonen A (1980) The effect of aging on the histochemically demonstrable catecholamines in the hypogastric (main pelvic) ganglion of the rat. Histochem J 12: 527-535

Partanen M, Waller SB, London ED, Hervonen A (1985) Indices of neurotransmitter synthesis and release in aging sympathetic nervous system. Neurobiol Aging 6: 227-232

Pelto-Huikko M, Hervonen A, Helen P, Linnoila I, Pickel VM, Miller RJ (1980) Localization of (Met⁵)- and (Leu⁵)-enkephalin in nerve terminals and SIF cells in adult human sympathetic ganglia. In: Eränkö O (ed) Histochemistry and cell biology of autonomic neurons, SIF cells, and paraneurons. Raven, New York, pp 379–383

Reis DJ, Ross RA, Joh TH (1977) Changes in the activity and amounts of enzymes synthesizing catecholamines and acetylcholine in brain, adrenal medulla, and sympathetic ganglia of aged rat and mouse. Brain Res 136: 465–474

Richards JG, Tranzer JP (1975) Localization of amine storage sites in the adrenergic cell body. J Ultrastruct Res 53: 204–216

Schmidt RE, Plurad SB, Modert CW (1983) Neuroaxonal dystrophy in the autonomic ganglia of aged rats. J Neuropathol Exp Neurol 42: 376–390

Schultzberg M, Hökfelt T, Terenius L, Elfvin L-G, Lundberg M, Brandt J, Elde RP, Goldstein M (1979) Enkephalin immunoreactive nerve fibers and cell bodies in sympathetic ganglia of the guinea-pig and rat. Neuroscience 4: 249–270

Taxi J, Derer M, Domich A (1983) Morphology and histophysiology of SIF cells in the autonomic ganglia. In: Elfvin L-G (ed) Autonomic ganglia. Wiley, Chichester, pp 67–95

Weinrich D (1985) Multiple sites of histamine storage in superior cervical ganglia. Exp Neurol 90: 36–43

Coexistence of Enkephalin-, VIP-, Somatostatin- and Substance P-Like Immunoreactivities in the Abdominal Paraganglia of Senescent Male Fisher-344 Rats

M. Partanen[1, 2], I. Linnoila[3], and A. Hervonen[1, 2]

[1] Section of Gerontology, Department of Public Health, University of Tampere, Medical School,
P.O.Box 607, 33101 Tampere, Finland
[2] Laboratory of Neurosciences, National Institute on Aging, NIH, Bethesda, MD 20205, USA
[3] National Cancer Institute, Navy Medical Oncology Branch, Naval Hospital, Bethesda, MD 20814, USA

Introduction

Paraganglia (PG) are clusters of extra-adrenal chromaffin cells, which contain catecholamines (Coupland 1965; Hervonen 1971). More recently, however, several neuropeptides have been demonstrated in the chromaffin tissue. Enkephalins first were demonstrated immuno-histochemically in the adrenal medulla of rat, guinea-pig and cat (Schultzberg 1978; Linnoila et al. 1980). Since then, multiple molecular forms of enkephalin-like immunoreactive material have been characterized (DiGuilio et al. 1978; Hanbauer et al. 1982). It is now established that opiate peptide material is co-stored with catecholamines in chromaffin granules of adrenal medulla (Hervonen et al. 1980; Viveros et al. 1983), and that the amine and peptide components can be co-secreted upon physiological stimulation (Cle-ment-Jones et al. 1980; Hexum et al. 1980; Hanbauer et al. 1982).

Partanen et al. (1981, 1984a) have demonstrated a marked, age-related increase of the volume of the extra-adrenal paraganglia in Fisher-344 rats. Subsequently, it was shown that hypertrophied para-ganglia are innervated, contain catecholamine synthesizing enzymes (Partanen et al. 1984b) and contain a dense network of fenestrated capillaries (Partanen et al. 1984c). The enlarged paraganglia in senescent rats showed bright catecholamine histofluorescence and the typical fine structure of the chromaffin cells (Partanen et al. 1984b,c).

The present study was performed to further characterize the endo-crine role of paraganglia in the aged rat. The coexistence of neuro-peptides was studied by immunohistochemical localization of peptides derived from different precursor molecules.

Methods

Male Fisher-344 rats, 30-34 months of age, were anesthetized and a para-aortic tissue block containing the retroperitoneal abdominal paraganglia was removed.

The paraganglia were traced within the tissue block by using form-aldehyde induced fluorescence (FIF) of catecholamines. The tissue block was frozen in liquid nitrogen, freeze dried and treated with vapour generated from paraformaldehyde powder, for 60 min at 80°C. The tissue was embedded in paraffin under vacuum and serially sec-tioned at a thickness of 15 um. Once paraganglia were identified,

Experimental Brain Research Series 16
© Springer-Verlag Berlin · Heidelberg 1987

paraffin was removed from consecutive sections and the sections were processed for immunohistochemistry.

Rabbit antisera against porcine synthetic vasoactive intestinal polypeptide (VIP), synthetic substance-P (SP) and somatostatin were purchased from Calbiochem (La Jolla, CA). Antisera against Met-enkephalin was purchased from Merseyside Laboratories (UK) or from Cambridge Research Biochemicals (UK). Leu-enkephalin antibody was a generous gift from Dr. R. Miller from the University of Chicago.

The peroxidase-antiperoxidase method of Sternberger (1974) was used. Incubation time for the primary antisera was 24 h at 4°C. Satisfactory staining was obtained with dilutions between 1:2000-1:5000. In order to assess the specificity of the reaction, serial sections were incubated with nonimmune rabbit serum and with antigen-inactivated antiserum (10 ug of the corresponding peptide in 1 ml of diluted antiserum).

Results

The abdominal paraganglia typically comprised cell clusters which exhibited a bright histofluorescence, specific for catecholamines (Figs. 1 and 6). These paraganglia were found close to sympathetic ganglia. Neurons were often located between the chromaffin cell clusters (Fig. 2). On the other hand, solitary small paraganglia (Fig. 6) commonly occurred within the larger prevertebral ganglia.

Somatostatin-like immunoreactivity was localized either in typical paraganglionic cords of cells or in single cells surrounded by nonreactive cells (Figs. 2 and 3a). In few, large PG all the cells were stained with antibody to somatostatin. The intensity of the staining varied from cell to cell (Fig. 3a). No staining with anti-somatostatin or with other antibodies used was seen following pre-incubation with corresponding antigens (Fig. 3b) or after use of normal rabbit serum.

The localization of SP-like immunoreactivity resembled that of somatostatin immunoreactivity. In some PG only selected groups of cells were labelled (Fig. 4), in others, large cell groups showed uniform labelling (Fig. 5). In some PG, both somatostatin and SP-like immunoreactivity could be demonstrated in the same area, suggesting the coexistence of these peptides within the same cells.

VIP-positive staining was found less frequently than was staining for somatostatin of SP. Typically, cells containing VIP immunoreactivity were arranged in small nests and anastomosing cords in close proximity to blood vessels (Fig. 7). Enkephalin-like immunoreactivity was localized mostly in single cells surrounded by unstained PG cells (Fig. 8). Met- and leu-enkephalin showed similar distributions. Occasionally, small groups of positive cells were found within the nonstained surroundings (Fig. 9).

Nerve fibers or nerve terminals with specific immunoreactivities with the antibodies could not be demonstrated within the paraganglia.

238

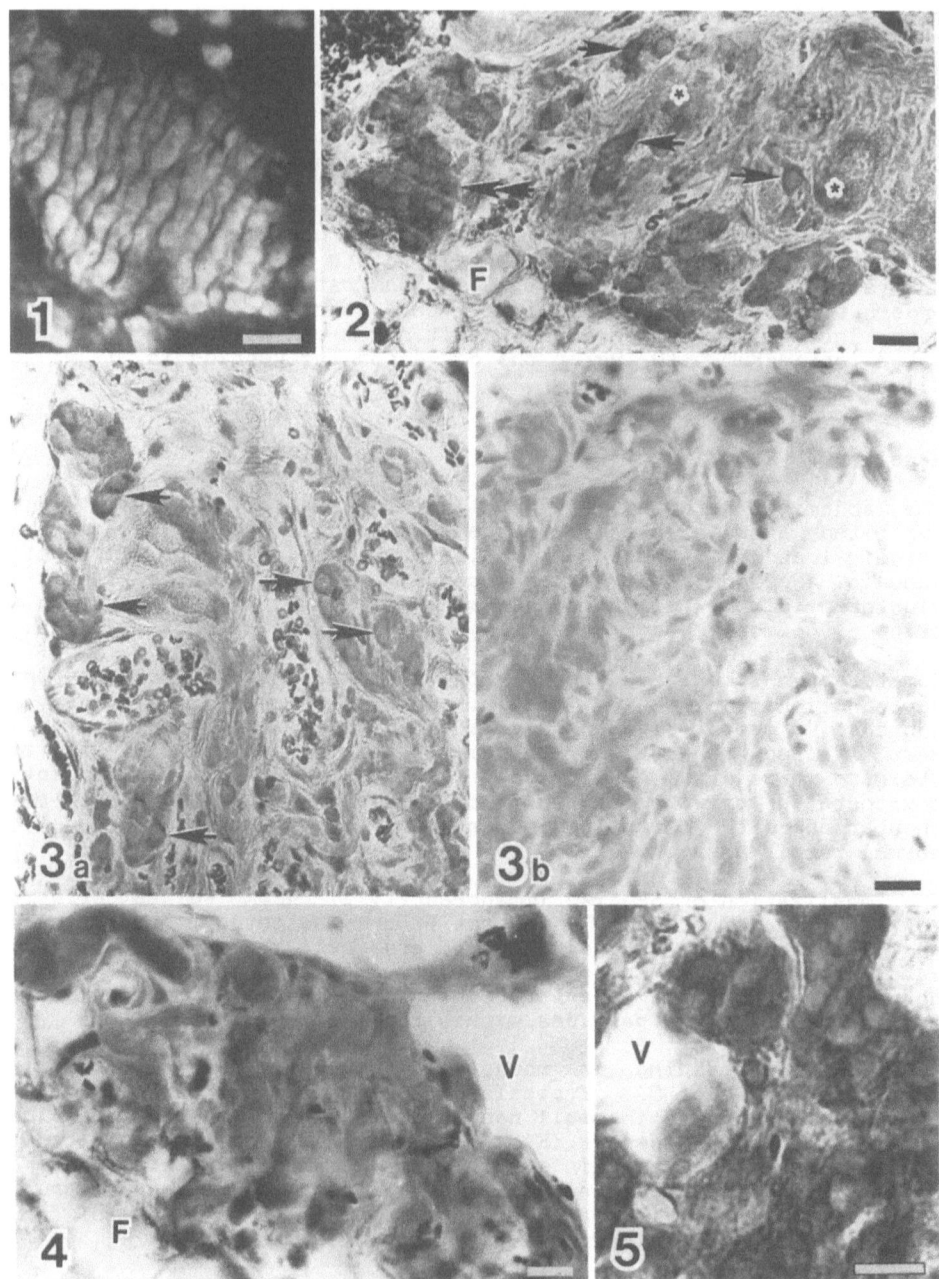

Fig. 1. Formaldehyde induced fluorescence in abdominal paraganglia of old rat. Bar indicates 50 um

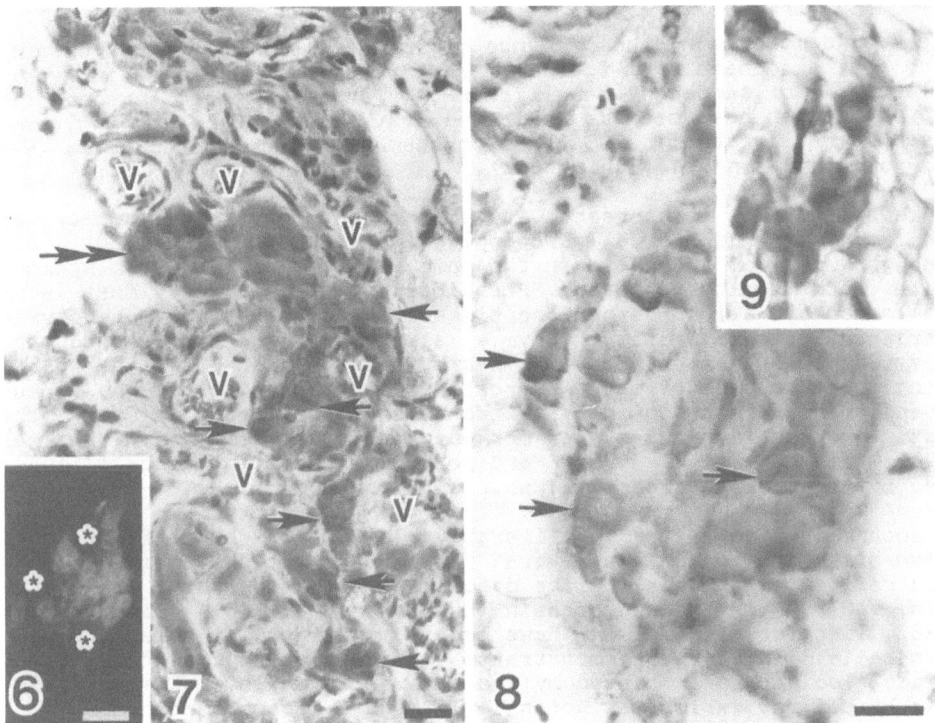

Fig. 6. Catecholamine histofluorescence of a paraganglion. Bar indicates 50 um

Fig. 7. VIP-like immunoreactivity of the same cell group as in Fig. 6. The labeled cells (arrows) are surrounded by numerous blood vessels (V). Bar indicates 20 um

Fig. 8. Enkephalin-like immunoreactivity in single cells scattered throughout a nonreactive paraganglion. Bar indicates 20 um

Fig. 9. Small solitary paraganglion consisting of only enkephalin-like immunoreactive cells

◄ **Fig. 2.** Somatostatin immunoreactivity in a paraganglion next to a sympathetic ganglion. Double arrow points to a cluster of labelled cells and arrowheads mark solitary stained catecholamine storing cells. F = fat cells, asterisk = sympathetic neurons. Bar indicates 20 um

Fig. 3. (a) Somatostatin-immunoreactive cells in a large abdominal paraganglion. Arrows point to a few strongly stained cells. (b) Consecutive control section incubated with preabsorbed antiserum. Bar indicates 20 um

Fig. 4. Substance P like immunoreactivity in a small solitary paraganglion. F = fat, V = vein. Bar indicates 20 um

Fig. 5. Part of a hypertrophied abdominal paraganglion shows immunoreaction with substance P antibody. V = vein. Bar indicates 20 um

Discussion

We have demonstrated the presence of immunoreactivity to neuropeptides that originate from four different precursor molecules in hyperplastic PG of senescent male Fisher-344 rats. Multiple neuroendocrine phenotypes have been demonstrated previously in pheochromocytomas
(see De Lellis et al. 1984) and in the human fetal paraganglia (see
Hervonen and Linnoila, this book).

Enkephalin-like immunoreactivity was present only in a minority of
cells, as in human fetal paraganglia (Hervonen et al. 1980), although most of the cells in the same clusters of PG were stained
with both somatostatin and SP antibodies. It is likely that some
cells contain all three neuropeptides. PG cells can show phenotypic
variance of antigenicity. Heym et al. (1984) recently showed the
coexistence of enkephalins with neurotensin or with SP in the paraganglionic tissue of cat, supporting the presence of multiple neuroendocrine phenotypes in the extra-adrenal chromaffin tissue. The
mature chromaffin cells show marked phenotypic plasticity (Unsicker
1983; Tischler et al. 1984), and genomic expression may be influenced by the microenvironmental shifts in the old animal.

Pheochromocytomas, the proliferative lesions of chromaffin tissue,
also are known to contain several regulatory peptides. Opioid peptides related to enkephalin-like dynorphin, beta-endorphin, somatostatin, ACTH, VIP and neurotensin (Lundberg et al. 1979; Eiden et al.
1982; Tischler et al. 1984) have been demonstrated in human chromaffin cells. This study demonstrates that several neuropeptide phenotypes also can be expressed by the chromaffin cells of the paraganglia of old rat, which may potentially become pheochromocytomas.

The cause of an age-related increase of extra-adrenal chromaffin
tissue of the rat is not known (Partanen et al. 1984a). The presence
of a large bulk of endocrine cells containing both catecholamines
and several neuropeptides may be of importance for the aged orga-
nism. Young and adult rats have minimal PG tissue (Coupland 1965;
Partanen 1984a).

References

Clement-Jones V, Lowry PJ, Rees LH, Besser GM (1980) Met-enkephalin
corculates in human plasma. Nature 283: 295-297

Coupland RE (1965) The natural history of chromaffin cell. Longmans,
London

De Lellis RA, Tischler AS, Wolfe HJ (1984) Multidirectional differentiation in neuroendocrine neoplasma. J Histochem Cytochem 32:
899-904

DiGuilio AM, Yang H-Y, Lutold B, Fratta W, Hong I, Costa E (1978)
Characterization of enkephalin-like material extracted from sympathetic ganglia. Neuropharmacology 17: 989-992

Eiden L, Giraud P, Hotchkiss A, Brownstein MJ (1982) Enkephalins and
VIP in human pheochromocytomas and bovine adrenal chromaffin cells.
In: Costa E, Trabucchi M (eds) Regulatory peptides: from molecular
biology to function. Raven, New York, pp 387-395

Hanbauer I, Kelly GD, Saiani L, Yang H-Y (1982) (Met5)-enkephalin-like peptides of the adrenal medulla: release by nerve stimulation and functional implications. Peptides 3: 469-473

Hervonen A (1971) Development of catecholamine-storing cells in the human fetal paraganglia and adrenal medulla. Acta Physiol Scand. Suppl 368: 1-94

Hervonen A, Linnoila I (1987) Coexistence of norepinephrine and neuropeptides originating from different precursor molecules in the human fetal paraganglia. (This volume)

Hervonen A, Pickel VM, Joh TH, Reis DJ, Linnoila I, Kanerva L, Miller RJ (1980) Immunocytochemical demonstration of catecholamine synthesizing enzymes and neuropeptides in the catecholamine storing cells of human fetal sympathetic nervous system. In: Eränkö O, Päivärinta H, Soinila S (eds) Histochemistry and cell biology of autonomic neurons, SIF-cells and paraneurons. Raven, New York, pp 373-378

Heym Ch, Reinecke M, Weihe E, Forssmann WG (1984) Dopamine-beta-hydroxylase-, neurotensin-, substance P-, vasoactive intestinal polypeptide- and enkephalin-immunohistochemistry of paravertebral and prevertebral ganglia in the cat. Cell Tissue Res 235: 411-418

Hexum TD, Hanbauer I, Govoni S, Yang H-Y, Costa E (1980) Secretion of enkephalin like peptides from canine adrenal gland following splanchnic nerve stimulation. Neuropeptides 1: 137-142

Linnoila I, DiAugustini RP, Hervonen A, Miller RJ (1980) Distribution of (met5)- and (leu5)-enkephalin, vasoactive intestinal polypeptide and substance-P-like immunoreactivities in human adrenal glands. Neuroscience 55: 2247-2259

Lundberg JM, Hamberger B, Schultzberg M, Hökfelt T, Granberg P-O. Efendic S, Terenius L, Goldstein M, Luft R (1979) Enkephalin- and somatostatin-like immunoreactivities in human adrenal medulla and pheochromocytoma. Proc Natl Acad Sci USA 76: 4079-4083

Partanen M, Chiueh CC, Rapoport SI (1981) Age-related increase in the catecholamine containing paraganglia in male Fischer-344 rats. Anat Rec 210: 563-566

Partanen M, Linnoila I, Hervonen A, Rapoport SI (1984a) The effect of aging on extra-adrenal catecholamine storing cells of the rat. Neurobiol Aging 5: 105-110

Partanen M, Rapoport SI, Reis DJ, Joh TH, Stolk JM, Linnoila I, Teitelman G, Hervonen A (1984b) Catecholamine synthesizing enzymes in paraganglia of aged Fischer 344 rats. Immunohistochemistry and fluorescence microscopy. Cell Tissue Res 238: 217-220

Partanen M, Hervonen A, Rapoport SI (1984c) The ultrastructure of hyperthrophied paraganglia in aged rats. J Anat (Lond) 139: 619-626

Schultzberg M, Hökfelt T, Terenius L, Elfvin L-G, Lundberg J, Brandt J, Elde RP, Goldstein M (1979) Enkephalin immunoreactive nerve fibers and cell bodies in sympathetic ganglia of the guinea pig and rat. Neuroscience 4: 249-270

Tischler A, Lee YC, Perlman RL, Costopoulos D, Slayton VW, Bloom SR (1984) Production of ectopic vasoactive intestinal polypeptide and

neurotensin like immunoreactivity in human pheochromocytoma ceₗl
cultures. J Neurosci 4: 1398-1404

Unsicker K (1983) Cell and tissue culture studies on the sympathoadrenal system. In: Elfvin L-G (ed) The autonomic ganglia. Wiley,
Chichester, pp 475-505

Viveros HO, Wilson S (1983) The adrenal chromaffin cell as a model
to study the co-secretion of enkephalins and catecholamines. J Auton
Nerv Syst 7: 41-58

Quantitative Studies on the Autonomic Innervation of the Human Detrusor Muscle: Effects of Age and Outlet Obstruction

J. S. Dixon and J. A. Gosling

Division of Anatomy, Department of Cell and Structural Biology, Medical School, University of Manchester, Manchester M13 9PT, Great Britain

Introduction

Very little is known about the effect of age on autonomic nerves and in particular those innervating the lower urinary tract. In the present study quantitative methods have been used to assess the density of autonomic innervation within the wall of the urinary bladder using biopsy samples obtained from a group of adult 'control' patients varying in age over a wide range. In addition, changes in innervation density due to the presence of outflow obstruction have been studied in bladder biopsies from a group of patients with urodynamically proven outflow obstruction. The latter is known to induce marked histological change within the bladder wall and, when artificially induced in animals, has been shown to affect the nervous control of bladder activity (Uvelius et al. 1984). However, the response of the autonomic innervation of the human detrusor to outflow obstruction has not previously been investigated.

Materials and Methods

'Control' Patients. 54 patients (10 male, 44 female) with an age range of 20-79 years (mean 47 years) were included in this group. Each patient was neurologically and urodynamically normal, with a stable cystometrogram and no evidence of bladder outlet obstruction or trabeculation. Bladder samples were obtained from the dome and lateral walls and processed for light and electron microscopy.

Obstructed Patients. 19 patients (17 male, 2 female) with an age range of 58-76 years (mean 69.5 years) were included in this group. Each patient received detailed clinical, urodynamic and cystoscopic evaluation. Those included in the study had a stable cystometrogram and no evidence of neurological disease but in each patient there was unequivocal urodynamic evidence of outflow obstruction, i.e. a detrusor pressure rise greater than 100 cm of water in association with a flow rate of less than 10 ml/sec accompanied by severe bladder trabeculation at cystoscopy. During endoscopy bladder biopsy samples were removed from the dome and/or lateral walls and processed for light and electron microscopy.

Light Microscopy. Each tissue sample was quickly frozen in 2-methyl butane cooled in liquid nitrogen. 15 um thick cryostat sections were stained either for routine histology using Masson's trichrome stain or to demonstrate acetylcholinesterase-containing nerves using a modification of the method due to Gomori (1952).

Experimental Brain Research Series 16
© Springer-Verlag Berlin · Heidelberg 1987

Electron Microscopy. Biopsy samples for electron microscopy were obtained from 12 patients in the control group (6 in the age range 25–35 years and 6 in the age range 60–72 years) and 7 patients in the obstructed group (age range 67–74 years, mean age 71 years). Each sample was double fixed in buffered glutaraldehyde and osmium tetroxide before embedding in epoxy resin.

Quantitative Methods for Light Microscopy. The total area of smooth muscle within each biopsy section was determined using a point counting technique and expressed as a percentage of detrusor tissue. This area was then divided by the number of cells present to determine muscle cell mean profile area for each patient. Sections stained to demonstrate acetylcholinesterase-containing nerves were examined using the same magnification and a similar point counting method to determine the amount of enzyme-positive nerve within the given area.

Quantitative Methods for Electron Microscopy. For each biopsy sample ten grid squares containing transversely-sectioned smooth muscle cells were selected and scanned at a magnification of x 24,000. Counts were then made of the total number of 'nerve profiles' present in these grid squares together with the number of smooth muscle cell profiles. Using a point counting technique on low magnification micrographs the total area of smooth muscle cell profiles was calculated and from this data a mean value for smooth muscle cell profile area was obtained.

The data obtained from both the control and obstructed groups were found to be normally distributed using Filliben's test. Thus, all statistical comparisons of data were performed using a two-tailed Student's t-test.

Results

A. 'Control' Patients. In 91 biopsies obtained from the 54 patients included in this study the amount of muscle per mm^2 of tissue varied from 63–77% (mean 69.0 $\pm$ 0.4%). No significant differences in the amount of nerve per mm^2 of muscle were detected between samples from males and females or between samples from the dome and lateral walls of an individual patient. Using the mean of all sites for each of the 54 patients (regardless of sex) the amount of nerve per mm^2 of muscle was shown to decrease linearly with the age of the patient using Pearson's correlation ($p < 0.01$) (Fig. 1).

Under e.m. mean values for smooth muscle cell profile areas were not significantly different between the 6 patients aged 25–35 years and the 6 patients aged 60–72 years.

The mean value for nerve profiles per mm^2 of smooth muscle in the younger group was 1,003 ($\pm$ 135 S.E.), significantly different from that for the elder group, 673 ($\pm$ 72 S.E.) ($p < 0.05$).

B. Obstructed Patients. The amount of smooth muscle per mm^2 of tissue in the biopsies from the 19 patients with outflow obstruction ranged from 44%–71% with a mean of 60% $\pm$ 2% (S.E.).

The amount of smooth muscle per mm^2 of tissue for each of the biopsies obtained from 17 age-matched 'control' patients ranged from

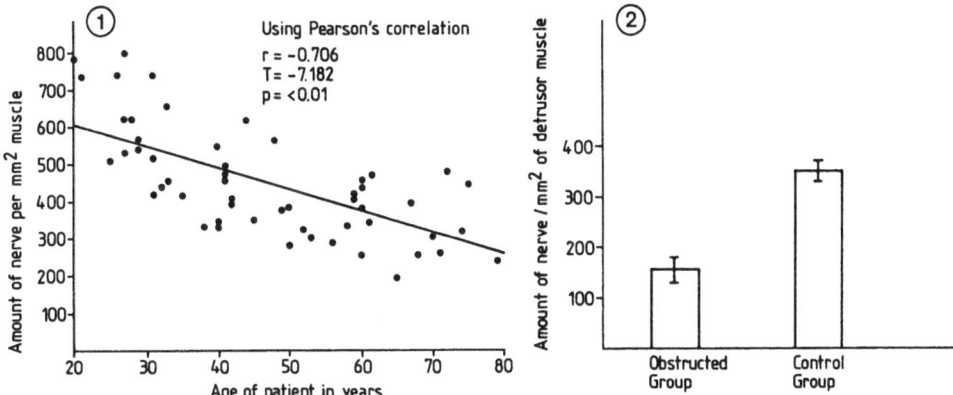

Fig. 1. The mean amount of nerve per mm² detrusor muscle (measured by light microscopy related to the age of each patient. Using Pearson's correlation there is a significant (p<0.01) negative correlation between amount of nerve and the age of the patient

Fig. 2. Effect of outflow obstruction on amount of nerve per mm² of detrusor muscle measured by light microscopy; significantly different from controls (p<0.01)

65-75% with a mean of 69% ± 1% (S.E.). When these two groups were compared, there was a significant decrease (p<0.01) in the amount of muscle per mm² of tissue in the obstructed patients, due to the infiltration of the detrusor muscle bundles by connective tissue. For each biopsy the figure for the percentage of smooth muscle within a given area was used to convert the value for the amount of nerve per mm² of tissue to the amount of nerve per mm² of muscle.

Values of smooth muscle cell mean profile area from individual obstructed patients were higher than those from the control group. Thus, for each biopsy from the obstructed group, an appropriate correction factor (the percentage increase in size) was applied when calculating the amount of nerve per mm² of muscle. The mean values of 'corrected' nerve per mm² of muscle for the obstructed group (154 ± 27 S.E.) were significantly lower than that of the control group (351 ± 22 S.E.) (p<0.01) (Fig. 2).

Under e.m. nerve profile counts for each biopsy from obstructed patients were adjusted by an appropriate factor to compensate for the increase in mean muscle cell profile area. The mean value obtained for the control group was 673 ± 72.6 (S.E.) nerve profiles per mm² of smooth muscle tissue, a value significantly different from the obstructed group (273 ± 90.2 S.E.) (p<0.01).

Discussion

The 'control' group included in the present study contained only those patients who, on the basis of clinical evaluation, were considered likely to possess a normal autonomic innervation; hence patients with neurological lesions or peripheral neuropathy were excluded. The results of the quantitative light microscopic study on 'control' biopsies demonstrated a significant linear decrease with age in the amount of acetylcholinesterase-containing nerve associa-

ted with the detrusor muscle. This technique relies on the histochemical localisation of acetylcholinesterase in order to visualise the
autonomic nerve fibres. In order to exclude the possibility of a
progressive loss of enzyme with advancing age rather than an actual
loss of nerve fibres, a parallel study was undertaken using the
electron microscope on biopsies from six young adults and six elderly patients. Counts of nerve axons (regardless of type) per unit
area of detrusor smooth muscle were carried out for these two groups
of patients and, when the results were compared, a statistically
significant decrease (p<0.05) was demonstrated in the nerve counts
from the older group, the amount of the decrease being comparable to
that demonstrated in the light microscope. The possibility that the
decrease in number of nerve axons per unit area of smooth muscle
from the older group might have been due to hypertrophy of individual smooth muscle cells was excluded since it was shown that there
was no significant difference between the size of smooth muscle
cells from the two groups of patients. Thus, in the present study
the fine structural results confirmed the light microscopic quantitative measurements of acetylcholinesterase-containing nerve fibres
and demonstrated a significant loss (of about one third) of nerve
associated with detrusor muscle between young adults and elderly
patients.

Previous studies have shown that bladder outflow obstruction is
associated with hypertrophy of detrusor smooth muscle cells and
infiltration of the muscularis by connective tissue (Brent and
Stephen 1975; Gosling and Dixon 1980; Susset 1983; Levin et al.
1984). In the present study both these changes were observed in the
biopsies from obstructed patients when compared with controls. Thus,
in the quantitation of autonomic nerves in the obstructed group,
allowances were made for the increased smooth muscle cell mean
profile area and intercellular connective tissue. When compared with
control patients a 56% reduction in the amount of enzyme-positive
nerves was demonstrated in the obstructed bladders. Although the
majority of the obstructed patients were male a comparison between
age-matched males and females from the control group showed that
there is no significant difference in the amount of nerve per mm^2 of
detrusor muscle between the sexes. Thus, the present results from
obstructed male patients cannot merely be explained as a sex difference in autonomic innervation.

As with the control patients it could be argued that the observed
reduction in the amount of nerve in bladders with outflow obstruction reflects decreased levels of enzyme activity and is not due to
an actual reduction in innervation. Thus, parallel studies using the
electron microscope were carried out using portions of the same
biopsy samples from selected patients in both groups. Counts of all
types of autonomic nerve profiles per mm^2 of smooth muscle (corrected for hypertrophy where applicable) were obtained for each biopsy
and when the two groups were compared, a significant reduction (of
about 60%) in nerve density was demonstrated in the obstructed
group. Thus, the fine structural results endorse those of the light
microscope study and confirm that the presence of outflow obstruction significantly reduces the amount of autonomic nerve supplying
human detrusor muscle.

The reduced innervation of the detrusor muscle in response to outflow obstruction may markedly impair the neuromuscular control of
the bladder. In this regard it is interesting to note that various
pharmacological changes compatible with this effect have been reported in experimentally induced bladder outflow obstruction. Furthermore recent unpublished studies by Brading and co-workers have

demonstrated "denervation supersensitivity" in samples of detrusor
muscle obtained from patients with outflow obstruction. Thus, the
changes in autonomic innervation presently demonstrated may play a
part in the aetiology of bladder instability which frequently accom-
panies outflow obstruction.

References

Brent L, Stephen FD (1975) The response of smooth muscle cells in
the rabbit bladder to outflow obstruction. Invest Urol 12: 494–502

Gomori G (1952) Microscopic histochemistry: principles and practice.
University of Chicago Press, Chicago

Gosling JA, Dixon JS (1980) Structure of trabeculated detrusor
smooth muscle in cases of prostatic hypertrophy. Urol Int 35: 351

Levin RM, High J, Wein AJ (1984) The effect of short-term obstruction on urinary bladder function in the rabbit. J Urol 132: 789

Susset JG (1983) Effects of ageing and prostatic obstruction on
detrusor morphology and function. In Hinman F (ed) Benign prostatic
hypertrophy. Springer, Berlin

Uvelius B, Ekstrom J, Larsson B, Mattiasson A (1984) Changes in the
nervous control of the rat urinary bladder subjected to infravesical
outflow obstruction. Proceedings of the International Continence
Society 14th Annual Meeting: 206

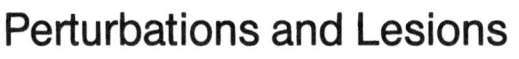
Perturbations and Lesions

Stress Effects on Central Autonomic Peptide Immunocytochemistry

J. Y. Jew and N. E. Hynes

Department of Anatomy, University of Iowa, Iowa City, IA 52242, USA

Introduction

The paraventricular nucleus (PVN) of the hypothalamus is classically regarded a portion of the magnocellular neurosecretory system, sending oxytocinergic and vasopressinergic projections to the posterior pituitary. However, recent morphologic findings regarding the PVN's multi-neuropeptide content, its different cytoarchitectonic subdivisions, and its afferent and efferent connections focus attention on other aspects of PVN function, particularly its role in autonomic functions. Vasopressinergic projections from the PVN have been localized in brain stem centers involved in cardiovascular regulation; it has been suggested that the connections between the PVN and cardiovascular control centers may function in the homeostatic regulation of blood pressure and blood volume. The PVN also sends vasopressinergic projections to the zona externa of the median eminence. While the function of these projections is unresolved, experiments showing the effects of adrenalectomy and glucocorticoids on vasopressin immunoreactivity (Stillman et al. 1977) have led some investigators to conclude that vasopressin in the median eminence may be a regulator of ACTH release. However, other investigators (Hashimoto et al. 1981) suggest that median eminence vasopressin is not involved in the feedback and acute stress mechanism of corticotropin releasing factor–ACTH secretion.

Biochemical studies carried out by Negro–Vilar and Saavedra (1980) showed striking increases in PVN vasopressin levels in spontaneously hypertensive rats following acute stress, whereas increases in PVN vasopressin levels in stressed normotensive rats were not statistically significant from those of nonstressed controls.

In view of reports as those cited above suggesting a role for PVN vasopressin in cardiovascular regulation and stress responses, the present study was carried out to determine the effects of chronic immobilization stress on vasopressinergic neurons of the PVN and whether these effects were detectable by immunohistochemistry.

Materials and Methods

Twelve week old male spontaneously hypertensive rats (SHR) and their normotensive controls, Wistar Kyoto (WKY) rats, were used in these experiments. The rats were housed two per cage with alternate 12 h periods of light and dark. Food and water were provided ad libitum.

Experimental Brain Research Series 16
© Springer-Verlag Berlin · Heidelberg 1987

The stress protocol was begun 2 weeks prior to the anticipated perfusion date and was carried out on two (one SHR and one WKY) of the four rats in each group. The two unstressed rats (one SHR and one WKY) in each group were brought into the lab but left unrestrained in their cages during the stress protocol. The chronic immobilization stress protocol consisted of strapping the rats down in the supine position for a period of 2 hours each day, for 10 out of 14 days. The last stress regimen was administered on day 14.

Approximately 20 minutes after the end of the last stress period, the rats were anesthetized (Nembutal, 40 mg/kg b.w., i.p.) and perfused according to a standard protocol (Williams and Jew 1975) with a fixative solution containing 4% paraformaldehyde, 0.05% glutaraldehyde and 15% saturated picric acid in 0.165 M phosphate buffer (pH 7.3). Tissue blocks were placed in fixative for an additional two hours and then sectioned at 30 um with a vibratome. All sections containing the PVN were collected in vials, washed overnight in cold phosphate buffered saline (PBS), and processed for light microscopic technique (Hsu et al. 1981). Control preparations for immunostaining consisted of tissue sections incubated in normal rabbit serum or preabsorbed primary antiserum rather than in primary antiserum.

After immunostaining, sections were mounted on chrome alum gelatin-coated slides, dehydrated and coverslipped with Permount. The PVN was surveyed throughout its rostral-caudal extent and photographed. Prints of the vasopressin-immunostained PVN were evaluated by these investigators and independently by a research assistant who was not informed of the treatment or animal strain; immunostaining was rated from + (very lightly stained) to ++++ (very heavily stained).

These experiments and evaluations were carried out for 10 groups of 4 animals per group.

Results

Vasopressin-immunoreactive perikarya were localized in the paraventricular, supraoptic and suprachiasmatic nuclei of the hypothalamus in both stressed and unstressed rats. Small groups of labelled perikarya were also encountered along the arc of the hypothalamohypophyseal pathway arising from the PVN. This presentation will be limited to our findings in the PVN.

The majority of vasopressin-immunostained perikarya were found in the posterior dorsolateral portion of the PVN, the posterior magnocellular subdivision (Figs. 1-4). Larger perikarya were concentrated in this subdivision whereas smaller perikarya were scattered across the medial subdivisions of the PVN (Figs. 1-4). The staining intensity varied somewhat among labelled cells with each PVN. Many labelled nerve fibers within the PVN were of large caliber and heavily stained; such fibers originated from perikarya within the PVN and were confined to the PVN, with some of these fibers extending to the ependyma of the ventricle. In contrast, labelled fibers leaving the nucleus to form the prominent hypothalamohypophyseal tract were of

Figs. 1-4. Light microscopic photographs taken from coronal sections ▶ through the PVN which have been immunostained for vasopressin. **Fig. 1.** WKY nonstressed. **Fig. 2.** WKY stressed. **Fig. 3.** SHR nonstressed. **Fig. 4.** SHR stressed. Magnification bars = 200 um

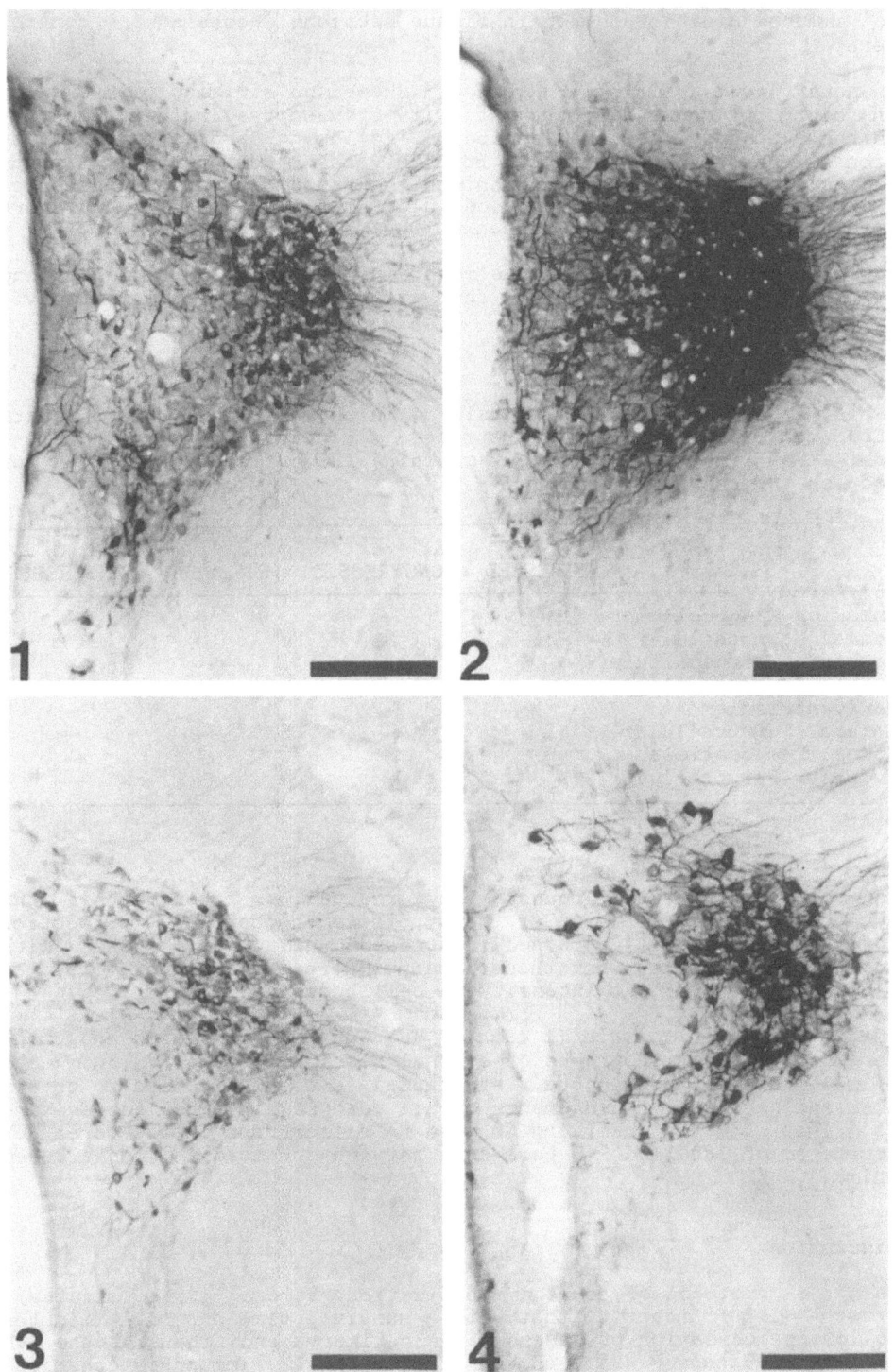

much finer caliber and beaded along the course of individual fibers. No immunostaining was seen in tissue sections incubated in control sera.

In nonstressed animals, SHR and WKY showed no striking differences in distribution or staining intensity of vasopressinergic perikarya in the PVN or of their nerve fibers. Following chronic immobiliza-tion stress, PVN perikarya in both SHR and WKY demonstrated increa-sed staining intensity, which was most apparent in the posterior magnocellular and medial parvocellular subdivisions (nomenclature according to Swanson and Sawchenko 1980).

Table 1 summarizes our findings regarding the relative intensity and quantity of cell body staining for vasopressin in different subdivi-sions of the PVN.

Table 1. Vasopressin immunostaining in the PVN of WKY and SHR rats after chronic immobilization stress. + = very light staining (0-3 perikarya); ++++ = very heavy staining (20 or more perikarya per section through PVN)

	WKY STRESSED	WKY NONSTRESSED	SHR STRESSED	SHR NONSTRESSED
Anterior Magnocellular	+	+	+	+
Medial Magnocellular	+	+	+	+
Posterior Magnocellular	++++	+++	++++	+++
Medial Parvocellular	+++	++	++++	++
Periventricular	+	+	+	+
Lateral Parvocellular	++	++	++	++
Dorsal Parvocellular	++	+	++	++
Anterior Parvocellular	++	++	++	++

In some experimental groups, the large caliber fibers within the nucleus as well as fibers of the hypothalamohypophyseal projection appeared more heavily stained in stressed animals as compared with nonstressed animals, but these findings were not so unequivocal as the increased staining intensity in cell bodies.

In one group of animals (1 each of WKY and SHR, stressed and non-stressed), photomicrographs of coronal sections of the PVN from each animal, taken at homologous rostrocaudal levels, were used to obtain area and perimeter measurements of all vasopressin-labelled perikar-ya. This limited sampling showed no differences in the area or perimeter or labelled PVN perikarya between stressed and nonstressed animals.

Discussion

Using a protocol of acute immobilization stress (five minutes), Krisch (1980) reported that with a survival time of five minutes following cessation of stress, the perikarya and fibers of the PVN in Wistar-Hannover rats showed increased LM immunohistochemical staining for vasopressin; but that with a survival period of 30 minutes, staining in the PVN was similar for stressed and nonstressed animals. She also found that the hypothalamohypophyseal tract stained more weakly in the stressed, 5 minute survival animal, but the 30 minute survival preparation resembled control preparations in stainability.

In comparison, Negro-Vilar and Saavedra (1980) found that acute immobilization stress significantly increased vasopressin content (measured by radioimmunoassay) in PVN of SHR but not of WKY (normotensive) rats.

Our experiments indicate that increased vasopressin-immunoreactive staining is observed in the PVN of chronically stressed animals. Whether the increased staining can be attributed to increased synthesis or decreased release of hormone will need to be addressed by ultrastructural immunocytochemical studies and is directly relevant to the question of the role of vasopressin in the stimulation of the adenohypophysial-adrenocortical system. Our use of a chronic stress model also enables us to examine the reported relationship between chronic stress and neuronal degeneration (Meehan and Naranjo 1980; Naranjo 1980); does stress induce degeneration in neurons and do peptidergic subsystems show variations in their vulnerability to stress?

Acknowledgements. This work was supported by BNS8216786, NS19578 and AM34986. We thank Evelyn Jew and Paul Reimann for their technical assistance.

References

Hashimoto K, Yunoki S, Kageyama J, Ohno N, Takahara J, Ofuji T (1981) Vasopressin and CRF-ACTH in adrenalectomized and dexamethasone-treated rats. Neuroendocrinology 32: 87-91

Hsu S-M, Raine L, Fanger H (1981) Use of avidin-biotin-peroxidase complex (ABC) in immunoperoxidase techniques: a comparison between ABC and unlabeled antibody (PAP) procedures. J Histochem Cytochem 29: 577-580

Krisch B (1980) Immunocytochemistry of neuroendocrine systems: vasopressin, somatostatin, luliberin. Progr Histochem Cytochem Vol 13/2

Meehan WP, Naranjo JN (1980) Behavioral and histological analysis of psychosocially stressed mice. Soc Neurosci Proc 10th Ann Mtg Cinn, Ohio, p 44

Naranjo JN (1980) Interaction between brain degeneration and psychosocial stress. Soc Neurosci Proc 10th Ann Mtg Cinn, Ohio, p 44

Negro-Vilar A, Saavedra JM (1980) Changes in brain somatostatin and vasopressin levels after stress in spontaneously hypertensive and Wistar-Kyoto rats. Brain Res Bull 5: 353-358

Stillman M, Recht L, Rosario SL, Seif SM, Robinson AG, Zimmerman EA (1977) The effects of adrenalectomy and glucocorticoid replacement on vasopressin and vasopressin-neurophysin in the zona externa of the rat. Endocrinology 101: 42-49

Williams TH, Jew JY (1975) An improved method for perfusion fixation of neural tissues for electron microscopy. Tiss Cell 7: 407-418

Cells in Sympathetic Ganglia After Stress

M. Mikulajova[1] and A. Kiss[2]

[1] Department of Anatomy, Medical School of Comenius University, Slovak Academy of Sciences, Sasinkova 2, 81108 Bratislava, CSSR
[2] Institute of Experimental Endocrinology, Center of Physiological Sciences, Slovak Academy of Sciences, 81108 Bratislava, CSSR

Introduction

Applying immobilization as a well introduced stress model (Kvetnan-sky and Mikulaj 1970; Heym and Addicks 1977), the ultrastructure of sympathetic neurons, satellite cells and small granule-containing (SGC) cells in the rat superior cervical ganglion (SCG) was investigated after acute and repeated immobilization stress.

Materials and Methods

Adult Wistar rats (200-300 g b.w.) kept under standard conditions were immobilized according to Kvetnansky and Mikulaj (1970) as follows: 1 x 150 min, 7 x 150 min, 8 x 150 min, and 14 x 150 min. Tissue was fixed by immersion in Milloning's fixative or by perfusion (3% glutaraldehyde in 0.1 M phosphate buffer at pH 7.4) and postfixed in 1% OsO_4 in 0.2 M phosphate buffer for 2 h. Tissue pieces were dehydrated and embedded in Durcupan ACM (Fluka). Ultrathin sections were stained with uranyl acetate and lead citrate and examined with a Tesla BS 513 electron microscope. Weibel's method (Weibel 1963) was used for morphometric evaluation of dense core vesicles after acute stress.

Results

After 1 x 150 min of immobilization, characteristic morphological changes in principal sympathetic neurons were observed: deep infoldings of the karyolemma, an enlarged dark nucleolus, vesicular type of the Golgi apparatus, a large number of dense mitochondria, an increased amount of lysosomes, nematosomes, free ribosomes, as well as expanded granular endoplasmic reticulum. The rough endoplasmic reticulum (RER) of repeatedly immobilized (IMO) rats was organized into several islets distributed throughout the cytoplasm (Fig. 1). The cytoplasm of sympathetic neurons revealed frequent occurrence of concentric membraneous bodies of 1.8-2.0 um in diameter, not exceeding a total number of 5 per cell. These bodies were located in close vicinity of the RER profiles (Fig. 2) surrounded with free ribosomes, Golgi apparatus and nematosomes. Their highest number was observed in cells following 7 or 8 x 150 min of immobilization. After 14 x 150 min IMO structures in similar locations consisted of centrally located dark granules and filaments lacking a concentric arrangement (Fig. 3).

The SCG satellite cells of acutely stressed rats did not bear any significant ultrastructural alterations compared to controls.

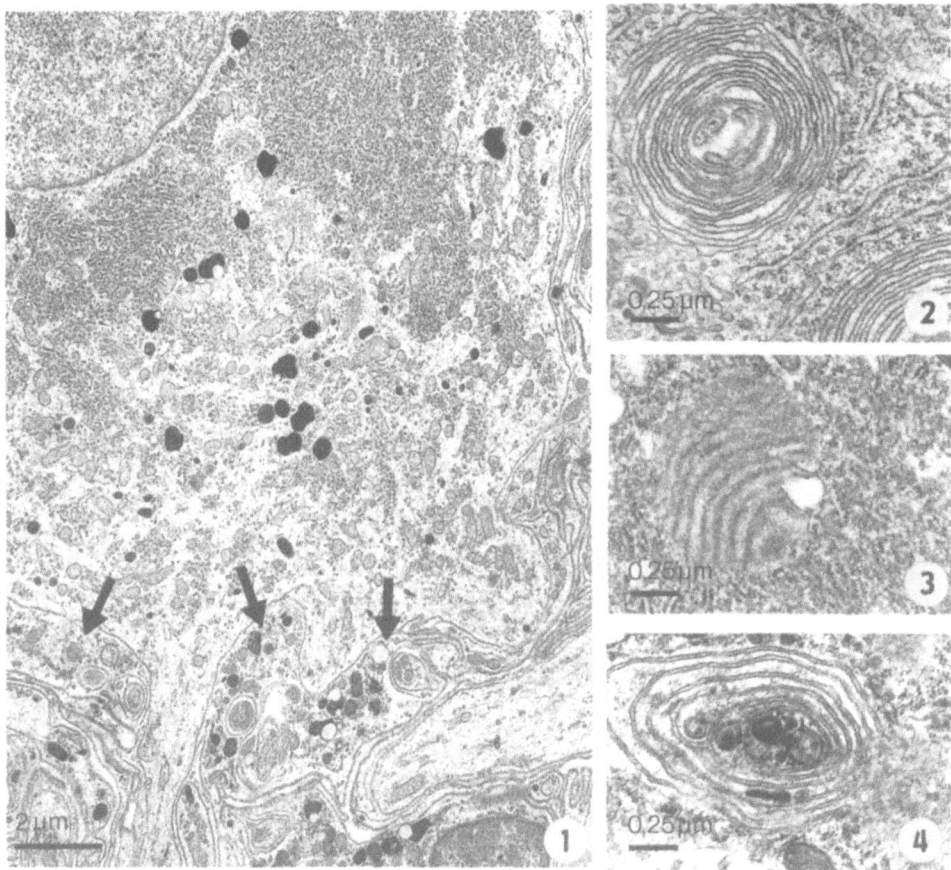

Fig. 1. Fine structure of sympathetic neuron and satellite cell (arrows) of the rat SCG after repeated immobilization stress (14 x 150 min)

Fig. 2. Concentric membraneous body in cytoplasm of a principal sympathetic neuron after repeated immobilization stress

Fig. 3. Concentric granular body in cytoplasm of a principal sympathetic neuron after repeated immobilization stress

Fig. 4. Cytoplasmic membrane of satellite cell enveloping disintegrated material in rats after repeated immobilization stress

Repeated immobilization, however, resulted in an increased number of lysosomes and lipofuscin granules in their cytoplasm. After 14 x 150 min of immobilization, disintegrated structures enveloped by cytoplasmic membranes of satellite cells were found (Fig. 4). After acute stress, SGC cells exhibited an increased number of dense-cored vesicles measuring 60-110 nm in diameter. On counting the rate of their relative volume, an increase from 20 to 38.2% in average was measured. However, the electron density of these vesicles was apparently decreased (Fig. 5). After repeated immobilization (7, 8, 14 x 150 min) the number of dense-cored vesicles appeared to have decreased (Fig. 6).

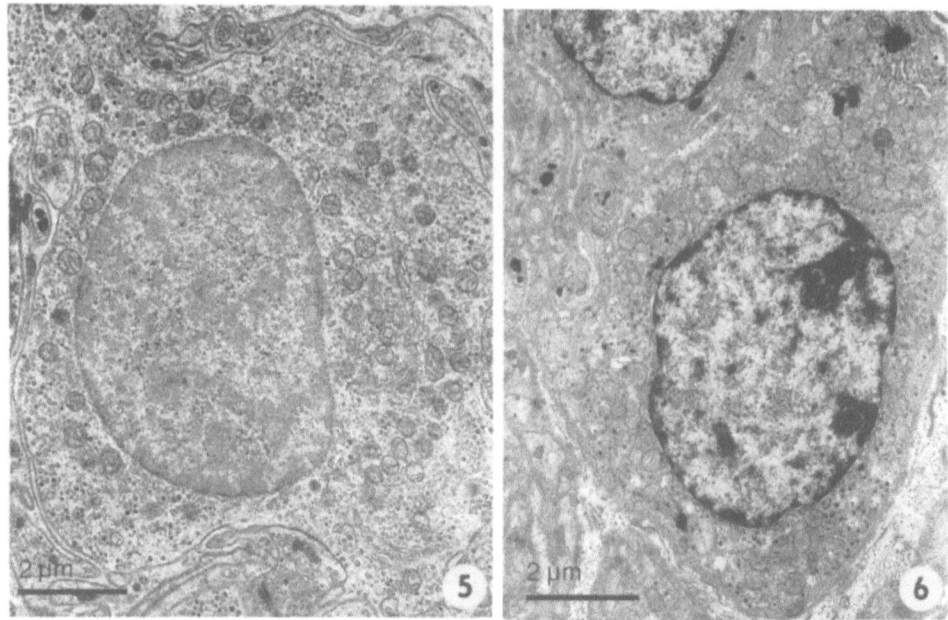

Fig. 5. SGC cell of the rat SCG after acute immobilization stress

Fig. 6. SGC cell of the rat SCG after repeated immobilization stress

Discussion

The ultrastructure of the sympathetic neuron after stress is considered to be a morphological manifestation of its increased functional activity. Our observations are in agreement with previous investigations (Heym et al. 1981; Heym and Addicks 1982; Mikulajova et al. 1984). The frequent occurrence of concentric membraneous bodies after repeated stress has been considered to be a morphological index of increased neuronal activity (Ford and Milks 1978). It seems that relationship of satellite cells to terminal and preterminal axons was also altered; structurally altered axons were manyfold enveloped by the cytoplasmic membrane of a satellite cell. The possibility that this phenomenon represents the segregation of disintegrated material of the satellite cell cannot be excluded. The observations on the SGC cells changes following acute stress confirm a participation of this cell type in stress response (Heym and Addicks 1977). On the basis of our observations, a time relation between the ultrastructural alterations of the observed cell types and the duration of stress is suggested. It may be regarded as a manifestation of adaptation processes in response to the immobilization stress.

References

Ford DH, Milks LC (1978) Smooth endoplasmic reticular whorls in neurons of the arcuate nucleus in male rats following adrenalectomy. Psychoneuroendocrinology 3: 65–83

Heym Ch, Addicks K (1977) Ultrastructure of the rat superior cervical ganglion following immobilization stress. Proc 18th Int Congr Neuroveg Res Tokyo, p 166

Heym Ch, Addicks K, Wedel C (1981) Effect of intermittent immobilization on the ultrastructure of the rat superior cervical ganglion. J Auton Nerv Syst 4: 283–296

Heym Ch, Addicks K (1982) Formation of perichromatin granules, nematosomes and concentric lamellar bodies in sympathetic postganglionic perikarya in response to immobilization stress. Anat Embryol 165: 281–290

Kvetnansky R, Mikulaj L (1970) Adrenal and urinary catecholamines in rats during adaptation to repeated immobilization stress. Endocrinology 87: 738–743

Mikulajova M, Kiss A, Mraz P, Polonyi J (1984) Morfologickévlastnosti granulárneho endoplazmatického retikula sympatikových neurónov po strese. Bratisl Lek Listy 81: 536–547

Weibel ER (1963) Principles and methods for the morphometric study of the lung and other organs. Lab Instr 12: 131–140

Ultrastructural Changes Induced in Rat Carotid Bodies and Superior Laryngeal Nerve Paraganglia by Chronic Hypoxia

D. M. McDonald[1] and A. Haskell[2]

[1] Cardiovascular Research Institute, University of California, San Francisco, CA 94143, USA
[2] Department of Anatomy, University of California, San Francisco, CA 94143, USA

Introduction

Small, catecholamine-containing cells (glomus cells) clustered next to tortuous capillaries constitute a distinctive feature of the carotid body. Glomus cells are sites of termination of sensory nerves that mediate the organ's chemoreceptive function. The carotid body's vasculature is thought to participate in this function because factors that regulate blood flow in the carotid body can modify its responsiveness to changes in the partial pressure of oxygen and carbon dioxide in blood. Paraganglia associated with branches of the vagus nerve have glomus cells and blood vessels with many similarities to those of the carotid body (McDonald and Haskell 1987).

Humans and other mammals living at high altitude have enlarged carotid bodies (Edwards et al. 1971; Arias-Stella and Valcarcel 1976). Carotid bodies of rats made hypoxic in the laboratory for several weeks are as much as eight times normal size (Smith et al. 1974; Pequignot and Hellström 1983; Dhillon et al. 1984; McGregor et al. 1984). Paraganglia associated with the recurrent laryngeal nerve are similarly enlarged (Dahlqvist et al. 1987). Most studies of carotid body enlargement have focused on glomus cell hypertrophy and hyperplasia, but marked vasodilatation is an equally prominent feature. Blessing and Wolff (1973) reported a 14-fold increase in the total volume of capillaries, which accounted for 43% of the enlargement of carotid bodies in rats made hypoxic for 99 days. However, there is little information on which portion of the carotid body's vasculature changes, whether the vascular circuitry is altered, whether the relationship between blood vessels and glomus cells is modified, and whether blood vessels of vagal paraganglia also change.

Materials and Methods

We used light and electron microscopic methods to study the plasticity of blood vessels supplying carotid bodies and superior laryngeal nerve (SLN) paraganglia of Long-Evans rats. We made observations on three groups of adult rats: normal rats, rats made hypoxic and hypercapnic for 10 min under anesthesia by ventilation with a gas mixture containing 10% oxygen and 10% carbon dioxide, and rats made hypoxic for 4 or 6 weeks by continuous exposure to an atmosphere containing 10% oxygen. The brief period of hypoxia and hypercapnia was used to maximally dilate blood vessels of normal carotid bodies (Seidl et al. 1977) for comparison to changes induced in the blood vessels by hypoxia of long duration. At the end of the period

Experimental Brain Research Series 16
© Springer-Verlag Berlin · Heidelberg 1987

of exposure, the rats (body weight 210–430 g) were anesthetized and then tissues were preserved by perfusion of glutaraldehyde through the heart at a pressure of 120 mmHg (McDonald and Larue 1983). Morphometric measurements and reconstructions of vascular circuits were made from serial sections 0.5 um in thickness of tissues embedded in epoxy resin. Alterations in the relationship between blood vessels and glomus cells were assessed by electron microscopy.

Results and Discussion

In carotid bodies of normal rats, the diameter of blood vessels ranged from 8 um to 25 um, and 20% of the tissue volume was occupied by blood vessels (Fig. 1). When vasodilatation was produced in the carotid body by a 10 min period of hypoxia and hypercapnia, this proportion increased to 28%. By comparison, after 6 weeks of hypoxia blood vessels composed 70% of the tissue volume, the diameter of blood vessels next to glomus cells had increased to as much as 80 um (Fig. 2), and the carotid bodies were enlarged. Blood vessels of SLN paraganglia showed similar changes in response to chronic hypoxia (Figs. 3, 4), but vessels of the superior cervical ganglion did not enlarge (Figs. 5, 6), even though this ganglion shares the arterial supply and venous drainage of the carotid body and SLN paraganglia.

Reconstructions of serial sections revealed that the enlarged blood vessels next to glomus cells provided the principal route for blood flow through the carotid body (Fig. 7). No arteriovenous anastomoses bypassing glomus tissue were found. In contrast to these findings, our previous studies (McDonald and Larue 1983; McDonald and Haskell 1987) revealed that terminal arterioles in carotid bodies and SLN paraganglia of normal rats divide near clusters of glomus cells. One branch enters the network of capillaries next to glomus cells, while the other branch bypasses glomus tissue by joining an arteriovenous anastomosis (Fig. 7).

Chronic hypoxia also changed the relationship between glomus cells and capillaries: the usual clusters of glomus cells were transformed into thin layers of cells bordering enormous blood vessels (Fig. 8). Sheath cells covered an abnormally small amount of the surface of such glomus cells, and expanses of glomus cell plasma membranes were exposed to and in some cases touching the vascular endothelium (Fig. 9).

Acker (1980) found that acute changes in the partial pressure of oxygen and carbon dioxide in blood can increase the total blood flow through the carotid body without having the same effect on the organ's local blood flow. Such measurements suggest that in normal carotid bodies a large proportion of the blood flow traverses arteriovenous anastomoses which bypass the capillaries next to glomus cell–sensory nerve complexes. Our morphological studies of the architecture of blood vessels in the normal carotid body are consistent with these observations (McDonald and Larue 1983). However, we conclude from the present studies that after several weeks of hypoxia the enlarged blood vessels next to glomus cells receive most of the blood flow in the carotid body and SLN paraganglia.

Acknowledgement. This research was supported in part by NIH Pulmonary Program Project Grant HL-24136 from the US Public Health Service.

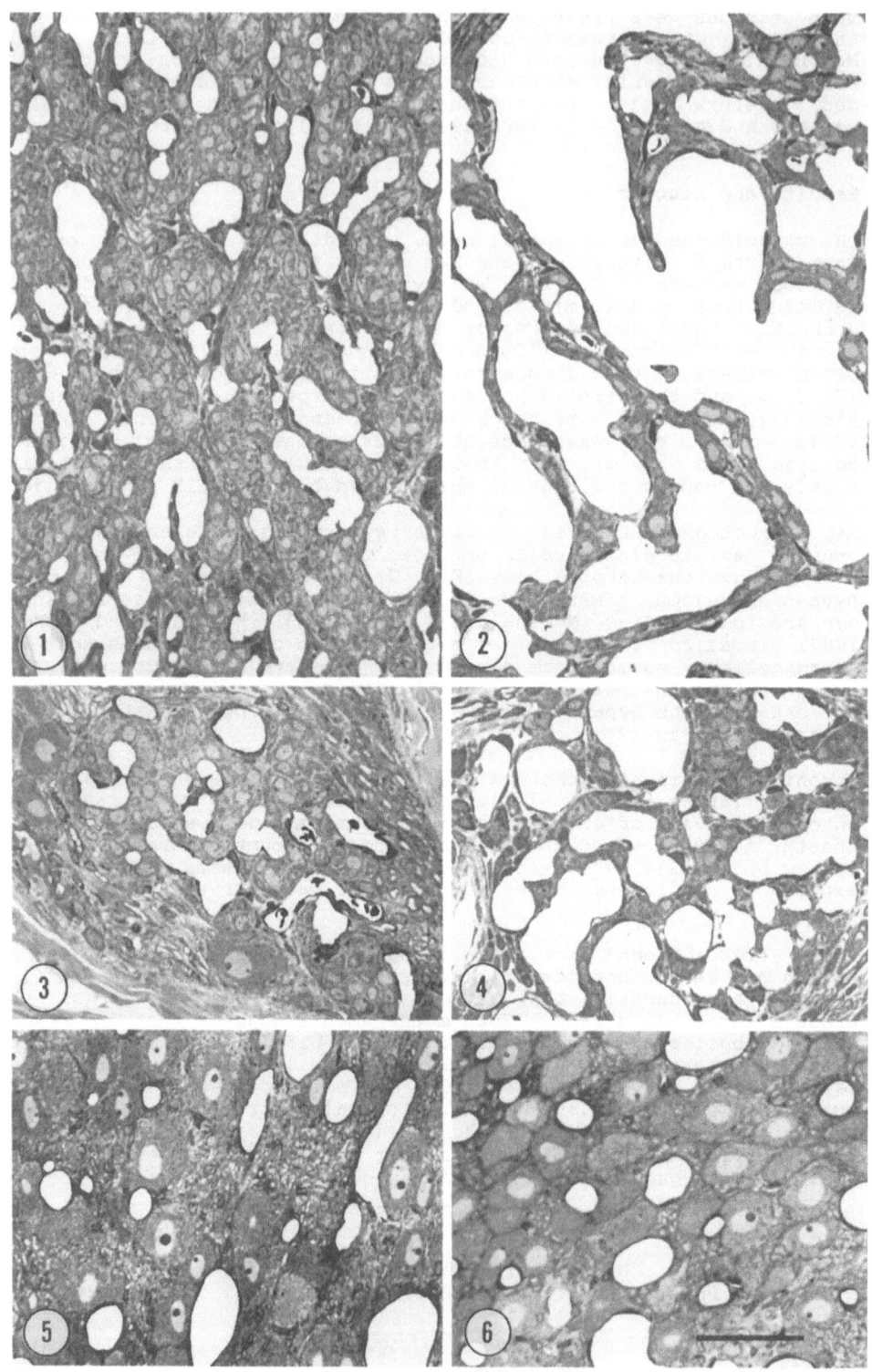

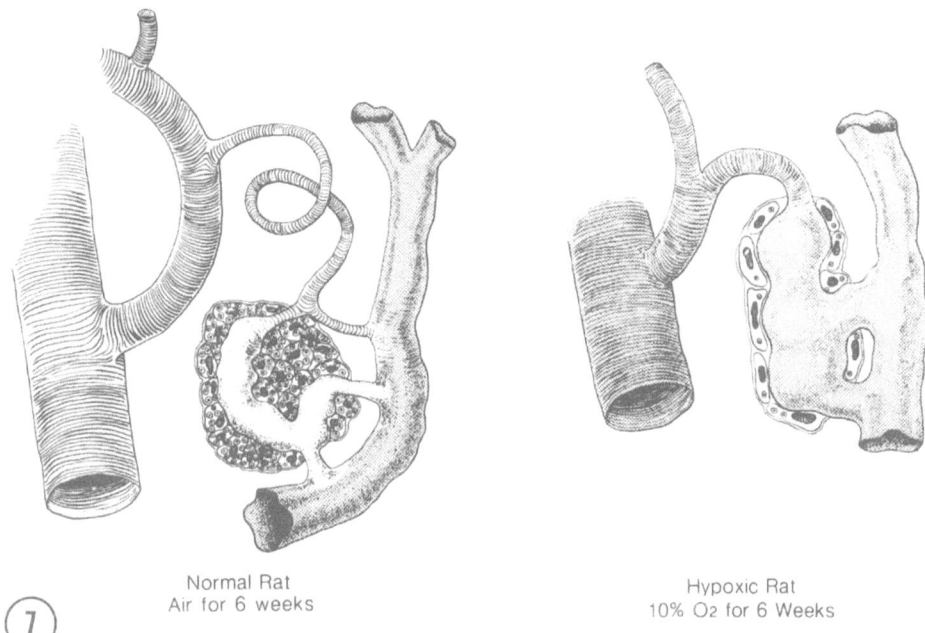

Normal Rat
Air for 6 weeks

Hypoxic Rat
10% O2 for 6 Weeks

(7)

Fig. 7. A comparison of connections of blood vessels in carotid bodies of normal rats with those in chronically hypoxic rats. In normal rats arterioles divide near their termination, with one branch supplying glomerular capillaries and the other branch joining an arteriovenous anastomosis. In chronically hypoxic rats, only one route for blood flow was identified: from arterioles to venules via enlarged, shunt-like vessels next to glomus cells

◀ **Figs. 1-6**. Light micrographs showing blood vessel profiles (emptied of blood by perfusion of fixative) in carotid bodies (Figs. 1, 2). SLN paraganglia (Figs. 3, 4) and superior cervical ganglia (Figs. 5, 6) of normal rats breathing air (Figs. 1, 3 ,5) and hypoxic rats breathing 10% oxygen for 4 or 6 weeks (Figs. 2, 4, 6). In the carotid body and SLN paraganglion of normal rats, glomus cells are clustered near capillaries, but in chronically hypoxic rats, glomus cells form sheets of tissue compressed by adjacent blood vessels. Blood vessels of the superior cervical ganglion do not exhibit these changes. Epoxy sections stained with toluidine blue. Scale marker = 50 um

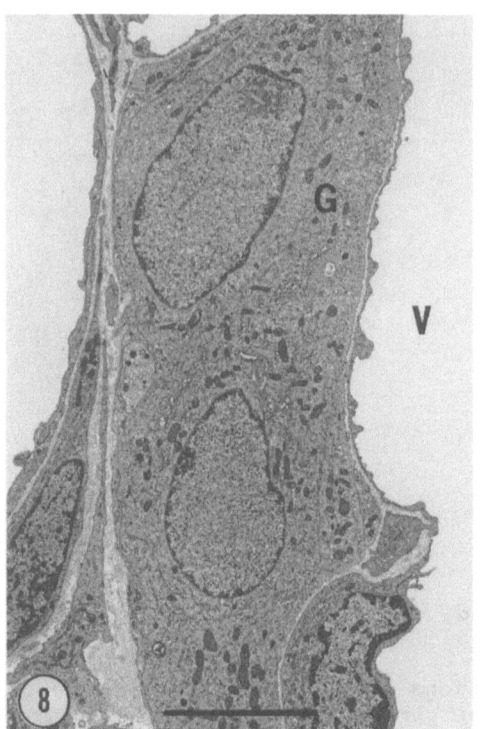 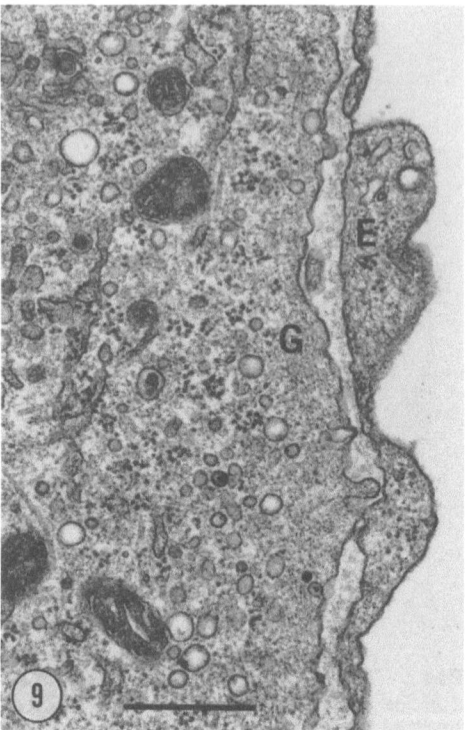

Figs. 8, 9. Electron micrographs of glomus cells (G) in a carotid body of a hypoxic rat (10% oxygen for 4 weeks). Figure 8 shows two glomus cells in a septum of tissue between blood vessels (V). Scale marker = 5 um. Figure 9 shows the proximity of one of these glomus cells to an endothelial cell (E). Only the basal lamina separates the two cells. Scale marker = 0.5 um

References

Arias-Stella J, Valcarcel J (1976) Chief cells hyperplasia in the human carotid body at high altitudes. Hum Pathol 7: 361–373

Blessing MH, Wolff H (1973) Befunde am Glomus caroticum der Ratte nach Aufenthalt in einer simulierten Höhe von 7500 m. Virch Arch (Pathol Anat) 360: 79–92

Dahlqvist A, Hellström S, Carlsöö B, Pequignot JM, Domeij S (1987) Morphological and biochemical characteristics of the laryngeal nerve paraganglia. In: Pallot DJ, Ribeiro JA (eds) 8th International Symposium on the Peripheral Arterial Chemoreceptors. Croom Helm Ltd, London (in press)

Dhillon DP, Barer GR, Walsh M (1984) The enlarged carotid body of the chronically hypoxic and chronically hypoxic and hypercapnic rat: a morphometric analysis. Quart J Exp Physiol 69: 301–317

Edwards C, Heath D, Harris P, Castillo Y, Kruger H, Arias-Stella J (1971) The carotid body in animals at high altitude. J Pathol 104: 231–238

McDonald DM, Haskell A (1987) Vascular geometry of arterial chemoreceptors: learning about the carotid body by studying paraganglia of
the superior laryngeal nerve. In: Pallot DJ, Ribeiro JA (eds) 8th
Internation Symposium on the Peripheral Arterial Chemoreceptors.
Croom Helm Ltd, London (in press)

McDonald DM, Larue DT (1983) The ultrastructure and connections of
blood vessels supplying the rat carotid body and carotid sinus. J
Neurocytol 12: 117-153

McGregor KH, Gil J, Lahiri S (1984) A morphometric study of the
carotid body in chronically hypoxic rats. J Appl Physiol: Respirat
Environ Exercise Physiol 57: 1430-1438

Pequignot J-M, Hellström S (1983) Intact and sympathectomized carotid bodies of long-term hypoxic rats. Virch Arch (Pathol Anat) 400:
235-243

Seidl E, Acker H, Heinrich R (1977) Variability of the diameter of
the vessels of the cat carotid body in dependence on arterial pO2.
Proc Int Union Physiol Sci XIII, paper 2027

Smith P, Moosavi H, Winson M, Heath D (1974) The influence of age
and sex on the response of the right ventricle, pulmonary vasculature and carotid bodies to hypoxia in rats. J Pathol 112: 11-18

Paraganglionic Cell Response to Chronic Imipramine: A Structural Model

J. C. Folan

Department of Anatomy, University College, Dublin 2, Ireland

Introduction

Deriving from the monoamine hypothesis for affective disorders (Schildkraut 1965) a role for dopamine in depression and in the mechanism of action of tricyclic antidepressant drugs was proposed (for ref. see Willner 1983). The objective of the present study was to use an ultrastructural approach to examine the effects of chronic administration of imipramine, the prototypic tricyclic antidepressant on the paraganglionic cells of the rat superior cervical ganglion. The main transmitter in these cells is dopamine (Björklund et al. 1970; Heym et al. 1981a), and they have been studied previously following various pharmacological manipulations (see Heym 1978).

Material and Methods

Male Wistar rats (90–100 g) were divided into three groups. The drug treated group of three animals were injected with imipramine (Tofranil, Ciba–Geigy), 10 mg/kg, intraperitoneally for 21 days. Two animals served as saline treated controls and a further three were unhandled and uninjected for the 21 day period. The animals were perfused with a phosphate buffered glutaraldehyde–PVP solution. Excised superior cervical ganglia were postfixed in osmium tetroxide and embedded in epon. Thin sections, contrasted with lead citrate and uranyl acetate were examined with a Phillips EM 301. Using a Kontron image analysis system paraganglionic cell profiles from the control, saline and drug treated groups were assessed for vesicle counts, vesicle density, vesicle morphology, cell morphology and plasticity. To determine if the chronic effect of imipramine had a measurable effect on the drug treated cells an analysis of variance was used. For variables yielding a significant 'F' value ($P < 0.05$) a studentised T range test was applied. This involved looking at the difference between control, saline and drug treated group means. If this exceeded a critical difference value the group comparisons were deemed significant ($P < 0.05$).

Results

The paraganglionic cell morphology showed no alteration when the control, saline and drug treated tissue were compared. The vesicle counts and vesicle morphology alterations are presented in Table 1. The number of vesicles per paraganglionic cell profile show a significant reduction for the imipramine treated compared to the saline and control groups. This is not accompanied by an alteration in the

Table 1. Mean measurements and group comparisons. The mean vesicle counts are from paraganglionic cell profiles from the control (22 cells), saline (22 cells) and drug (27 cells) treated groups. The vesicle morphology measurements are from an average of 90 dense cored vesicles in each of the control, saline and drug treated groups. The group comparisons are derived from the ANOVA and the studentised T range method for the comparison of group means. C = control group. S = saline group. D = drug group. * = significant group comparison (P<0.05). NS = non significant group comparison

VESICLE COUNTS	VESICLE MORPHOLOGY			
	External rim		Internal core	
	Area (nm2)	d-max (nm)	Area (nm2)	d-max (nm)
CONTROL 107.2+/-43.2	7000+/-1780	97.9+/-11.9	3560+/-1370	70.4+/-14.1
SALINE 107.5+/-58.8	6309+/-1974	94.2+/-17.1	2940+/-1316	65.4+/-11.3
DRUG 62.6+/-31.9	6304+/-1870	90.8+/-14.0	2729+/- 998	60.3+/-11.3
GROUP COMPARISONS				
C vs S NS	NS	NS	NS	NS
C vs D *	NS	*	NS	*
S vs D *	NS	NS	NS	NS

vesicle density estimated from over 1000 vesicles per group as 24/um³ for control, 27/um³ for saline and 26/um³ for drug treated cells. The external rim and internal core diameter of the dense cored vesicles are significantly reduced for the drug treated group compared to the saline and the control, although there is no alteration in the area of the vesicles. The interrelationships of the paraganglionic cells to surrounding satellite cells, other paraganglionic cells, neural processes and the basal lamina shows a significant alteration with regard to the contact with neural processes. In the control tissue 14% of the cell surface contact neural processes. In the saline and the drug treated tissue the contact area is significantly increased to 27% and 25% respectively.

Discussion

The vesicle alterations in the paraganglionic cells of the rat may be due to the strong antimuscarinic action of imipramine (Snyder and Yamamura 1977; Hall and Ogren 1981) demonstrated in cat superior cervical ganglion (Weinstock and Cohen 1976). Another possibility is that imipramine inhibits the reuptake of dopamine resulting in a change in vesicle morphology. Ahonen and Pentilla (1971) demonstrated uptake inhibition of dopamine in Paneth cells of imipramine treated mice. The alterations in the interrelationships of the cells to their surrounding neural processes, evident in both the saline and drug treated tissue, may reflect stress alterations following a daily intraperitoneal injection for a 21 day period. Alterations in paraganglionic cells following stress have been reported (Heym et al. 1981a). This preliminary study using the paraganglionic cells as a model provides morphological evidence for a possible involvement of the tricyclic antidepressant imipramine with dopaminergic transmission.

References

Ahonen A, Pentilla A (1971) Effect of some drugs on the uptake of monoamines and their precursors in the mouse Paneth cells. Acta Physiol Scand 82: 59–69

Björklund A, Cegrell L, Falck B, Ritzen M, Rosengren E (1970) Dopamine containing cells in sympathetic ganglia. Acta Physiol Scand 78: 334–338

Hall H, Ogren SD (1981) Effects of antidepressant drugs on different receptors in the brain. Eur J Pharmacol 70: 393–407

Heym Ch (1978) Functional morphology of monoamine storing cells in the rat superior cervical ganglion. In: Coupland RE, Forssmann WG (eds) Peripheral neuroendocrine interaction. Springer, Berlin, pp 70–79

Heym Ch, Addicks K, Wedel Ch (1981a) Effect of intermittent immobilisation on the ultrastructure of the rat superior cervical ganglion. J Auton Nerv Syst 4: 283–296

Heym Ch, König R, Schröder H, Gerold N (1981b) Immunofluorescent and biochemical studies on dopamine–beta–hydroxylase and catecholamines in SIF cells of the superior cervical ganglion. Acta Histochem Suppl XXIV: 123–129

Schildkraut JJ (1965) The catecholamine hypothesis of affective disorders. Amer J Psychiat 122: 509–522

Snyder SH, Yamamura HI (1977) Antidepressants and the muscarinic acetylcholine receptor. Arch Gen Psychiat 34: 236–239

Weinstock M, Cohen D (1976) Tricyclic antidepressant drugs as antagonists of muscarinic receptors in sympathetic ganglia. Eur J Pharmacol 40: 321–328

Willner P (1983) Dopamine and depression: a review of recent evidence. III. The effects of antidepressant treatments. Brain Res Rev 6: 237–246

Expression and Regulation of Catecholaminergic Phenotypic Traits in Autonomic Primary Sensory Neurons

D.M.Katz, J.E.Adler, and I.B.Black

Division of Developmental Neurology, Cornell University Medical College, 515 East 71st Street, New York, NY 10021, USA

Introduction

In the peripheral nervous system, catecholaminergic (CA) neurotransmitter traits have classically been associated with cells of the sympathoadrenal axis. Accordingly, it has largely been assumed that CA phenotypic expression in the periphery is restricted to derivatives of the embryonic neural crest and that catecholamines function as sympathetic motor transmitters and adrenal hormones. However, recent studies challenge these assumptions and indicate a much wider role for catecholamines than previously recognized. Specifically, the discovery of catecholamine-containing <u>primary sensory neurons</u> indicates that the ability to express CA traits is shared by diverse subsets of peripheral neurons, and suggests that catechoamines function as sensory, as well as motor neurotransmitters (Gaudin-Chazal et al. 1983; Katz et al. 1983; Price and Mudge 1983; Kai-Kai and Keen 1985; Katz and Black 1986). In addition, studies of sensory CA regulation indicate that despite similarities among diverse CA populations, expression of the CA phenotype in the periphery is not necessarily associated with an obligatory repertoire of regulatory mechanisms (Katz et al. 1986).

Expression of CA Traits by Autonomic Primary Sensory Neurons

We have been studying expression and regulation of CA phenotypic traits in primary sensory neurons of the nodose (vagal) and petrosal (glossopharyngeal) cranial nerve ganglia (NPG) of the rat. Unlike sympathetic neurons, which are of neural crest origin, NPG cells are believed to be derived from ectodermal placodes (LeDouarin et al. 1986), and therefore differ both functionally and embryologically from classical peripheral CA neurons. Initial studies in our laboratory demonstrated that approximately 5-10% of NPG neurons in the unmanipulated adult rat express functional CA traits, including catalytically active tyrosine hydroxylase (TH), the rate-limiting enzyme in CA biosynthesis, formaldehyde-induced catecholamine fluorescence, and increased catecholamine levels in response to monoamine oxidase inhibition (Katz et al. 1983). These cells express morphologic traits typical of primary sensory neurons (Fig. 1), and are readily distinguished from small intensely fluorescent (SIF) cells by size, shape and TH staining intensity. Moreover, CA NPG neurons are insensitive to neonatal treatment with 6-hydroxydopamine, thereby distinguishing them from ectopic sympathetic neurons. The cells appear to be dopaminergic, as they do not exhibit immunoreactivity to dopamine-beta-hydroxylase, the norepinephrine-synthesizing enzyme.

Experimental Brain Research Series 16
© Springer-Verlag Berlin · Heidelberg 1987

270

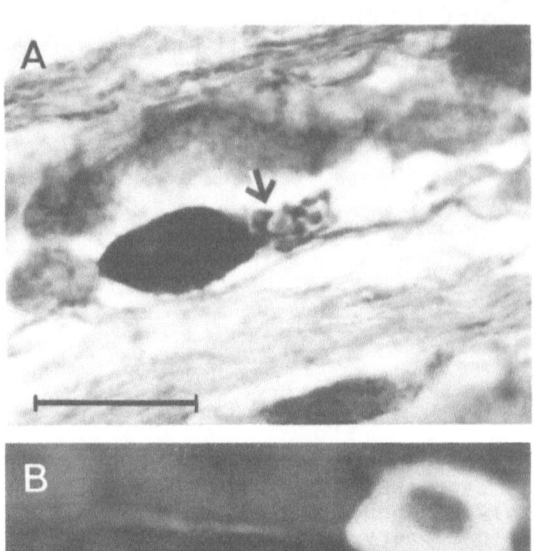

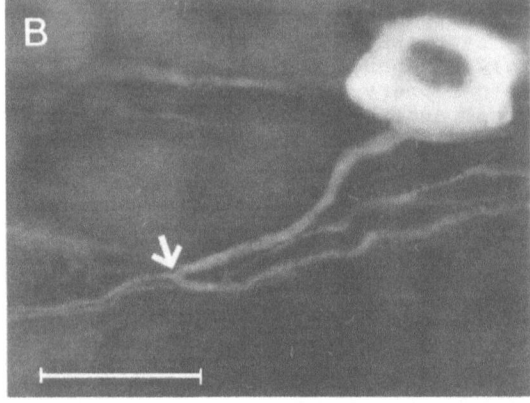

Fig. 1. (A) Bright-field photomicrograph of a TH-immunoreactive NG
cell, showing the initial axon glomerulus (arrow), a specific morphologic trait of primary sensory neurons. Peroxidase-anti-peroxi-
dase stained preparation; bar indicates 50 um. (B) Fluorescence
micrograph of a TH-immunoreactive NG cell, showing the bifurcating
neuritic process (arrow). Fluorescein isothiocyanate stained preparation; bar indicates 50 um

Peripheral Targets Innervated by CA NPG Neurons

NPG sensory and sympathetic neurons differ markedly in embryonic
origin and adult function, yet share the ability to express common
CA traits. This finding suggests that developmental lineage and
adult function alone are not determinative of CA phenotypic expression in peripheral neurons. We therefore sought to determine whether
other factors, such as the pattern of peripheral target innervation,
might correlate with sensory CA expression. In fact, retrograde
labeling studies, combined with TH immunocytochemistry, demonstrated
that 80-90% of CA sensory neurons in the petrosal ganglion project
selectively in the carotid sinus nerve (CSN), a branch of the glossopharyngeal nerve (Katz and Black 1986). Moreover, most of these
cells innervate a single target, the carotid body, revealing a
striking correlation between sensory CA expression and the pattern
of target innervation.

Regulation of Sensory CA Traits

The fact that CA expression occurs in diverse neuronal populations raises the question of whether or not regulatory mechanisms are also shared. Indeed, the apparent structural identity of TH in sensory and sympathetic neurons suggested that mechanisms of regulation might also be similar. Sensory and sympathetic TH, for example, exhibit immunoreactivity to the same highly specific antiserum, and display similar catalytic activities under identical assay conditions (Katz et al. 1983). In view of the well-documented role for target innervation in regulation of sympathetic CA traits (Black 1978), initial studies examined the effects of peripheral axotomy, and separation from peripheral targets on TH in the petrosal ganglion (Katz and Black 1986). The CSN was unilaterally transected, and sham-operated ganglia from age-matched animals served as controls. TH catalytic activity decreased by approximately 50% by 1 week post-operatively; total protein was unchanged, indicating that TH specific activity was reduced. However, activity returned to control levels within three weeks, coincident with regrowth of axons to the periphery. Parallel changes in TH immunocytochemical staining of NPG perikarya were observed. These data indicate that peripheral target innervation may play a role in regulating adult sensory CA traits, as in sympathetic neurons. Moreover, colchicine blockade of axonal transport in the CSN mimicked the effect of axotomy, suggesting that transport mechanisms may mediate interactions between CA sensory neurons and the periphery (Katz and Black 1986). In contrast, transection of ganglion cell projections to the central nervous system did not alter enzyme levels, suggesting that central projections do not normally play a major role in regulation of TH in the petrosal ganglion.

Molecular Mechanisms of Sensory CA Regulation

Nerve Growth Factor (NGF), synthesized in peripheral tissues and transported in a retrograde fashion to sympathetic perikarya, plays a critical role in target regulation of sympathetic CA traits (Thoenen and Barde 1980). To determine whether sensory CA traits are similarly regulated, we examined the effects of NGF on TH activity in explant cultures of adult NPG neurons. Cultured NPG neurons express catalytically active and immunoreactive TH, as in vivo (Katz et al. 1986).

NPG explants from adult rats were grown for up to three weeks on collagen-coated tissue culture dishes in serum-containing medium (Katz et al. 1986). Parallel cultures were grown in the presence of various concentrations of NGF or in the presence of an antiserum against the beta-subunit of NGF to remove endogenous NGF activity. TH activity was the same in all groups. In contrast, adult sympathetic ganglia, grown under identical conditions, exhibited a 5-10-fold rise in enzyme activity in response to a 10-fold increase in NGF. Thus, despite similarities in phenotypic expression, CA sensory and sympathetic traits are not necessarily regulated by common extracellular factors, such as NGF.

TH activity in ganglion neurons was, however, regulated by other extracellular stimuli. Specific TH activity increased 2-3-fold, for example, in the presence of depolarizing concentrations of potassium (Katz et al. 1986). Similar effects of depolarizing stimuli have been noted in classical CA populations as well (Zigmond and Ben-Ari 1977; Hefti et al. 1982; Dreyfus et al. 1986). Thus, some mechanisms of CA regulation may be ubiquitous, whereas others are not. These findings suggest, therefore, that CA phenotypic expression in the periphery is not necessarily associated with an obligatory repertoire of regulatory mechanisms.

Acknowledgements. This work was supported by a Grant-in-Aid from the American Heart Association with funds contributed in part by the New York Heart Association, the Dysautonomia Foundation, Inc. and NIH grant NS 10259. I.B.B. is the recipient of a McKnight Research Project Award.

References

Black IB (1978) Regulation of autonomic development. Ann Rev Neurosci 1: 183-214

Dreyfus CF, Friedman WF, Markey K, Black IB (1986) Depolarizing stimuli increase tyrosine hydroxylase in the mouse locus coeruleus in culture. Brain Res (in press)

Gaudin-Chazal G, Portalier P, Barrit MC, Puizillout JJ (1983) Serotonin-like immunoreactivity in paraffin-sections of the nodose ganglia of the cat. Neurosci Lett 33: 169-172

Hefti F, Gnahn H, Schwab ME, Thoenen H (1982) Induction of tyrosine hydroxylase by nerve growth factor and by elevated K concentrations in cultures of dissociated sympathetic neurons. J Neurosci 2: 1554-1566

Kai-Kai MA, Keen P (1985) Localization of 5-hydroxytryptamine to neurons and endoneurial mast cells in rat sensory ganglia. J Neurocytol 14: 63-78

Katz DM, Black IB (1986) Expression and regulation of tyrosine hydroxylase in primary sensory neurons: relationship to target innervation in vivo. J Neurosci 6: 983-989

Katz DM, Adler JE, Black IB (1986) Expression and regulation of tyrosine hydroxylase in adult sensory neurons in culture: effects of elevated potassium and nerve growth factor. Brain Res (in press)

Katz DM, Markey KA, Goldstein M, Black IB (1983) Expression of catecholaminergic characteristics by primary sensory neurons in the normal adult rat in vivo. Proc Natl Acad Sci USA 80: 3526-3530

LeDouarin NM, Fontaine-Perus J, Couly G (1986) Cephalic ectodermal placodes and neurogenesis. Trends in Neurosci 9: 175-180

Price J, Mudge AW (1983) A subpopulation of rat dorsal root ganglion neurones is catecholaminergic. Nature 301: 241-243

Thoenen H, Barde Y-A (1980) Physiology of nerve growth factor. Physiol Rev 60: 1284-1385

Zigmond RE, Ben-Ari Y (1977) Electrical stimulation of preganglionic nerve increases tyrosine hydroxylase activity in sympathetic ganglia. Proc Natl Acad Sci USA 74: 3078-3080

"Immunotech" and Computer-Assisted Image Analysis Used Together for Studying Peptide Transmitter Perturbations

T. H. Williams, J. Y. Jew, J. B. Crabtree, and M.-Q. Zhang

Department of Anatomy, University of Iowa, Iowa City, IA 52242, USA

Introduction

This report addresses two major technical advances that are completely revolutionizing studies using morphology for cell and neurobiology research – the development and applications of the microcomputer and the introduction and refining of immunohistochemical techniques – and in particular how a <u>combination</u> of these technologies make it possible to address many kinds of questions using a neuroplasticity experimental paradigm.

Need for Quantification. A longstanding presumption of anatomists has been that structure predominantly reflects function; but studies focusing on changes in the structure in consequence of experimental or other factors cannot be studied adequately by qualitative anatomical descriptions alone. However, anatomists have only recently accepted quantification as an essential need. Introduction of potent microcomputers and digitizing systems has made it possible to collect quantitative data more rapidly and accurately, and mathematical and statistical calculations can be performed automatically.

We have used two types of computer–based morphometric systems: digitizing tablet and image analyzer. The digitizing tablet surface is electrically referenced to a pen and is mapped out in Cartesian coordinates. The digitizer generates a signal representing the x,y coordinate position of the pen as it is moved. The advantage of image analyzers is their ability to reproduce a digital image of a specimen that can then be manipulated and analyzed quantitatively. The image analyzer was particularly suitable for looking at density and distribution of peptide localization in our light microscopic (LM) preparations. However, electron microscopic (EM) investigations of immunolabelled profiles require pattern recognition tasks that have not been performed very successfully by automatic image analyzers. With such a situation existing, the digitizing tablet was used to obtain quantitative EM data.

The Model. The central nucleus of the amygdala (CNA) offers a number of advantages as a model for studying neuroplastic responses in central peptidergic pathways: (1) numerous neuropeptides have been identified within the CNA; (2) the CNA has 2 major inputs which are accessible to surgical deafferentation; (3) the size and shape of the CNA make it possible to use transverse sections of the whole nucleus for quantitative LM and EM studies; (4) and last, but not least, marked changes were observable in immunostaining of several peptides in the CNA after lesioning its medial input.

Experimental Brain Research Series 16
© Springer-Verlag Berlin · Heidelberg 1987

Materials and Methods

Surgery. Male Sprague Dawley rats were used for these studies and received either no lesion or unilateral lesions. Unilateral lesions of the medial amygdala input were made with a glass microknife.

After survival periods of 3, 10, 30–45, 70–77 and 230–256 days, perfusion-fixed lesioned and nonlesioned animals' heads were mounted in a Kopf stereotaxic frame and tissue blocks cut out with a scalpel blade inserted into the frame's needle holder. Tissue blocks containing the CNA were post-fixed, then washed and placed in 0.1 M PO_4 buffered 10% sucrose at 4°C overnight.

Immunohistochemistry. We have directed extensive effort towards developing a standardized protocol for immunohistochemical staining (Jew et al. 1986, manuscript in review). After sectioning followed by PBS washes, immunostaining was carried out with a modification of the ABC immunoperoxidase technique (Hsu and Fanger 1981) using an antiserum to somatostatin (SS) (Immunonuclear) or angiotensin II/III (a gift from Professor Detlev Ganten). Note that the latter antiserum cross-reacts with both AII and AIII, but will be referred to subsequently simply as AII. For immunostaining controls, tissue sections were incubated in normal rabbit serum or preabsorbed antisera rather than in primary antiserum.

For EM studies, tissue sections were postfixed in cold 1% OsO_4 solution, dehydrated in graded ethanol solutions and propylene oxide, and embedded in epon between two sheets of Aclar plastic film. Areas of the tissue sections containing the CNA were cut out with a razor blade and mounted on plastic blocks for ultrathin sectioning. Ultrathin sections were mounted on Formvar coated grids and viewed with a Philips 201 electronmicroscope.

Quantitative Analysis. For LM preparations, the CNA was examined and photographed with a Nikon microscope. With darkfield illumination, an image is produced according to the light scattering properties of the opaque reaction product. Examination of darkfield images has revealed that the principal light scattering elements are immunoreactive terminals and preterminal swellings (Williams et al. 1981; Cassell et al. 1982). We have hypothesized that, under darkfield illumination, the degree of brightness of areas of immunocytochemically labeled sections of the CNA is a function of the density of labeled elements in these areas (Williams et al. 1981; Cassell et al. 1982). From darkfield photomicrographs, the EyeCom II PDP11/34 was used to (1) calculate the area of the lesioned and nonlesioned CNA; (2) generate grey level frequency histograms of the SS- and AII-immunostained CNA from lesioned and unlesioned sides of the same tissue section; and (3) calculate the mean grey level (i.e. the average brightness of the immunostained area) of the lesioned vs. nonlesioned CNA.

For EM, thin sections were mounted on 100-mesh copper grids, and synapses and axon profiles were counted directly from the electron microscope. A Zeiss MOP-3 digitizing tablet was used to measure the areas of individual axonal and axon terminal profiles.

Results

Somatostatin. In the nonlesioned CNA, bright and darkfield LM revealed that the densest distribution of SS-immunostained presumed terminals is within the central lateral subdivision, with some extension ventralwards. Three days post-lesion, LM and grey level frequency histograms indicated no significant change in SS-immuno-reactivity, mean grey level or the histogram shape, in lesioned vs. nonlesioned CNA. However, 10 days post-lesion, SS-immunoreactivity was considerably increased on the lesioned side. Further increases were elicited at 30 and 90 days, which persisted even in the 256 days post-lesion animals. The histograms generated by image analysis were shifted for the lesioned CNA, indicating higher densities of presumed SS-positive terminals on the deafferentated side.

Our ultrastructural quantitative study showed increased numbers of SS-immunostained terminals in the 10 to 256 day lesioned CNA. Further details of this and other data reported including synapse numbers and area statistics of axonal profiles (to be reported elsewhere) support the hypothesis that SS axons sprout in the lesio-ned CNA.

Angiotensin II. AII-like immunoreactivity in the nonlesioned CNA was heaviest in the central lateral division, followed by the central ventral division and then the central medial division (nomenclature after McDonald 1982 and Cassell et al. 1986). Both the measured area of the CNA and AII-immunostaining were greatest in sections taken from the more rostral and caudal ends of the CNA. Likewise, the area and staining of the central lateral and central medial divisions were greatest at the very rostral and very caudal parts of the CNA.

Grey-level frequency histograms (Fig. 1) of the CNA (all CNA subdi-visions) showed that grey level distribution was similar on the two sides in control animals. Thirty days after unilateral medial deaf-ferentation of the CNA, there was a shift of the histogram to the left on the lesioned side, indicating less immunoperoxidase reaction product. By 230 days post-lesion the histograms were again similar for lesioned and nonlesioned sides. From these preliminary results, it is inferred that the deafferentation resulted in a reduction, followed by a restoration in AII-like immunoreactivity. The time frame for both these changes was very slow.

EM surveys have been initiated to investigate changes in AII-posi-tive structures as well as other components of the microenvironment after various post-lesion survival times. EM localization of AII-like peptide in nonlesioned rat CNA is the first stage of plans for ultrastructural quantitative studies.

Conclusions

Our findings indicate that both AII and SS neural elements in the CNA appear to have reorganized after deafferentation of the medial input, but in different ways. The mechanism of the effect of this lesion on SS-positive elements is not completely elucidated, but the changes revealed by image analysis of LM preparations coupled with quantification of structures at the EM level fit the hypothesis that medial deafferentation generated somatostatinergic sprouting. We propose that the normal medial CNA input may act to inhibit the formation of additional SS terminals, giving these afferents a hypothesized role as a SS terminal regulator. In the case of AII,

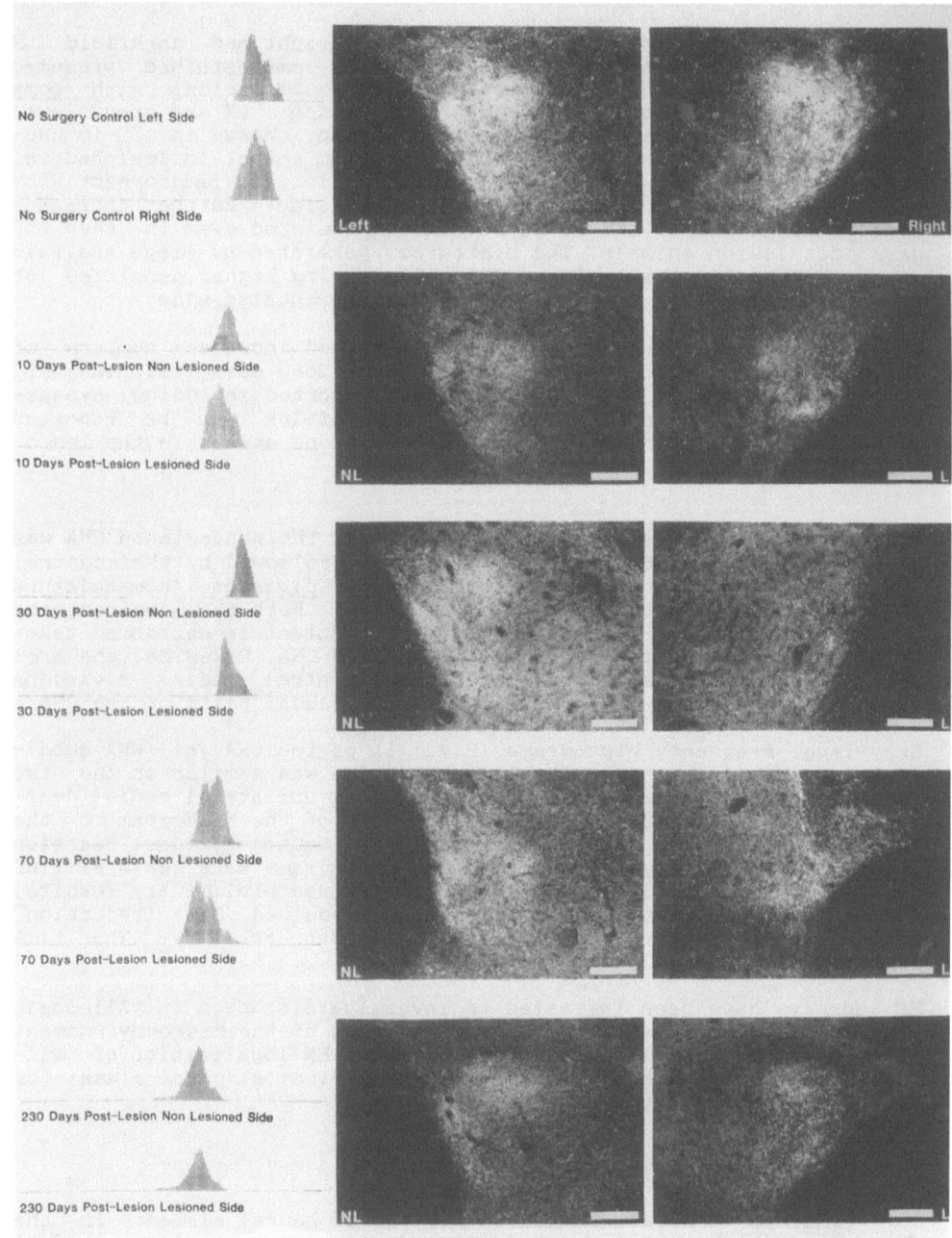

Fig. 1. Grey level frequency histograms shown on the left were generated from the accompanying darkfield images of AII-immunostained left and right CNA, in tissue sections of comparable rostrocaudal levels from various survival periods. Thirty days after a medial unilateral deafferentation of the CNA there is a shift to the left in the lesioned grey level histogram, reflecting reduced immunoreaction product on the lesioned side. Histograms of the 230 days post-lesion CNA are similar for nonlesioned (NL) and lesioned sides (L). Bar scale = 0.2 mm

preliminary studies show that medial deafferentation results in a gradual decrease followed by a gradual restoration of AII-positive elements. The mechanism(s) whereby these changes occur (e.g., why is AII loss so gradual?) need to be addressed.

Acknowledgements. This work was supported by NS 19578, BNS 8216786, and AM 34986. We thank Paul Reimann and Evelyn Jew for technical assistance.

References

Cassell MD, Gray TS, Kiss JZ (1986) Neuronal architecture in the rat central nucleus of the amygdala: a cytological, hodological, and immunocytochemical study. J Comp Neurol 246: 479–499

Cassell MD, Mankovich NJ, Gray TS, Williams TH (1982) Computer-assisted image analysis of the distributions of peptidergic terminals in the central nucleus of the amygdala: a preliminary study. Peptides 3: 283–290

Hsu SM, Raine R, Fanger H (1981) Use of avidin–biotin–peroxidase complex (ABC) in immunoperoxidase techniques: a comparison between ABC and unlabeled antibody (PAP) procedures. J Histochem Cytochem 26: 577–580

McDonald AJ (1982) Cytoarchitecture of the central amygdaloid nucleus of the rat. J Comp Neurol 208: 401–418

Williams TH, Cassell MD, Mankovich NJ, Hwang HK, Gray TS (1981) The distribution of neuropeptides in the central nucleus of the amygdala: a quantitative approach using image analysis. In: Ben-Ari Y (ed) The amygdaloid complex. INSERM Symposium No. 20. Elsevier, Amsterdam, pp 197–202

Morphometric Analysis of Alkaline Phosphatase Stained Capillaries in Rat Sympathetic Ganglia: Changes During Development and Ageing

R. M. Santer and U. Kabeer

Department of Anatomy, University College, P.O. Box 78, Cardiff CF1 1XL, Great Britain

Introduction

Morphometric analyses of the microvasculature of selected brain cortical regions have been made using either image analysis (Hunziker et al. 1974; Bar 1978) or alkaline phosphatase (AP) staining of endothelial cells (Bell and Ball 1981; Bell and Scarrow 1984) usually with specific reference to ageing or pathological conditions. Arterioles and capillaries stain intensely for AP whilst venules are virtually unstained. Thus the distinction between arterioles and capillaries must be made according to diameter. Morphometric analyses of the microvasculature of the carotid body (McDonald 1983) and of the sciatic nerve (Bell and Weddell 1984) are available but few studies of sympathetic ganglionic vasculature exist (Baker and Santer 1985). In the present report the AP method is utilised in the quantitative investigation of sympathetic ganglion microvasculature in rats throughout their lifespan.

Materials and Methods

Superior cervical (SCG) and coeliac-superior mesenteric ganglia (CSMG) were removed from all animals under ether anaesthesia and frozen in cryomountant onto cryostat chucks. 30 u frozen sections, cut longitudinally, were mounted on polylysine coated glass slides and incubated for 30 min at room temperature for the demonstration of AP in the following medium: sodium-B-glycerophosphate (Sigma), 1.2 g, $CaCl_2.2H_2O$, 1.2 g, $Mg_2SO_4.6H_2O$, 0.8 g, sodium barbitone, 2.0 g in 200 ml of distilled water. Following incubation, the slides were washed, placed in 2% cobalt nitrate for 5 min, fixed in 10% buffered formalin for 5 min and dipped in 0.1% ammonium sulphide for 30 s. After haematoxylin counterstaining the sections were conventionally processed. Control sections from all stages were incubated in a substrate-free medium and processed identically. Ganglia were removed from rats at various ages between E18 and 24 months.

Morphometrics. For all age groups n = 5, and for each animal five random fields from both ganglia were photographed, giving a sample of 25 fields per ganglion per age group. Both negatives and prints were calibrated in order to assess magnification. Total capillary length per field was measured in cm with a calibrated Combi 2 map measurer but finally expressed in mm. Capillary diameters intersecting a superimposed 1 cm grid were measured on each photograph to the nearest 0.5 mm at their point of intersection with the grid. Mean capillary diameters were finally expressed in um. The volume of tissue sampled per ganglion was calculated from the dimensions of the photographs, their magnification, number of photographs and the section thickness. Capillary density was expressed as "total measured length/sample volume mm/mm". It was possible to derive certain parameters from the mean capillary diameters and total capillary lengths which gave approximate values based on the concept of the capillary bed as a continuous tube of known diameter. Thus capillary volume (V) can be derived from the equation $\pi r^2 \times 1$ (mm³) and capillary surface area (S) from the equation $2\pi r \times 1$ (mm²). The ratio of the capillary surface area to the capillary volume (S/V) was also calculated as estimate of the surface area available for gas exchange for a given volume of blood. The results indicate only relative values indicating trends in vascular measurements between the two ganglion types. For this reason statistical significances have not been sought.

Results

AP-stained endothelial cells were first observed at 24 hours after birth in both ganglia (Fig. 1). These few lightly stained capillaries were located mainly beneath the capsule but some were seen deeper into the ganglia amongst the neuroblasts. By day 6, however, the AP staining of capillary endothelial cells was apparent throughout the ganglia and of an increased intensity which rendered the capillary profiles extremely sharp. It was therefore possible to commence quantitative analyses of the capillary bed at day 6. The appearance and AP stain intensity did not alter in any of the subsequent age stages up to 24 months (Fig. 2).

The results of the morphometric analyses are summarised in Tables 1 (SCG) and 2 (CSMG) and present a similar picture for both ganglia. After the first month there is an increase in average capillary diameter which is maintained into old age. However, there is a decrease in CSMG capillary diameter between 12 and 24 months. The density of the capillary bed rises rapidly to a peak at day 21, decreases to approximately half the peak value by day 33 and remains

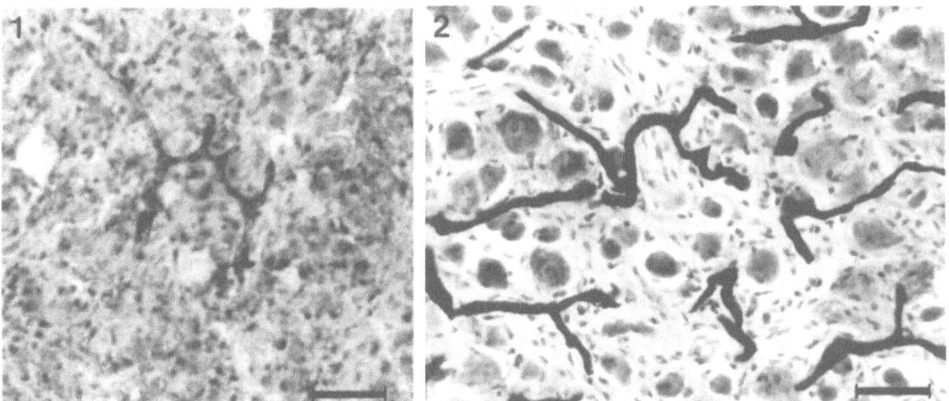

Fig. 1. Lightly stained capillaries in the coeliac—superior mesenteric ganglion 24 hours post-partum. Bar indicates 50 um

Fig. 2. Capillary bed of a 24 months coeliac—superior mesenteric ganglion showing intense alkaline phosphatase activity. Bar indicates 50 um

relatively constant thereafter. Of the derived parameters, S/V is of functional significance as it provides information on the exchange potential of the capillary bed. From its peak at day 33 S/V declines throughout life in both ganglia but shows a slight increase in the CSMG between 12 and 24 months.

Discussion

Variable changes in cortical capillary lumen area with age have been reported including an increase in the frontal cortex capillaries (Hicks et al. 1983) and this has been attributed to passive expansion of the capillaries due to an age-associated neuronal fall-out. The increase in capillary diameter with age observed in both ganglia in this study may also be a consequence of neuronal fall-out which has been reported to occur in both SCG and CSMG (Partanen et al. 1982). However, the increase in luminal diameter in the CSMG between 12 and 24 months which results in an increased exchange potential (S/V) may reflext the much greater neurone loss which occurs in this ganglion from 12 months of age (D.M. Baker, personal communication) which will increase diffusion distances.

In both ganglia the capillary density peaks towards the end of the first month post-partum and it is likely that this represents a stage at which maturation of sympathetic neurones is incomplete and subsequent growth in neurone size, dendritic proliferation and the development of interstitial elements (Davies 1978) will serve to decrease the capillary density. The high capillary exchange potential during the latter part of the first month is also probably a reflection of the high metabolic demands of the maturing ganglion cells.

References

Baker DM, Santer RM (1985) Quantitative changes in the microvasculature of the rat coeliac-superior mesenteric ganglion (CSMG). J Anat 140: 523-524

Bar T (1978) Morphometric evaluation of capillaries in different laminae of rat cerebral cortex by automatic image analysis: changes during development and aging. Adv Neurol 20: 1-9

Bell MA, Ball MJ (1981) Morphometric comparison of hippocampal microvasculature in ageing and demented people: diameters and densities. Acta Neuropathol (Berl) 53: 299-318

Bell MA, Scarrow WG (1984) Staining for microvascular alkaline phosphatase in thick celloidin sections of nervous tissue: morphometric and pathological applications. Microvasc Res 27: 189-203

Bell MA, Weddell AGM (1984) A morphometric study of intrafascicular vessels of mammalian sciatic nerve. Musc Nerv 7: 524-534

Davies DC (1978) Neuronal numbers in the superior cervical ganglion of the neonatal rat. J Anat 127: 43-51

Hicks P, Rosten C, Brizzee D, Samorajski T (1983) Age-related changes in rat brain capillaries. Neurobiol Aging 4: 69-75

Hunziker O, Frey H, Schulz U (1974) Morphometric investigations of capillaries in the brain cortex of the cat. Brain Res 65: 1-11

McDonald DM (1983) A morphometric analysis of blood vessels and perivascular nerves in the rat carotid body. J Neurocytol 12: 155–199

Partanen M, London ED, Rapoport SI (1982) Glucose utilization in sympathetic ganglia of male Fischer-344 rats at different ages. J Auton Nerv Syst 5: 391–398

Cytofluorimetric Scanning Studies on Axonal Transport in Reserpinized Adrenergic Nerves

P.-A. Larsson[1], S. Bööj[1], K. Lundmark[1], M. Goldstein, and A. Dahlström[1]

[1] Institute of Neurobiology, University of Göteborg, P.O. Box 33031, 40033 Göteborg, Sweden
[2] Department of Psychiatry, New York University, New York, NY, USA

Introduction

Earlier results obtained by biochemical assay of noradrenaline (NA) have indicated that the proximo–distal transport of NA–storage granules in rat sciatic nerve is supranormal between the 2nd and 4th day after reserpine (Dahlström and Häggendal 1969) (Fig. 1). The present cytofluorimetric scanning (CFS) study was undertaken to investigate if also other known components of axonal amine granules, viz. dopamine–beta–hydroxylase (DBH) and neuropeptide Y (NPY), were transported in increased amounts after reserpine. Surprisingly, the results obtained indicate that this may not be the case, but rather suggest that the biochemical composition of the amine granules in transit distally in the axons may be altered after reserpine.

Material and Methods

Male Sprague–Dawley rats (150–200 g) were given reserpine (10 mg/kg i.p.) or the vehicle. At various times (6 hours to 7 days) thereafter the sciatic nerves were crush-operated bilaterally. Twelve hours later the rats were anaesthetized with Nembutal and one nerve was dissected out, frozen and processed for NA-fluorescence histochemistry according to the Hillarp–Falck method (Corrodi and Jonsson 1966). The rats were then perfusion fixed with 4% paraform-aldehyde in PBS, and the remaining nerve was, after postfixation and rinsing, cryostat-sectioned and processed for indirect immunofluo-rescence using antisera to tyrosine hydroxylase (TH), DBH (Goldstein et al. 1972) or NPY (Amersham Internat. RPN 1702). Longitudinal sections were investigated using CFS which gives a graphical record of the amounts of immunoreactive material accumulated relative to the crush (Fig. 2) (Dahlström et al. 1982; Larsson et al. 1984). Values from 5–12 sections per nerve were estimated, and used to calculate mean $\pm$ S.E. for 4–7 nerves, expressed in % of control.

Results

As seen in Fig. 3A, NA-accumulations were initially very low (because of the NA-depletion caused by reserpine) but supranormal (160% of control) on day 4, as in the previous study (Fig. 1) (Dahlström and Häggendal 1969). DBH-like immunoreactivity (LI) (Fig. 3B), NPY-LI (Fig. 3C) and TH-LI (Fig. 3D) were decreased by approx. 50% during day 1 but reached normal levels of accumulation on day 2 after reserpine. On day 4 DBH-LI reached supranormal levels (140% of normal) (Fig. 3B) while NPY-LI and TH-LI were in the normal range

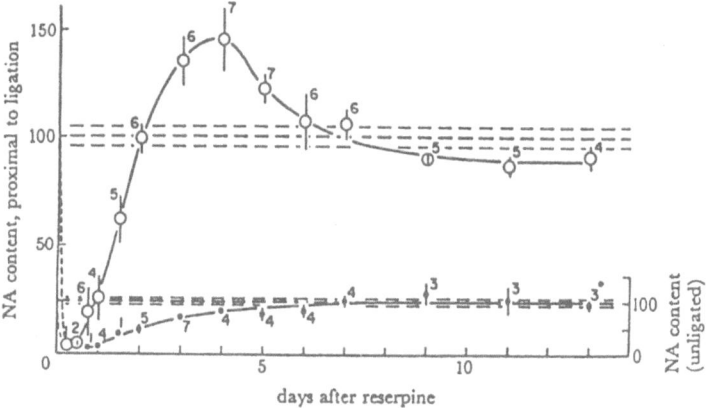

Fig. 1. The axonal transport of NA in rat sciatic nerve after reserpine. The amount accumulated NA proximal to a 6 h crush (o—o) is
expressed in % of the NA accumulated in control nerves. The lower
curve shows NA in intact nerves (●—●). (From Dahlström and Häggendal 1969). Small numbers represent number of animals. In each nerve
5-12 sections were scanned. Vertical bars show S.E. of the mean of
the individual animals. *** = p<0.001

(Fig. 3C,D). At 7 days NA, DBH-LI and TH-LI accumulated in normal
amounts while NPY-LI appeared to accumulate in subnormal amounts.

Discussion

During the recovery phase (7 days) after reserpine, anterogradely
transported NA initially decreased to virtually zero, became norma-
lized at day 2, and was supranormal at day 4 after reserpine. Thus,
the results registered in this study were identical to results
obtained earlier with biochemical methods (Dahlström and Häggendal
1969).

This, together with previous results showing a good parallelism
between biochemical results and CFS (Dahlström et al. 1982), shows
that the CFS technique can indeed be used for studies on axonal
transport of enzyme molecules, transmitters and neuropeptides. Also,
the model experiments using various concentrations of synthetic NPY
in gelatine slabs, sectioned and treated for indirect immunofluorescence, showed that cytofluorimetric readings are proportionally
related to the concentration of NPY (Larsson 1985).

The decrease at day 1 in amounts of transported TH-LI, DBH-LI and
NPY-LI was about 50% of normal. This initial decrease in the proxi-
mo-distally transported amounts of immunoreactive material is probably caused by the lowered body temperature (Δ 1-3°C) initially
after reserpine (Larsson et al. 1986), which may influence the
protein synthesis as well as the rate of axonal transport.

On day 4 supranormal levels of both NA and DBH-LI were registered,
and in 3 out of 6 rats also TH-LI was supranormal (133, 151 and 141%
of control), although the average of 6 rats was 118 $\pm$ 11.8% (see
Larsson et al. 1986). Thus, the adrenergic components (NA, DBH-LI
and in some rats also TH-LI) were increased, but not the peptide. It

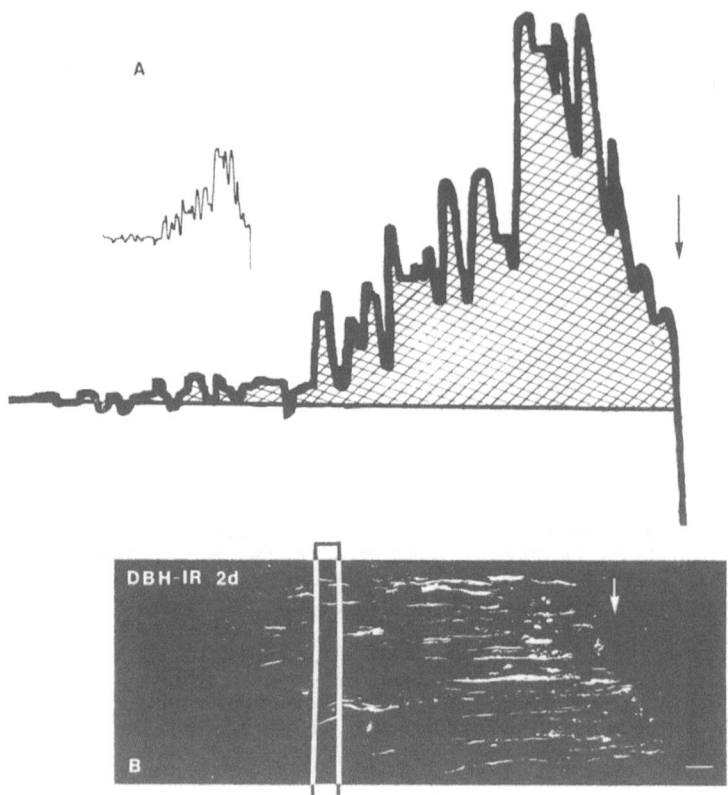

Fig. 2. Schematic illustration of the CFS—method. A nerve section is slowly moved under a measuring slit (white rectangle in B); the fluorescence intensity passing through the slit is registered by a photomultiplier, and the signal fed into a graphical recorder. The resulting graph (top) gives a quantitative estimate, expressed in arbitrary units, of the immunofluorescence (proportional to the amount of immunoreactive material) along the section. This section of a sciatic nerve is from a reserpine treated (2 days) rat, and was incubated with antiserum against DBH. (For further explanation see Larsson 1985)

is known from several studies on the adrenal gland (Black et al. 1971), sympathetic ganglia (Thoenen et al. 1969; Molinoff et al. 1970) and brain (Levin 1981) that reserpine causes an induction of the adrenergic enzymes DBH and TH, probably via an increased presynaptic activity. Cholinoceptor agonists induce a similar increase in TH in ganglion cells (cf. Zigmond 1985).

The present results indicate that although NPY appears to be a component of axonal amine granules, as well as DBH (Fried et al. 1985), NPY is not transsynaptically regulated in parallel with DBH. These observations are in contrast to those of Lundberg et al. (1985) who found a doubling of the amounts of axonally transported NPY-LI (measured by RIA) in guinea pig sciatic nerve on day 4 after reserpine. We suggest that species differences in the reaction to

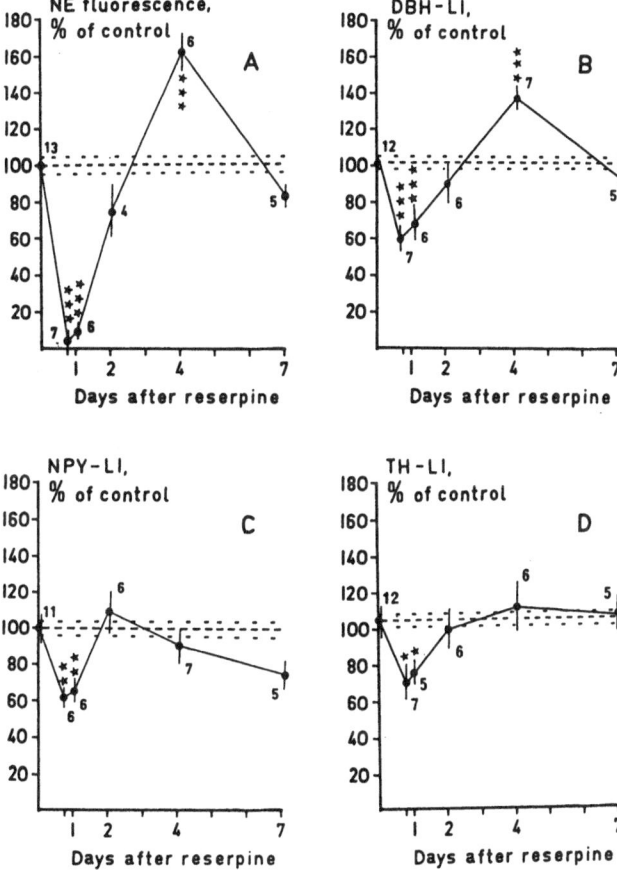

Fig. 3. A. Accumulations of NA in rat sciatic nerves 12 h after crushing, performed at various times after reserpine (10 mg/kg). The formalin-induced NA-fluorescence was registered with the CFS method (see text and Fig. 2). B-D. Accumulations of DBH-LI (B), NPY-LI (C) and TH-LI (D) in rat sciatic nerve proximal to a 12 h crush 18 h to 7 days after reserpine treatment. CFS values are given in arbitrary units in % of levels in 12 h crushed control nerves. Mean ± SEM (vertical bars) are given. Small numerals represent number of animals. In each nerve 3-16 sections were measured. (* = p<.05; ** = p<.01; *** = p<.001)

reserpine may explain this difference, and this will be investigated further.

However, our results are in analogy with the observation that after, e.g. reserpine, mRNA coding for TH is increased, but not mRNA coding for preproenkephalin, the precursor molecule for enkephalin which coexists with catecholamines in adrenal medullary granules (La Gamma et al. 1985). Consequently, amine granules in transit distally in adrenergic axons on day 4 after reserpine probably have an altered biochemical composition, compared to the normal state (Larsson et al. 1986), at least in lumbar sympathetic adrenergic nerves of the rat. Alternatively, NPY and DBH may exist in separate populations of granules, with similar size and density. The relative proportion of these 2 populations may be altered after reserpine. Double immunolabelling experiments at the EM level may solve this question.

Acknowledgements. Supported by grants from the Swedish MRC (2207), the Medical Faculty, University of Göteborg, and the Medical Society, Göteborg. The technical assistance of Anne Johnsson and Margareta Ogenholm is gratefully acknowledged.

References

Black IB, Hendry IA, Iversen LL (1971) Differences in the regulations of tyrosine hydroxylase and DOPA decarboxylase in sympathetic ganglion and adrenal. Nature (New Biology) 231: 27–29

Corrodi H, Jonsson G (1967) The formaldehyde fluorescence method for the histochemical demonstration of biogenic monoamines. A review on the methodology. J Histochem Cytochem 15: 65–78

Dahlström A, Häggendal J (1969) Recovery of noradrenaline in adrenergic axons of rat sciatic nerves after reserpine treatment. J Pharm Pharmacol 21: 633–638

Dahlström A, Bööj S, Goldstein M, Larsson P-A (1982) Cytofluorimetric scanning: a tool for studying axonal transport in monoaminergic neurons. Brain Res Bull 9: 61–68

Fried G, Lundberg JM, Theodorsson-Norheim E (1985) Subcellular storage and axonal transport of neuropeptide Y (NPY) in relation to catecholamines in the cat. Acta Physiol Scand 125: 145–154

Goldstein M, Fuxe K, Hökfelt T (1972) Characterization and tissue localization of catecholamine synthesizing enzymes. Pharmacol Rev 24: 293–309

LaGamma EF, White JD, Adler JE, Krausea JE, McKelvy JF, Black IB (1985) Depolarization regulates adrenal preproenkephaline mRNA. Proc Natl Acad Sci USA 82: 8252–8255

Larsson P-A (1985) Axonal transport of amine storage granules in sympathetic adrenergic neurons. Thesis, The Medical Faculty, Göteborg. ISBN 91-7222-930-6

Larsson P-A, Bööj S, Lundmark K, Goldstein M, Dahlström A (1986) Reserpine-induced effects in the adrenergic neuron as studied with cytofluorimetric scanning. Brain Res Bull 16: 67–74

Larsson P-A, Goldstein M, Dahlström A (1984) A new methodological approach for studying axonal transport: cytofluorimetric scanning of nerves. J Histochem Cytochem 32: 7–16

Levin BE (1981) Reserpine effect on axonal transport of dopamine-beta-hydroxylase and tyrosine hydroxylase in rat brain. Exp Neurol 72: 99–112

Lundberg JM, Saria A, Franco-Cereceda A, Hökfelt T, Terenius L, Goldstein M (1985) Differential effects of reserpine and 6-hydroxydopamine on neuropeptide Y (NPY) and noradrenaline in peripheral neurons. Naunyn-Schmiedebergs Arch Pharmacol 328: 331–340

Molinoff P, Brimijoin S, Weinshilboum R, Axelrod J (1970) Neurally
mediated increase in dopamine-beta-hydroxylase activity. Proc Natl
Acad Sci USA 66: 453-458

Thoenen H, Mueller RA, Axelrod J (1969) Increased tyrosine hydroxylase activity after drug-induced alteration of sympathetic transmission. Nature 221: 1264

Zigmond RE (1985) Biochemical consequences of synaptic stimulations:
the regulation of tyrosine hydroxylase activity by multiple transmitters. Trends in Neurosci 8: 63-69

A Morphometric and Biochemical Study on the Effect of Unilateral Adrenalectomy on the Adrenal Medulla in Neonatal and Adult Rats

A. Tomlinson, J. Durbin, and R. E. Coupland

The Medical School, Queen's Medical Centre, University of Nottingham, Nottingham, Great Britain

Introduction

From the observations of Wislocki and Crowe (1924) and Coupland (1965), it is apparent that bilateral adrenal medullectomy does not result in hypertrophy of extra-adrenal chromaffin tissue and it has been generally accepted that in the adult, unilateral adrenalectomy is not followed by compensatory growth of the intact medulla. These findings are in keeping with the apparent lack of specific adrenal medullar trophic hormone and the origin of chromaffin cells from neural crest as well as their structural and functional associations with post-ganglionic sympathetic neurons.

Unlike nerve cells, however, mature chromaffin cells replicate throughout life as evidenced by mitotic activity and uptake of tritiated thymidine (Monkhouse 1985). In neonatal animals the labelling index is increased by the administration of glucocorticoid (Monkhouse and Coupland 1985).

Recently (Verhofstad et al. 1985), functional dimensions have been added to earlier morphological studies on the developing adrenal medulla (e.g., Coupland and Weakley 1970a,b; El-Haghraby and Lever 1980; Millar and Unsicker 1981). Chromaffin cells mature only slowly during foetal life. During the first week after birth the definitive cell forms, i.e. adrenaline (A) or noradrenaline (NA) storing cells can be identified and these possess appropriate synthetic enzymes. In the rat during the later part of the first week of post-natal life evidence of functional innervation of chromaffin tissue is first obtained (see Slotkin et al. 1980) and this approaches the adult state by post-natal day 14 (Gripois et al. 1986).

In the present work, the response of adrenal chromaffin cells of neonatal and adult Wistar rats to unilateral adrenalectomy has been observed by making complementary morphometric and biochemical observations.

Materials and Methods

Wistar strain rats were maintained under standard conditions of temperature and lighting (12 h light, 12 h darkness), 41B modified rat pellets (Heygates Ltd., Northampton) and water were available ad libitum.

Experimental Brain Research Series 16

Neonatal Animals

Three day old rats were anaesthetised by hypothermia and the left adrenal gland removed through lumbar approach: the skin and body wall were sutured with 8 x 0 non-absorbent silk. The animals were returned to the mother until aged 26 days when they were anaesthetised by intraperitoneal sodium pentobarbitone (60 mg/kg i.p. Sagatal, May and Baker Ltd., Dagenham).

From the experimental animals, the remaining (right) adrenal gland was removed and weighed and processed for catecholamine or choline acetyltransferase assay, or morphometry. In control animals both glands were similarly processed.

Adults

Using animals weighing 250 ± 10 g left adrenalectomy was performed. Both operated and non-operated control animals were used. After periods of 5 or 10 weeks, the animals were anaesthetised with Sagatal, as above, and the right adrenal gland (experimental) or both (controls) were processed.

Morphometry

Electron Microscopy. Using the method described previously (Tomlinson et al. 1986) adrenal fixation was achieved with 2% glutaraldehyde in 0.11 M cacodylate buffer (pH 7.2, 550 m Osmols) for 15 minutes. Parallel slices of glands (cut on random axes) were fixed for a further 3 hours at room temperature, washed in buffer, postfixed in 1% Millonig's osmium tetroxide before embedding in Araldite. Photographs of thin seections of randomly selected slices were analysed by a point counting method.

Light Microscopy. A prewash perfusion was followed by 2% paraformaldehyde with 1% glutaraldehyde in 0.1 M cacodylate buffer. Buffer-washed adrenal glands were immersed for 8 hours in 2% aqueous potassium dichromate, washed and embedded in JB4 plastic. Every fifth 5 um section was mounted and stained with Mayers haematoxylin. Medullary areas were determined using a Kontron MOP AMO3 image analyser (Zeiss, Oberkochen, West Germany) and a microscope with an attached drawing tube.

Catecholamine Determination. Assay was performed using the high pressure liquid chromatography with electrochemical detection procedure described previously (Tomlinson et al. 1986).

Choline Acetyltransferase Activity (ChAT). This was measured by a radiochemical method based upon conversion of ^{3}H-acetyl CoA to ^{3}H-acetylcholine and purification of the latter by ion pair partition. Duplicate assays were performed on each sample. For details of the method see Fonnum (1975) and Tomlinson et al. (1984).

Results

Morphometry

As expected, unilateral adrenalectomy results in compensatory growth of the remaining gland as evidenced by increase in adrenal weight and volume. In animals unilaterally adrenalectomised on day 3, the results were control glands (n = 16) 7.3 ± 0.2 (SEM) mg and 6.24 ± 0.16 mm³: experimental group (n = 15) 8.2 ± 0.2 mg and 7.03 ± 0.19 mm³. The t-test significance is p = 0.006.

Medullary volumes were compared. In the neonatal group controls of the total medullary area was 6.83 ± 0.35 (SEM) mm²: in experimental animals it was 7.8 ± 0.60 mm². This was not significant (t-test p = 0.17).

In adult animals after 5 weeks control and experimental adrenal weights were 16.5 ± 1.8 (20) and 22.8 ± 3.1 (14) respectively; after 10 weeks they were 18.8 ± 1.3 (19) and 24.7 ± 2.8 (13). Hence both showed a significant increase. Medullary volumes showed no significant change.

Medullary Components. The volume densities of medullary components showed no significant difference in control and experimental groups (Table 1). Between 26 days of age and adult life the ratio of NA- to A-storing cells falls from 28-30 percent to 17-19 percent. Chromaffin cells form some 55 percent of medullary volume. Nervous tissue in the adult adrenal medulla is proportionally c 50 percent less than in the 26 day old animal.

Volume Densities of Cytoplasmic Components of A and NA Cells. No significant differences were observed between the experimental and control groups (Table 2).

Chromaffin Cell and Nuclear Volume. No changes in cell volumes or nuclear diameters (Table 3) occurred in the 26 day old neonates or in adults after 5 weeks. However, 10 weeks after unilateral adrenalectomy there was an increase in both nuclear diameter and cell volume in both A and NA cells that was associated with a decrease in the proportion of interstitial tissue (Table 1).

Biochemical Assays

Choline Acetyltransferase Activity. There was no difference between control (3.51 ± 0.21 nmol Ach/h/gland) and experimental (3.30 ± 0.23 nmol Ach/h/gland) groups in adult animals. In young animals there was a slight increase in ChAT activity (controls 3.59 ± 0.35, experimental 4.69 ± 0.38 nmol Ach/h/gland) significance p = 0.06.

Catecholamine Content of Adrenal Glands. NA and A contents (Table 4) of adrenal glands of 26 day old experimental animals decreased (p = 0.11 and 0.076 respectively). Dopamine content was unchanged.

In adult animals there was no significant difference recorded between control and experimental groups.

Table 1. Volume densities of medullary components of controls and unilaterally adrenalectomised animals. Results expressed as means ± SEM. Number of animals in parentheses. A = adrenaline-storing cell; NA = noradrenaline-storing cell

	A cells	NA cells	Vasculature	Neuronal tissue	Interstitial tissue
Controls					
Neonates 26 days (8)	0.397 ± 0.037	0.169 ± 0.031	0.214 ± 0.022	0.106 ± 0.005	0.139 ± 0.013
Adults 5 weeks (9)	0.406 ± 0.029	0.096 ± 0.011	0.276 ± 0.024	0.050 ± 0.005	0.173 ± 0.033
Adults 10 weeks (9)	0.480 ± 0.036	0.078 ± 0.002	0.235 ± 0.085	0.078 ± 0.028	0.128 ± 0.022
Experimental					
Neonates 26 days (9)	0.394 ± 0.037	0.150 ± 0.025	0.211 ± 0.024	0.113 ± 0.012	0.130 ± 0.016
Adults 5 weeks (5)	0.512 ± 0.047	0.110 ± 0.008	0.211 ± 0.037	0.049 ± 0.010	0.120 ± 0.016
Adults 10 weeks (5)	0.514 ± 0.032	0.107 ± 0.026	0.228 ± 0.028	0.055 ± 0.005	0.100 ± 0.010

Table 2. Volume densities of cytoplasmic components. Results expressed as means ± SEM. Number of animals in parentheses.
A = adrenaline-storing cell; NA = noradrenaline-storing cell

| | Chromaffin granules | | | Mitochondria | | Cytosolic | |
| | Whole | | Cores | | | | |
	A	NA		A	NA	A	NA
Controls							
Neonates 26 days (8)	0.240 ± 0.011	0.301 ± 0.011		0.062 ± 0.005	0.072 ± 0.003	0.698 ± 0.009	0.627 ± 0.010
Adults 5 weeks (9)	0.308 ± 0.013	0.358 ± 0.011	0.118 ± 0.009	0.060 ± 0.006	0.062 ± 0.005	0.632 ± 0.012	0.579 ± 0.010
Adults 10 weeks (9)	0.307 ± 0.011	0.365 ± 0.013	0.126 ± 0.007	0.055 ± 0.003	0.057 ± 0.002	0.642 ± 0.011	0.577 ± 0.012
Experimental							
Neonates 26 days (8)	0.259 ± 0.016	0.309 ± 0.014		0.067 ± 0.010	0.070 ± 0.007	0.673 ± 0.015	0.621 ± 0.011
Adults 5 weeks (5)	0.302 ± 0.014	0.378 ± 0.013	0.126 ± 0.015	0.068 ± 0.007	0.066 ± 0.006	0.630 ± 0.013	0.556 ± 0.011
Adults 10 weeks (5)	0.299 ± 0.020	0.305 ± 0.014	0.110 ± 0.007	0.051 ± 0.009	0.066 ± 0.004	0.650 ± 0.013	0.630 ± 0.016

Table 3. Mean nuclear diameters and mean cell volumes. Mean nuclear diameters are corrected according to Abercrombie. A = adrenaline-storing cell; NA = noradrenaline-storing cell

	NA		A	
	Nucleus (μm)	Cell (μm³)	Nucleus (μm)	Cell (μm³)
Control				
Neonates 26 days	6.3	650	6.9	900
Adults 5 weeks	6.6	950	7.2	1450
Adults 10 weeks	6.8	1050	7.3	1600
Experimental				
Neonates 26 days	6.6	700	6.9	950
Adults 5 weeks	6.5	1100	7.4	1500
Adults 10 weeks	7.0	1350	8.1	2150

Table 4. Catecholamine content of adrenal glands of control and experimental animals. Results expressed as means ± SEM. Number of animals in parentheses

	Noradrenaline nMoles/adrenal	Adrenaline nMoles/adrenal	Dopamine nMoles/adrenal
Controls			
Neonates 26 days (11)	6.43 ± 0.54	17.81 ± 1.00	0.33 ± 0.03
Adults 5 weeks (6)	21.11 ± 1.26	112.55 ± 4.70	0.51 ± 0.05
Adults 10 weeks (6)	26.08 ± 1.76	139.00 ± 3.27	0.91 ± 0.06
Experimental			
Neonates 26 days (8)	3.84 ± 0.53	13.98 ± 2.15	0.34 ± 0.02
Adults 5 weeks (8)	22.16 ± 1.51	119.31 ± 4.14	0.93 ± 0.16
Adults 10 weeks (8)	22.70 ± 1.16	133.90 ± 4.11	0.89 ± 0.11

The proportion of catecholamine formed by NA was $\underline{c}$ 25 percent in 26 day controls, 22.5 percent in 250 g rats and $\underline{c}$ 15 percent in the 290-350 g group.

Discussion

Recently, Monkhouse and Tomlinson (1986) reported significant increases in the labelling index and in adrenal medullary volume in animals adrenalectomised on day 3 and killed on day 8 post-natal. In the present work the survival period of the neonatal group was increased to 26 days. By 26 days no evidence of medullary hypertrophy or hyperplasia was apparent and the proportions of medullary components and of chromaffin granules, mitochondria and cytosol in chromaffin cells were unchanged. Hence, the effect noted by Monkhouse and Tomlinson appears to be transitory and eliminated or masked by subsequent normal growth changes. In the light of previous evidence that 11-oxycorticosteroids increase the labelling index of chromaffin cells in young mice during the first week after birth but not in more mature animals (Monkhouse and Coupland 1985), it is possible that the transient effect was the result of a local increase in concentration of adrenocortical steroids consequent upon the increased secretion of ACTH and the compensatory growth of the adrenal cortex of the remaining gland - as evidenced by adrenal weights and adrenal medullary volumes.

A slight increase in choline acetyltransferase activity was observed in 26 day old animals that may reflect increased cholinergic nervous activity. However, this was not accompanied by any change in the neural tissues content of the medulla. In this group, concentrations of A and NA in the adrenal glands showed slight decreases, possibly reflecting increased secretory activity in the single gland.

The increase in cell volume and nuclear diameter of chromaffin cells in adult animals 10 weeks after unilateral adrenalectomy was not accompanied by any change in catecholamine content of the glands or in volume densities of cytoplasmic components and hence it is possible that this represents an osmotic effect during tissue fixation and processing, possibly consequent upon changes in functional activity. This explanation would be in keeping with the apparent decrease in volume density of interstitial tissue in this group.

In conclusion, the present findings indicate that persistent compensatory growth of chromaffin tissue does not follow unilateral adrenalectomy in either neonatal or adult animals.

References

Coupland RE (1965) The natural history of the chromaffin cell. Longmans, London

Coupland RE, Weakley BS (1970a) Developing chromaffin tissue in the rabbit: an electron microscopic study. J Anat 102: 425-455

Coupland RE, Weakley BS (1970b) Electron microscopic observations on the adrenal medulla and extra-adrenal chromaffin tissue of the postnatal rabbit. J Anat 106: 213-231

El-Maghraby M, Lever JD (1980) Typification and differentiation of medullary cells in the developing rat adrenal. A histochemical and electron microscopic study. J Anat 131: 103-120

Fonnum F (1975) A rapid radiochemical method for the determination
of choline acetyl transferase. J Neurochem 24: 407-409

Gripois D, Valens M, Diarra A (1986) Adrenal medullary responses to
insulin-induced hypoglycaemia in the young rat. J Auton Nerv Syst
15: 165-178

Millar TJ, Unsicker K (1981) Catecholamine-storing cells in the
adrenal medulla of the pre- and postnatal rat. Cell Tissue Res 217:
155-170

Monkhouse WS (1986) The effect of in vivo hydrocortisone administration on the labelling index and size of chromaffin tissue in the
postnatal and adult mouse. J Anat 144: 133-144

Monkhouse WS, Coupland RE (1985) The effect of in vitro hydrocortisone administration on the labelling index and size of the intra-
and extra-adrenal chromaffin tissue of the fetal and perinatal
mouse. J Anat 140: 679-696

Monkhouse WS, Tomlinson A (1986) The effect of unilateral adrenalectomy on the labelling index and size of intra-adrenal chromaffin
tissue in newborn rats. J Anat (in press)

Slotkin TA, Smith PG, Lau C, Bareis DL (1980) Functional aspects of
development of catecholamine biosynthesis and release in the sympathetic nervous system. In: Parvez H, Parvez S (eds) Biogenic amines
in development. Elsevier, Amsterdam, pp 29-48

Tomlinson A, Durbin J, Coupland RE (1986) A quantitative analysis of
rat adrenal chromaffin tissue: morphometric analysis at tissue and
cellular level correlated with catecholamine content. Neuroscience
(in press)

Tomlinson DR, Moriarty RJ, Mayer JH (1984) Prevention and reversal
of defective axonal transport and motor nerve conduction velocity in
rats with experimental diabetes by treatment with the aldose reduct-
ase inhibitor Sorbinil. Diabetes 33: 470-476

Verhofstad AAJ, Coupland RE, Parker TL, Goldstein M (1985) Immunohistochemical and biochemical study on the development of the nor-
adrenaline- and adrenaline-storing cells of the adrenal medulla of
the rat. Cell Tissue Res 242: 233-243

Wislocki GB, Crowe SJ (1924) Experimental observations on the adrenals and the chromaffin system. Johns Hopk Hosp Bull 35: 187-193

Experiments on the Connectivity of Axotomized Rat Sympathetic Neurones

M. R. Matthews and G. J. Clowry

Department of Human Anatomy, South Parks Road, Oxford OX1 3QX, Great Britain

Introduction

In sympathetic neurones, as in somatic motoneurons, axotomy causes a
reduced responsiveness to synaptic activation (Brown and Pascoe
1954; Matthews and Nelson 1975; Purves 1975) accompanied by detachment of their various synaptic inputs and loss of post-synaptic
membrane densities (Matthews and Nelson 1975; Purves 1975; Case and
Matthews 1986). These observations suggest a change in neuronal
receptor characteristics which might during recovery lead to an
altered tolerance for 'foreign' or heterotypic connexions. To test
this hypothesis in the rat superior cervical sympathetic ganglion
(SCG) Matthews and Ramsay (1980) sutured a somatic motor nerve, the
hypoglossal nerve, to the ganglionic stump of the divided cervical
sympathetic trunk, the central end of which was prevented from
regenerating. Significantly more synapses per sympathetic neurone
were formed by the hypoglossal nerve in ganglia after postganglionic
nerve crush than in ganglia with intact postganglionic nerves. The
incidence of intrinsic aminergic synapses (Ramsay and Matthews 1985)
also increased dramatically in these ganglia, significantly more so
in axotomized ganglia.

Up to 50% of sympathetic neurones may however die after axotomy
(Acheson and Remolina 1955; Purves 1975; Smolen 1983), so it might
be argued that the hypoglossal nerve could more readily innervate a
reduced population of neurones in axotomized ganglia. We have therefore tested the effect of amputating approximately one third of the
SCG on the outcome of its innervation by the hypoglossal nerve. We
have also tested the effect of ganglioside treatment, a procedure
reported to cause expansion of somatic motor neurone territories
during regeneration (Gorio et al. 1983). The results of these experiments suggest that in the longer term (9-12 months) the extent of
hypoglossal innervation which persists in the ganglion may not be
inconsistent with the size of the surviving neurone population, but
that in the shorter term (up to 6 months) an excess of synapses is
formed in axotomized ganglia which may disappear as the injured
neurones re-establish their peripheral connexions.

Material and Methods

Young male albino rats (120-180 g), in littermate groups, were used.
Under intraperitoneal chloral hydrate anaesthesia the left SCG was
preganglionically denervated by cutting the cervical sympathetic
trunk (CST) and suturing its central end into the sternomastoid
muscle to prevent regeneration. The left hypoglossal nerve was cut
near its entry to the tongue and its central end was anastomosed

Experimental Brain Research Series 16

either a) to the ganglionic stump of the left CST (HD animals), or b) directly to the partially amputated left SCG (HDX animals), by a single transfixing atraumatic suture of 10/0 nylon. Partial amputation of the SCG was achieved by excising approximately its caudal one third, containing 15-30% of the ganglionic neurones (as determined by cell counting in excised pieces). Care was taken to preserve the major blood supply which the ganglion receives close to the carotid bifurcation (McDonald and Larue 1983; personal observations), and this limited the amount of ganglion tissue which could be removed. In littermates of the above series, ganglionic denervation and hypoglossal anastomosis were performed as above but at the same time the left internal and external carotid nerves were crushed with smooth-tipped watchmakers' forceps, to axotomize the majority of ganglionic neurones (HDA and HDXA animals, respectively). All the above animals were allowed to survive for between 9 and 16 months.

In a second set of experiments, the operative procedures were as in the HD and HDA animals but post-operatively half the rats in each group were treated for 30 d with daily intraperitoneal injections of 5-10 mg per kg body weight of a ganglioside preparation, in normal saline (Gorio et al. 1983), and the other half with isovolumetric injections of saline, beginning immediately following operation (HD-G, HD-S; HDA-G, HDA-S animals). These animals were allowed to survive from 6 weeks to 6 months. Gangliosides were prepared by ion exchange chromatrographic separation (Leeden et al. 1973) of glycolipids extracted from bovine brain by the method of Tettamanti et al. (1973). The ganglioside preparation exhibited 108% (91-126%) of the sialic acid concentration of a bovine brain ganglioside preparation supplied by Sigma, and gave similar silica gel thin layer chromatograms.

At regular intervals post-operatively the recovery of crushed post-ganglionic nerves and the development of hypoglossal nerve influence were assessed by observing the palpebral fissures of each rat under conditions of calm, agitation and feeding.

Experiments were terminated by re-anaesthetizing the rats, as above, and surgically removing the left and right SCGs, with associated nerve trunks. The rats were then killed. To label adrenergic elements ganglia (divided into cranial, middle and caudal blocks) were incubated for 30 min at 37°C with 5-hydroxydopamine (5-OHDA, Sigma), as described by Ramsay and Matthews (1985), before rinsing, osmication, dehydration and embedding in Araldite. Ultrathin sections of the middle block from two levels at least 100 um apart, were collected on mesh grids (with 100 um squares) and stained with uranyl acetate and lead citrate. In grid squares free of major defects, large vessels or nerve tracts and containing 2 to 6 (usually 3 or 4) nucleated neuronal profiles with surrounding neuropil, counts were made from the electron microscope screen of all vesicle-containing nerve terminal profiles (VCPs) and synapses. The VCPs were classified according to their content of vesicles and other organelles (cf. Ramsay and Matthews 1985). A maximum of ten grid squares was counted (five from each level) from each of three left SCGs in each experimental group. Right SCGs and ganglia from normal littermates were used for comparison.

Results

Functional Effects of Hypoglossal Innervation

In normal rats, and on the right side of the operated rats, the palpebral fissure tends to narrow slightly during feeding but opening and closing of the eye bear no strict relationship to the processes of eating, such as chewing. However, for the rats with hypoglossal innervation of the SCG, we observed a widening of the left palpebral fissure during feeding, an effect which began during chewing and was maximal during, or just after swallowing. Of the rats subjected to hypoglossal-CST anastomosis without axotomy and treated with either ganglioside (HD-G) or saline (HD-S), by the end of the 30 day injection period most, and by 60 days all, showed this effect in slight to moderate degree. There was no consistent difference between -G and -S animals in the strength of this effect or in its post-operative time of onset. After additional axotomy, the effect was not seen in most of the HDA-S animals until 2.5-3 months post-operatively, but it appeared in the HDA-G rats after 1.5-2 months, suggesting that ganglioside may facilitate reinnervation of the orbital smooth muscle by ganglionic neurones. In the longer-term experiments the effect was still evident in all animals which were tested, including those with partial amputation of the ganglion.

Ultrastructural Findings

In all the left SCGs post-operatively the preganglionic cholinergic type of nerve ending, containing small regular clear vesicles and a varying proportion of larger cored vesicles (Fig. 1a) was absent or found in very small numbers, attributable to intrinsic cholinergic

Table 1. Incidences of nerve endings and synapses of hypoglossal (H-) type

Experiment	Survival (months)	H-type endings per NNP[a]	H-type synapses per NNP[b]	H-type synapses per ending[b]
HD	9-12	0.18 (0-0.25)	0.05	0.24
HDX	9-16	0.22 (0-0.33)	0.06	0.20
HDA	9-12	0.35 (0.20-0.67)	0.12	0.28
HDXA	9-12	0.50 (0.29-0.75)	0 25	0.38
HD-S	2-6	0.29 (0.15-0.40)	0.12	0.22
HD-G	2-6	0.20 (0-0.33)	0.12	0.52
HDA-S	4-6	0.83 (0.33-1.14)	0.44	0.48
HDA-G	1.5-4	0.90 (0.50-1.33)	0.21	0.21

[a], median value per grid square + 95% confidence limits
[b], calculated from pooled data
*, P <0.05 (Mann-Whitney U test, two-tailed). For further explanation see text.

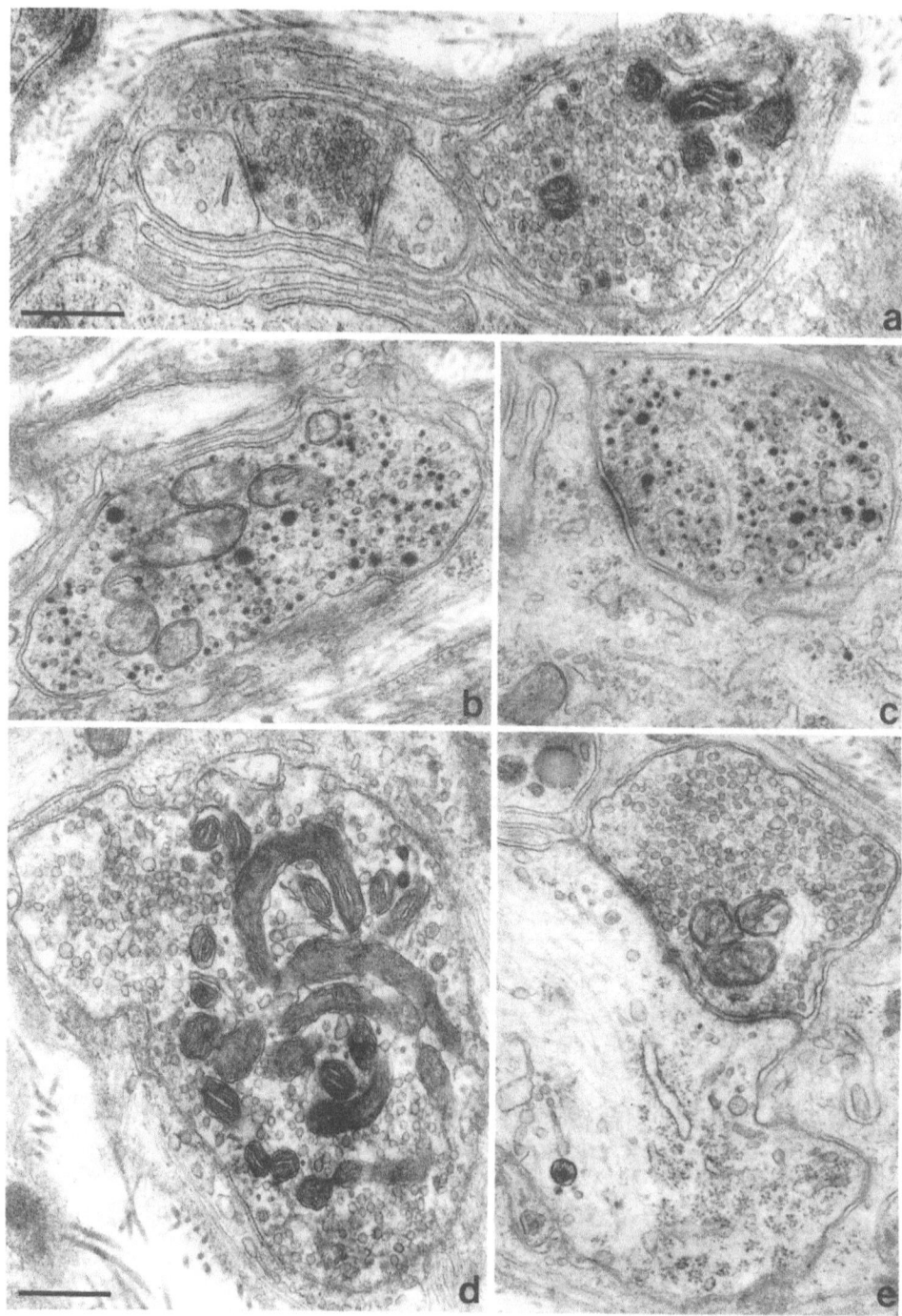

neurones (Ramsay and Matthews 1985). The vesicles of these endings did not become labelled with 5-OHDA. These ganglia however contained numerous nerve terminals filled with dense-cored vesicles, mostly small but some of larger size, which became strongly labelled (Fig. 1b,c). These nerve endings formed synapses upon the ganglionic neurones, as in denervated ganglia (cf. Ramsay and Matthews 1985). In addition, however, there were present nerve endings filled with clear vesicles of irregular shape and size, of a type not seen in normal or denervated ganglia (Fig. 1d). These terminals are typically large and often contain many mitochondria and bundles of neurofilaments; they are identifiable as of hypoglossal origin (Matthews 1983). They were seen to form synapses, mostly axodendritic or axospinous (Fig. 1e) but occasionally axosomatic, upon the ganglionic neurones.

The counts of these nerve endings of hypoglossal type (H-type) are summarized in Figure 2 and in Table 1. The number of the endings or synapses in each grid square was divided by the number of nucleated neuronal profiles (NNP) contained in the same square, to give an incidence related to the local neurone population. Figure 2 shows the distributions of the incidences of all H-type nerve endings seen in every grid square sampled; each histogram represents between 23 and 30 squares. Data for all the nerve endings are included, irrespective of whether or not they showed a synapse in the plane of section. The incidence of synapses is much lower and more variable, reflecting the random chance of a synapse coinciding with the section plane; there was however a well-marked correlation over all groups between the number of endings and the number of synapses ($r = 0.781$), suggesting that the presence of an ending is usually indicative of the presence of a synapse.

The histograms illustrate that the incidences of H-type nerve endings in all HD groups of experiments (left-hand histograms) differ from those in all HDA groups (right-hand histograms). All the HDA groups show higher medians (arrows) and a greater scatter of incidences toward the higher values. To test the significance of these differences, observations from pairs of consecutively counted grid squares were summed before the calculation of incidences (to reduce the frequency of ties); the results of statistical comparisons are shown in Table 1. The incidences per NNP of H-type nerve endings for HD and HDA groups were found to differ significantly ($P < 0.05$) in both the shorter (HD-S, HDA-S) and longer-term experiments. However, there were no significant differences between the incidences for the following consecutive pairs of groups in Table 1: HD versus HDX, HDA versus HDXA, HD-S versus HD-G and HDA-S versus HDA-G.

Table 1 shows also the frequency of hypoglossal-type synapses in relation both to NNP and to H-type nerve endings. Their lower overall numbers precluded the level of statistical analysis applied to the incidence of nerve endings, and therefore their incidences are calculated from pooled counts for each experimental group. For each

◄ **Fig. 1.** (a) two nerve terminal profiles of preganglionic type, one synaptic and the other non-synaptic, from a littermate control rat. Scale = 0.5 um. (b-e) nerve endings from ganglia following hypoglossal nerve anastomosis. Scale = 0.5 um. (b) adrenergic (intrinsic) nerve ending, HDA-G, survival 118 days; (c) adrenergic synapse on the base of a somatic spur, HDA-S, survival 176 days; (d) hypoglossal-type nerve ending, HDA-G, survival 48 days; (e) hypoglossal-type axo-dendritic synapse, HDA-S, survival 178 days. All the above are from ganglia incubated with 5-OHDA

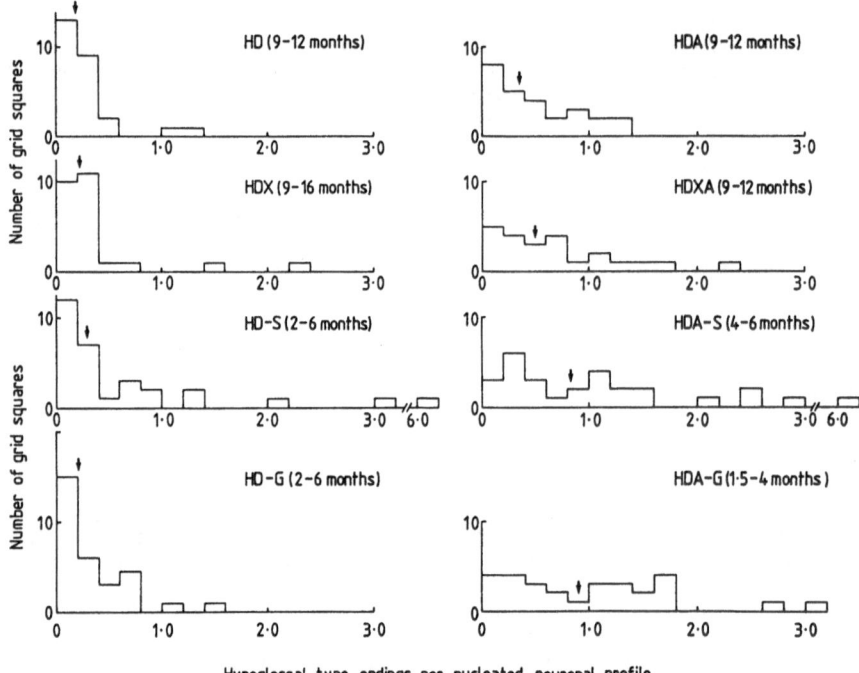

Hypoglossal type endings per nucleated neuronal profile

Fig. 2. Histograms to show the incidence of nerve endings of hypoglossal type, per nucleated neuronal profile, for each group of experiments. Arrows indicate median values

of the groups involving axotomy (HDA groups) the incidence of H-type synapses thus calculated is greater than in the corresponding non-axotomized (HD) group, as was found also for the H-type nerve endings. Higher incidences of synapses are also seen in general in the shorter-term than in the longer-term experiments. These values are however still low in comparison with normal ganglia, in which the incidence of synapses is between 3 and 5 per NNP.

It should be noted that the HDA–G and HDA–S groups are not strictly comparable in terms of survival, as each series was unfortunately interrupted by intercurrent infections. Further experiments are in progress to extend the series.

Summary and Conclusions

These experiments have confirmed that in the shorter term, axotomy of sympathetic neurones leads to a significant, approximately three-fold increase in the incidence of hypoglossal nerve endings which innervate them following hypoglossal-sympathetic anastomosis, as compared with non-axotomized neurones (cf. Matthews and Ramsay 1980, and in preparation). This is a greater increase than might be expected if a finite number of hypoglossal endings were redistributed among a population of neurones reduced even as much as 50% by axotomy. They have further indicated that ganglioside treatment, if it does increase the number of hypoglossal nerve fibres innervating the ganglion, still may not significantly increase the number of nerve endings present, in either axotomized or non-axotomized ganglia.

These results suggest that axotomized neurones, temporarily at least, either stimulate growth of, or are better able to accept, or are less likely to reject or be rejected by, regenerating foreign nerve fibres, and that this situation enhances the formation of nerve endings and synapses. In the longer term, although the incidences of hypoglossal-type nerve endings in axotomized and non-axotomized ganglia still differ significantly, the difference is now no more than two-fold, and this could be explained by a 50% loss of neurones and redistribution of available nerve terminals. Results from the two groups of amputated ganglia show no demonstrably significant differences from the corresponding groups of whole ganglia, but the median incidences of endings per neurone are greater in both series and would not be incompatible with predictions based upon a reduced population of neurones. The results of synapse counts on the whole mirror those for nerve endings. It appears that the incidence of hypoglossal synapses declines in both axotomized and non-axotomized ganglia between 6 months and 9-12 months, but that they may still be increasing in axotomized ganglia between 1 and 4 months post-operatively (compared HDA-G and HDA-S groups). An alternative, and interesting, possible explanation of the latter might however be that ganglioside treatment, by facilitating recovery of the axotomized sympathetic neurones, for which there is some functional evidence, has the effect of reducing their affinity for the foreign nerve fibres, thereby suppressing synapse formation (although not the formation of nerve endings). Toffano et al. (1985) have shown that ganglioside may enhance regeneration of catecholaminergic neurones in the central nervous system.

It is concluded that there is substantial if not yet unequivocal evidence that axotomy of sympathetic neurones leads to an increase in their affinity for, or acceptability to, a class of foreign nerve fibre, the somatic motor axon, to which the uninjured neurone appears highly unattractive. However, the decrease in endings and synapses in the long term, particularly in the axotomized ganglia, suggests a decrease of mutual affinity with time which may in the case of the axotomized sympathetic neurones be linked with re-differentiation following recovery of their peripheral target innervation.

Acknowledgements. We thank Mr. P. Belk for technical assistance, Messrs. B. Archer and A. Barclay for photographic work and Miss G. Davies for secretarial assistance. GC acknowledges an MRC studentship.

References

Acheson GH, Remolina J (1951) The temporal course of the effects of postganglionic axotomy on the inferior mesenteric ganglion of the cat. J Physiol (Lond) 127: 603-616

Brown GL, Pascoe JE (1954) The effect of degenerative section of ganglionic axons on transmission through the ganglion. J Physiol (Lond) 123: 565-573

Case CP, Matthews MR (1985) A quantitative study of structural features, synapses and nearest-neighbour relationships of small granule-containing cells in the rat superior cervical sympathetic ganglion at various adult stages. Neuroscience 15: 237-282

Case CP, Matthews MR (1986) Outgoing synapses of small granule-containing cells in the rat superior cervical ganglion after post-ganglionic axomoty. J Physiol (Lond) 374: 1–32

Gorio A, Marini P, Zanoni R (1983) Muscle reinnervation. III. Moto-neuron sprouting capacity, enhancement by exogenous gangliosides. Neuroscience 8: 417–429

Leeden RW, Yu RK, Eng LF (1973) Gangliosides of human myelin: sialo-galactosylceramide (G7) as a major component. J Neurochem 21: 829–839

Matthews MR (1983) The ultrastructure of junctions in sympathetic ganglia of mammals. In: Elfvin L-G (ed) Autonomic ganglia. Wiley, Chichester

Matthews MR, Nelson VH (1975) Detachment of structurally intact nerve endings from chromatolytic neurones of rat superior cervical ganglion during the depression of synaptic transmission induced by post-ganglionic axotomy. J Physiol (Lond) 245: 91–135

Matthews MR, Ramsay DA (1980) Evidence that chromatolysis enhances heterotypic innervation of sympathetic neurones. J Physiol (Lond) 307: 15–16P

McDonald DM, Larue DT (1983) The ultrastructure and connections of blood vessels supplying the rat carotid body and carotid sinus. J Neurocytol 12: 117–153

Purves D (1975) Functional and structural changes in mammalian sympathetic neurones following interruption of their axons. J Physiol (Lond) 252: 429–463

Ramsay DA, Matthews MR (1985) Denervation-induced formation of adrenergic synapses in the superior cervical sympathetic ganglion of the rat and the enhancement of this effect by postganglionic axotomy. Neuroscience 16: 997–1026

Smolen AJ (1983) Retrograde transneuronal regulation of the afferent innervation of the rat superior cervical sympathetic ganglion. J Neurocytol 12: 27–45

Tettamanti G, Bonali M, Marchesini S, Zambotti V (1973) A new procedure for the extraction, purification and fractionation of brain gangliosides. Biochem Biophys Acta 296: 160

Toffano G, Agnati LF, Aldino G, Consolazione A, Valenti G, Savioni G (1985) Effects of GM1 ganglioside treatment on the recovery of dopaminergic nigrostriatal lesions after different types of lesion. Acta Physiol Scand 122: 313–321

The Effect of Hydrocortisone on the Number of 5-Hydroxytryptamine- and Glutamic Acid Decarboxylase-Immunoreactive Cells in the Superior Cervical Ganglion of the Rat

H. Päivärinta[1], L. Eränkö[1], O. Häppölä[1], S. Soinila[1], H. Steinbusch[2], J.-Y. Wu[3], and P. Panula[1]

[1] Department of Anatomy, University of Helsinki, Siltavuorenpenger 20A, 00170 Helsinki, Finland
[2] Department of Pharmacology, Faculty of Medicine, Free University, Van der Boechorststraat 7, 1081 BT Amsterdam, The Netherlands
[3] Department of Physiology, Pennsylvania State University, Milton S. Hershey Medical Center, Hershey, PA 170033, USA

Introduction

The development of sympathetic cells has been most thoroughly studied in the superior cervical ganglion (SCG) of the rat. The cells of the sympathetic ganglia are derivatives of the neural crest. Catecholamine synthesis is first observed around the 12th embryonic day at the site of the developing SCG. As revealed by formaldehyde-induced catecholamine fluorescence, a wide range of cells can be seen during prenatal development: the size of the developing sympathetic cells varies considerably and the fluorescence intensity in them alternates from weak to very bright. First in the newborn ganglia two distinct cell types can be observed: the principal nerve (PN) cell and the small intensely fluorescent (SIF) cell. In the ganglia of newborn animals a few intermediate forms between PN cells and SIF cells can be observed but they disappear during the first postnatal week.

Glucocorticoids and the nerve growth factor have opposite effects on the developing sympathetic precursor cells (Landis and Patterson 1981), the former favouring differentiation into the SIF cell type and the latter into the PN cell type. Glucocorticoids cause approximately a 10-fold increase in the number of SIF cells when injected into newborn rats (Eränkö and Eränkö 1972). The effect of the nerve growth factor on immature sympathetic precursor cells was dramatically demonstrated when a sympathetic ganglion was observed at the site of the adrenal medulla in rats injected pre- and postnatally with the nerve growth factor (Aloe and Levi-Montalcini 1979).

In addition to the increase in the number of SIF cells by glucocorticoids, it also affects the amount of catecholamines in the rat SCG. Glucocorticoids injected into newborn rats cause a large increase in adrenaline content and in the activity of phenylethanolamine N-methyltransferase (PNMT), the enzyme synthesizing adrenaline from noradrenaline. On the other hand, during the early postnatal period glucocorticoids cause only minor changes in the amount of dopamine and noradrenaline (Koslow et al. 1975).

During the last few years a number of neuroactive substances other than catecholamines have been described in sympathetic cells (for references, see Eränkö et al. 1986; Heym and Lang 1986). 5-hydroxytryptamine (5-HT) immunoreactivity has been described in SIF cells of adult rats (Verhofstad et al. 1981) and also transiently in neurons during the early postnatal development (Häppölä et al. 1986). Furthermore, immunoreactivity to glutamic acid decarboxylase (GAD) (Häppölä et al. 1987), the enzyme necessary in the synthesis

Experimental Brain Research Series 16
© Springer-Verlag Berlin · Heidelberg 1987

of gamma-aminobutyric acid, has been described both in SIF cells and PN cells of adult rats.

The purpose of the present study was to determine if glucocorticoids have an effect in the expression of 5-HT or GAD in sympathetic cells.

Materials and Methods

Newborn Sprague-Dawley rats were subcutaneously injected daily with 20 ug/g body weight hydrocortisone acetate (Korti^R, Lääke) on postnatal days 2-6. Control rats were injected with saline.

On the day after the last injection, on the 7th postnatal day, the rats were perfused transcardially with 4% paraformaldehyde solution. The SCGs were dissected out and immersed in the same fixative. They were then transferred to 20% sucrose overnight. 10 um cryostat sections were cut and 5-HT and GAD immunoreactivities were shown using the indirect immunohistochemical method of Coons (1958) with fluorescein isothiocyanate (FITC) as a fluorescent label of the secondary antibody. For further details on the demonstration of 5-HT and GAD immunoreactivities or specificity of the antisera, see Häppölä et al. (1986, 1987). Even in ganglia from hydrocortisone-treated rats, 5-HT (1000 uM), when added to the 5-HT antiserum 24 h before incubation, completely abolished the specific staining. The fluorescence was not abolished with preabsorption of 1000 uM adrenaline, noradrenaline or dopamine.

Specimens were examined with a Leitz Dialux 20 fluorescence microscope equipped with epi-illumination and a specific filter block 1 2 for FITC. The photographs were taken on Kodak Tri-X film using an automatic Vario-Orthomat microscope camera.

Results and Discussion

In 7-day-old saline-injected control rats 5-HT-immunoreactive SIF cells were observed (Fig. 1). They were seen in clusters or singly scattered among PN cells. Also some 5-HT-immunoreactive PN cells were observed (Fig. 1) and sometimes neuronal processes were seen to emanate from these cells. In sections from 7-day-old control rats, stained with GAD antiserum, no immunoreactivity was observed. In 7-day-old rats injected with hydrocortisone, a dramatic increase in 5-HT-immunoreactive small cells was observed (Fig. 2), and also a large number of GAD-immunoreactive small cells appeared in ganglia of glucocorticoid-treated animals (Fig. 4).

Both 5-HT (Liuzzi et al. 1977) and gamma-aminobutyric acid (GABA) (Bertilsson et al. 1977) as well as the GABA-synthesizing enzyme GAD (Kanazawa et al. 1976) have been shown biochemically to be present in sympathetic ganglia of adult animals. In the SCG of adult rats the 5-HT immunoreactivity has been observed only in SIF cells and transiently also in PN cells in early postnatal rats (Häppölä et al. 1986), as observed also in this study. These findings indicate that in the normal development, 5-HT is expressed early postnatally both in PN cells and SIF cells but, in the adult rat, only in SIF cells. On the other hand, GAD immunoreactivity has been shown to be present immunohistochemically both in SIF cells and PN cells in the SCG of adult rats (Fig. 3) (Häppölä et al. 1987). In the present study, no GAD immunoreactivity was observed in 7-day-old control rats. The possibility must be considered that GAD is present in the ganglia of

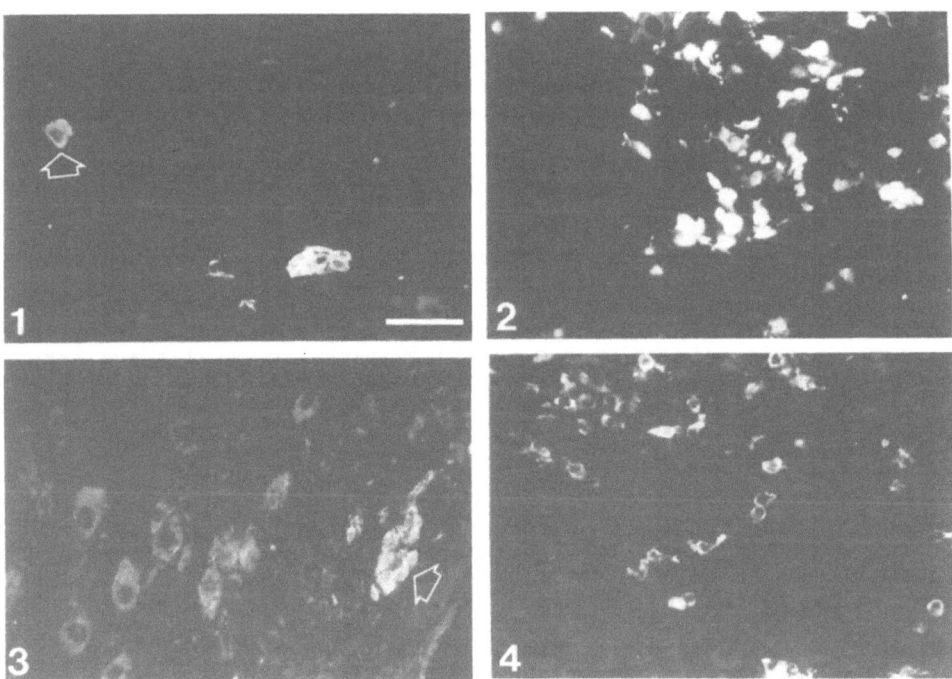

Figs. 1–4. Fluorescence micrographs of cells in superior cervical ganglia of the rat. Magnification in all micrographs is the same. Bar in Fig. 1 represents 50 um. **Fig. 1.** A cluster of 5-HT-immunoreactive SIF cells and a 5-HT-immunoreactive PN cell (open arrow) of a 7-day-old control rat. **Fig. 2.** 5-HT-immunoreactive SIF cells of a 7-day-old hydrocortisone-treated rat. **Fig. 3.** GAD-immunoreactive PN cells and a SIF cell cluster (open arrow) of an adult rat. **Fig. 4.** GAD-immunoreactive SIF cells of a 7-day-old hydrocortisone-treated rat

early postnatal rats in a concentration lower than the detection limit of the present immunohistochemical method. It seems that GAD, a marker of the GABAergic neurons, and 5-HT are differently expressed in normal development.

Since glucocorticoids, injected in early postnatal rats, cause only minute changes in the amount of dopamine and noradrenaline in the SCG, but a dramatic increase in the amount of adrenaline (Koslow et al. 1975), the synthesis of different neuroactive substances seems to be specifically regulated by glucocorticoids. The present study shows that glucocorticoids cause the appearance of GAD-immunoreactive cells and an increase in the number of 5-HT-immunoreactive cells in ganglia of early postnatal rats, as shown earlier for 5-HT in vitro (Eränkö and Eränkö 1981). Also nerve growth factor (NGF) has been reported to increase the amount of 5-HT as well as the activity of PNMT in the rat SCG. Evidence has been presented that NGF-induced increase in PNMT activity occurs at least partly in neurons (Saavedra and Liuzzi 1976), but the NGF-stimulated increase in the amount of 5-HT is obscure (Liuzzi et al. 1977). In the present study, on the basis of the small size and high nucleus/cytoplasm ratio, the glucocorticoid-induced 5-HT- and GAD-immunoreactive cells were regarded to be SIF cells. It is suggested that the glucocorticoid-induced SIF cells also attain GABAergic and serotoninergic features.

References

Aloe L, Levi-Montalcini R (1979) Nerve growth factor-induced trans-formation of immature chromaffin cells in vivo into sympathetic neurons: effect of antiserum to nerve growth factor. Proc Natl Acad Sci USA 76: 1246-1250

Bertilsson L, Suria A, Costa E (1976) Gamma-aminobutyric acid in rat superior cervical ganglion. Nature 260: 540-541

Coons AH (1958) Fluorescence antibody methods. In: Danielli AH (ed) General cytochemical methods. Academic Press, New York, pp 399-422

Eränkö L, Eränkö O (1972) Effect of hydrocortisone on histochemical-ly demonstrable catecholamines in the sympathetic ganglia and extra-adrenal chromaffin tissue of the rat. Acta Physiol Scand 84: 125-133

Eränkö L, Päivärinta P, Soinila S, Häppölä O (1986) Transmitters and modulators in the superior cervical ganglion of the rat. Med Biol 64: 75-83

Eränkö O, Eränkö L (1981) Catecholamine storage and synthesis sites in the sympathetic system: histochemical aspects. In: Stoward PJ, Polak JM (eds) Histochemistry, the widening horizons of its applica-tion in the biochemical sciences. Wiley, Chichester, pp 31-46

Häppölä O, Päivärinta H, Soinila S, Steinbusch H (1986) Pre- and postnatal development of 5-hydroxytryptamine-immunoreactive cells in the superior cervical ganglion of the rat. J Auton Nerv Syst 15: 21-31

Häppölä O, Päivärinta H, Soinila S, Wu J-Y, Panula P (1987) Locali-zation of L-glutamic acid decarboxylase (GAD) and GABA transaminase (GABA-T) immunoreactivity in the sympathetic ganglia of the rat. Neuroscience (in press)

Heym Ch, Lang R (1986) Transmitters in sympathetic postganglionic neurons. In: Panula P, Päivärinta H, Soinila S (eds) Neurohistoche-mistry: modern methods and applications. Alan R Liss, New York, pp 493-526

Kanazawa I, Iversen LL, Kelly JS (1976) Glutamate decarboxylase activity in the rat posterior pituitary, pineal gland, dorsal root ganglion and superior cervical ganglion. J Neurochem 27: 1267-1269

Koslow SH, Bjegovic M, Costa E (1975) Catecholamines in sympathetic ganglia of the rat: effects of dexamethasone and reserpine. J Neuro-chem 24: 277-281

Landis SC, Patterson PH (1981) Neural crest cell lineages. Trends in Neurosci 4: 172-174

Liuzzi A, Foppen FH, Saavedra JM, Levi-Montalcini R, Kopin IJ (1977) Gas chromatographic-mass spectrometric assay of serotonin in rat superior cervical ganglia. Effects of nerve growth factor and 6-hydroxytryptamine. Brain Res 133: 354-357

Saavedra JM, Liuzzi A (1976) Nerve growth factor: effects on 5-hydroxytryptamine and phenylethanolamine-N-methyltransferase in the superior cervical ganglion of the rat. In: Eränkö O (ed) SIF cells. Structure and function of the small intensely fluorescent sympathe-

tic cells. DHEW Publication No 30 (NIH), U.S. Government Printing
Office, Washington D.C., pp 124-130

Verhofstad AAJ, Steinbusch HWM, Penke B, Varga J, Joosten HWJ (1981)
Serotonin-immunoreactive cells in the superior cervical ganglion of
the rat. Evidence for the existence of separate serotonin- and
catecholamine-containing small ganglionic cells. Brain Res 212: 39-
49

The Influence of Hypothyroidism on the Rat Adrenergic Nerve Plexus: Quantitative Image Analysis of Plexus Density in Auerbach's Plexus and the Right Atrium

D. M. Baker

William Osler House, John Radcliffe Hospital, Oxford University, Oxford, Great Britain

Introduction

Heterogeneity within the sympathetic nervous system is demonstrated experimentally by the differing responses of various adrenergic neurone populations to various hormones, drugs and the effects of ageing (see Baker 1986). Clinically, heterogeneity amongst sympathetic neurones is demonstrated by the effect of the thyroid hormone status on the cardiovascular (CVS) and gastrointestinal systems. Thyroid hormones have a sympathomimetic effect on the CVS such that in the hyperthyroid patient there is an increase in heart rate, cardiac output and pulse pressure (Skelton 1982). The influence of thyroid hormones on the gut, however, is antagonistic to sympathetic activity such that hyperthyroidism in man and animal (Johansson 1966) increases gut motility and hypothyroidism leads to constipation, ileus and even pseudo-obstruction (Duret and Bastenie 1971). The exact nature of the interaction between thyroid hormones and the sympathetic activity in either the CVS or the gut is unknown. This study determines, in the rat, the influence of hypothyroidism on the density of the adrenergic nerve plexuses in the right atrium and in Auerbach's plexus of the jejunum.

Materials and Methods

Seven post-weaned male Wistar albino rats were rendered hypothyroid by addition of 0.1% methimazole to their drinking water. After four months the rats were sacrificed and the tissues immediately removed and processed to reveal the adrenergic axons by glyoxylic acid-induced fluorescence. In the right atrium the SPG method was used on 10 um sections (de la Torre and Surgeon 1976) and in the jejunum lamina stretch preparation, and processing was as described elsewhere (Baker 1986). Six standard areas of 1.4×10^5 um^2 from preparations of each tissue from each rat were randomly selected and photographed on a Leitz Laborlux 12 UV microscope using Ilford XP1 400 film. From micrographs, plexus density was quantified using the Quantimet 800 image analyser. This determined the ratios of fluorescent areas and perimeters to total frame area (Aa% and Pa). From micrographs of Auerbach's plexus taken at a magnification of x 350 the number of varicosities per 100 um of axon was determined by measuring 2,500 um of clearly visible axon in each animal. Results were statiscally compared, using the two sample t-test, with results similarly obtained from control rats supplied with normal drinking water.

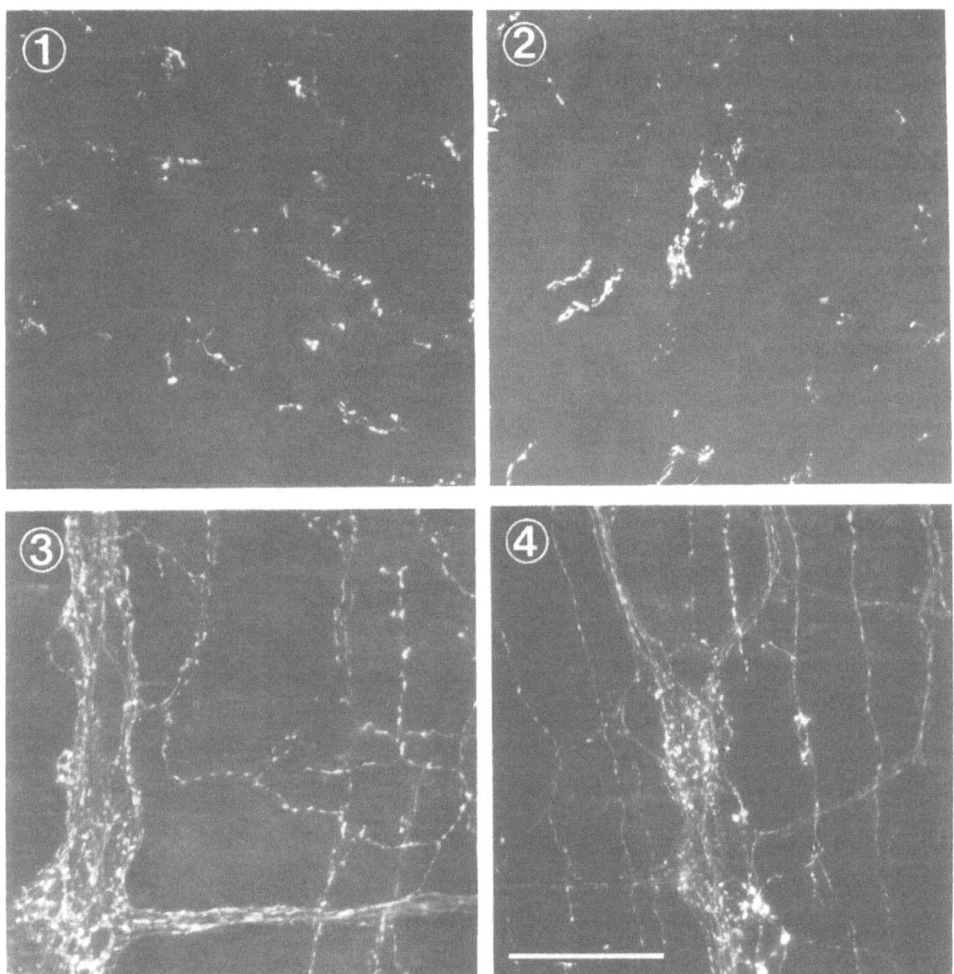

Figs. 1–4. The influence of hypothyroidism on the glyoxylic acid induced fluorescence adrenergic plexus (scale = 100 um). **Fig. 1.** Right atrium of an euthyroid rat. **Fig. 2.** Right atrium of hypothyroid rat. **Fig. 3.** Auerbach's plexus from the jejunum of an euthyroid rat. In the rat, Auerbach's plexus can be divided into three parts: a primary plexus of circularly running ganglia of neurone somata and interlinking large nerve bundles; a secondary plexus of large nerve bundles running longitudinally between the primary plexus and a tertiary plexus of single fibres running between the other two parts. **Fig. 4.** Auerbach's plexus from the jejunum of a hypothyroid rat

Results

Qualitatively, suppression of thyroid hormone production did not alter the structure, density or fluorescence intensity of the adrenergic fibres in either the heart or Auerbach's plexus (Figs. 1–4). Statistical analysis of quantitative results are shown in Table 1.

Table 1. Results and analysis of quantitative data from image analysis of Auerbach's plexus and right atrium from hypothyroid (Methimazole treated, n = 7)) and euthyroid (control, n = 7) rats. Aa% = percentage of fluorescent area to frame area, Pa = total fluorescent area perimeter to frame area. 1 + 2 = primary and secondary components and 3 = tertiary component of Auerbach's plexus

Plexus	Methimazole treated rats (mean + SD)	Control rats (mean + SD)	t-value	p-value
Auerbach's Aa%				
1 + 2	8.350 (1.260)	8.310 (1.380)	0.065	0.950
3	2.777 (0.526)	2.563 (0.658)	0.673	0.515
Pa				
1 + 2	0.123 (0.029)	0.109 (0.026)	0.862	0.407
3	0.132 (0.250)	0.136 (0.034)	-0.233	0.820
varicosity per 100 um	14.54 (1.231)	13.31 (0.546)	2.427	0.041
Atrium Aa%	1.879 (0.288)	1.686 (0.470)	0.925	0.379
Pa	0.050 (0.009)	0.047 (0.012)	0.502	0.626

Aa% is proportional to the total number or length of individual nerve axons and Pa to the total number or length of nerve bundles, both therefore indicating relative plexus density (Baker 1986).

Discussion

Thyroid hormones influence the maturing sympathetic nervous systems, in utero and in the neonate, at receptor, neurotransmitter and structural levels (Slotkin et al. 1980). However, insufficient data are available to determine the levels at which thyroid hormones exert their heterogeneous influence over the mature sympathetic nervous system.

In the heart, thyroid hormones enhance adrenergic receptor sensitivity as well as increase receptor number (Skelton 1982), but the effect of thyroid hormones on gut adrenergic receptors is unknown.

Data concerning thyroid hormones and catecholamine levels are contradictory. Even though in hypothyroidism plasma catecholamine levels increase (Christensen 1972), thyroid hormones stimulate catecholamine production in the heart (Skelton 1982) and thyroid hormones increase noradrenaline release from the inferior mesenteric ganglion following electrical stimulation of the lumbar splanchnic or hypogastric nerves (Bulygin et al. 1972). Despite this, as the qualitative assessment of fluorescence intensity is a good indicator of the quantity of catecholamines within nerve fibres (Fuxe and Jonsson 1973), from this study it is likely that thyroid hormones do not influence fibre catecholamine levels in either the heart or Auerbach's plexus.

Thyroidectomy, in the rat, causes no structural change to the somata of the superior cervical ganglion or abdominal ganglia (Atech 1962). Chemically induced hypothyroidism does not affect the adrenergic fibre density in the heart or Auerbach's plexus, but does cause an increase in the varicosity number per 100 um of axon in Auerbach's plexus (p<0.05). As all varicosities of an axon discharge neurotransmitter during axon depolarisation (Furness and Costa 1974), an increase in varicosity number per 100 um of axon would explain the sympathomimetic effects observed in the hypothyroid gut better than the previous hypothesis that myxoedema causes such severe smooth muscle atrophy as to result in decreased gut motility (Duret and Bastinie 1971). This latter hypothesis is opposed by the observation that the gut of most myodematous patients responds normally to parasympathomimetic stimulation (Duret and Bastinie 1971).

In conclusion, although this study suggests that hypothyroidism exerts its effect on the gut by increasing varicosity numbers per 100 um of adrenergic axon, the effect of hypothyroidism on varicosity numbers in the heart remains unknown. The mechanism by which the thyroid hormones exert their heterogeneous influence over the sympathetic nervous system is yet to be determined. Consideration of the ultrastructure of sympathetic neurones in animals in various thyroid states may well help resolve this.

Acknowledgements. The assistance of CA Edwards (Department of Medicine, University of Wales College of Medicine, Cardiff) and the use of Dr. RM Santer's laboratory (Department of Anatomy, University College, Cardiff) are gratefully acknowledged.

References

Atech YL (1962) Reactions structurales des neurones des ganglions de la chaine sympathique du rat thyroidectomise a l'hormone thyrotrope hypophysaire. C R Soc Biol Paris 156: 1322-1324

Baker DM (1986) Quantitative morphological and histochemical studies on autonomic neurones. PhD thesis. University of Wales

Bulygin IA, Petrov VI, Reprintseva VM (1982) Hormone regulation of adrenaline and noradrenaline release in the inferior mesenteric ganglion of the dog. J Auton Nerv Syst 6: 55-64

Christensen NJ (1972) Increased levels of plasma noradrenaline in hypothyroidism. J Clin Endocrin metab 35: 359-363

de la Torre JC, Surgeon JW (1976) A methodological approach to rapid and sensitive monoamine histofluorescence using a modified glyoxylic acid technique: the SPG method. Histochemistry 49: 81-93

Duret RL, Bastinie PA (1971) Intestinal disorders in hypothyroidism. Clinical and manometric study. Am J Dig Dis 16: 723-727

Furness J, Costa M (1974) The adrenergic innervation of the gastrointestinal tract. Ergebn Physiol 69: 1-51

Fuxe K, Jonsson G (1973) The histochemical fluorescence method for demonstration of catecholamines. J Histochem Cytochem 21: 293–311

Johansson H (1966) Gastrointestinal motility function related to thyroid activity. Acta Chir Scand Suppl 359: 1–88

Skelton LC (1982) The heart and hyperthyroidism. New Eng J Med 307: 1206–1208

Slotkin TA, Smith PG, Lau C, Bareis DL (1980) Functional aspects of development of catecholamine biosynthesis and release in the sympathetic nervous system. In: Parvez H, Parvez S (eds) Biogenic amines in development. Elsevier, Amsterdam, pp 29–48

Myenteric Ganglia: Size and Number of Neurones in Different Animal Species and in the Hypertrophic Intestine

G. Gabella

Department of Anatomy and Embryology, University College London, Glower Street, London WC1E 6BT, Great Britain

Introduction

The question of the number of neurones in the enteric plexuses has been studied by many authors. Neuronal counts are available for several mammalian species: guinea-pig, mouse, rat, rabbit, cat, dog, and monkey (reviewed in Gabella 1971). However, only in the cat and the rat the total number of myenteric neurones has been estimated. Yet, number of neurones, neuronal spatial density and other quantitative aspects of the enteric ganglia (such as nerve cell size and extent of the territory of innervation) are important parameters for understanding the morphogenesis of the autonomic ganglia and the interaction between peripheral nervous tissue and target tissues.

Materials and Methods

In the present experiments, the neurones of the myenteric plexus of the small intestine were studied quantitatively in mice, guinea-pigs and sheep. The neurones were stained electively in the intestinal wall in toto, with a histochemical method (Gabella 1971) to detect B-nicotinamide adenine diaphorase activity, applied to the full length of the small intestine. Constant conditions of distension were used, and diameter and full length of the intestine were measured after staining. Selected areas of the intestine were dissected and delaminated, and large laminae of the muscle coat with the myenteric plexus (the muscle cells being virtually unstained) were prepared.

Results

Mouse. The myenteric neurones are gathered in groups of variable size, narrow and elongated, of ill-defined extent, and spreading, sometimes as cord-like structures, in the direction of the circular musculature (Fig. 1A). In 7 preparations the average neuronal density was 10,600/cm². The length of the small intestine was about 33 cm and the circumference averaged 11.5 mm. It was therefore calculated that the serosal surface area of the entire small intestine is about 38 cm² and the total number of myenteric neurones in the small intestine about 403,000 (Table 1).

Guinea-Pig. The pattern of the plexus is regular and uniform from one segment of the small intestine to another. The ganglia are long and thin, usually not more than three or four neurones in width, and their long axis run almost invariably parallel to the circumference

Table 1. Summary of neuronal cell counts in myenteric plexus of the small intestine of mice, guinea-pigs and sheep

animal species	spatial density of myenteric neurons	length of small intestine	circumference of small intestine	total serosal surface	total number of myenteric neurons
MOUSE	10,600 .cm^{-2}	33 cm	11.5 mm	38.0 cm^2	403,000
GUINEA-PIG	8,600 .cm^{-2}	145 cm	22 mm	319 cm^2	2,750,000
SHEEP	2,500 .cm^{-2}	2,100 cm	60 mm	12,600 cm^2	31,500,000

of the intestine (Fig. 1B). The connecting strands of the plexus, in contrast, run mainly parallel to the length of the gut. The neurones are densely packed, but in the conditions of distension used there is only a small degree of overlap between adjacent neurones. The limits between ganglia and connecting strands are sometimes difficult to define and it is often arbitrary to define what constitutes a ganglion. In 7 laminar preparations measuring between 20 and 35 mm^2 the average neuronal density was 8,600/cm^2. The average length of the small intestine was 145 cm and the circumference averaged 22 mm. It was therefore calculated that the serosal surface area in the guinea-pig small intestine is about 319 cm^2 and the total number of myenteric neurones in the small intestine about 2,750,000 (Table 1).

Sheep. The myenteric neurones are gathered in very large groups. Some of these contain well over a hundred neurones, but the boundaries between ganglia and between ganglia and connecting strands are often uncertain (Fig. 1C). The pattern of the plexus is more variable and less geometrical than in the guinea-pig and mouse. The average neuronal density in 7 laminae was 2,500/cm^2. The length of the

Fig. 1. Micrographs of preparations of the muscle coat and the myenteric plexus from the small intestine of a mouse (A), a guinea-pig (B) and a sheep (C). The long axis of the intestine lies vertical in all three micrographs, which are at the same magnification. Calibration bar: 100 um

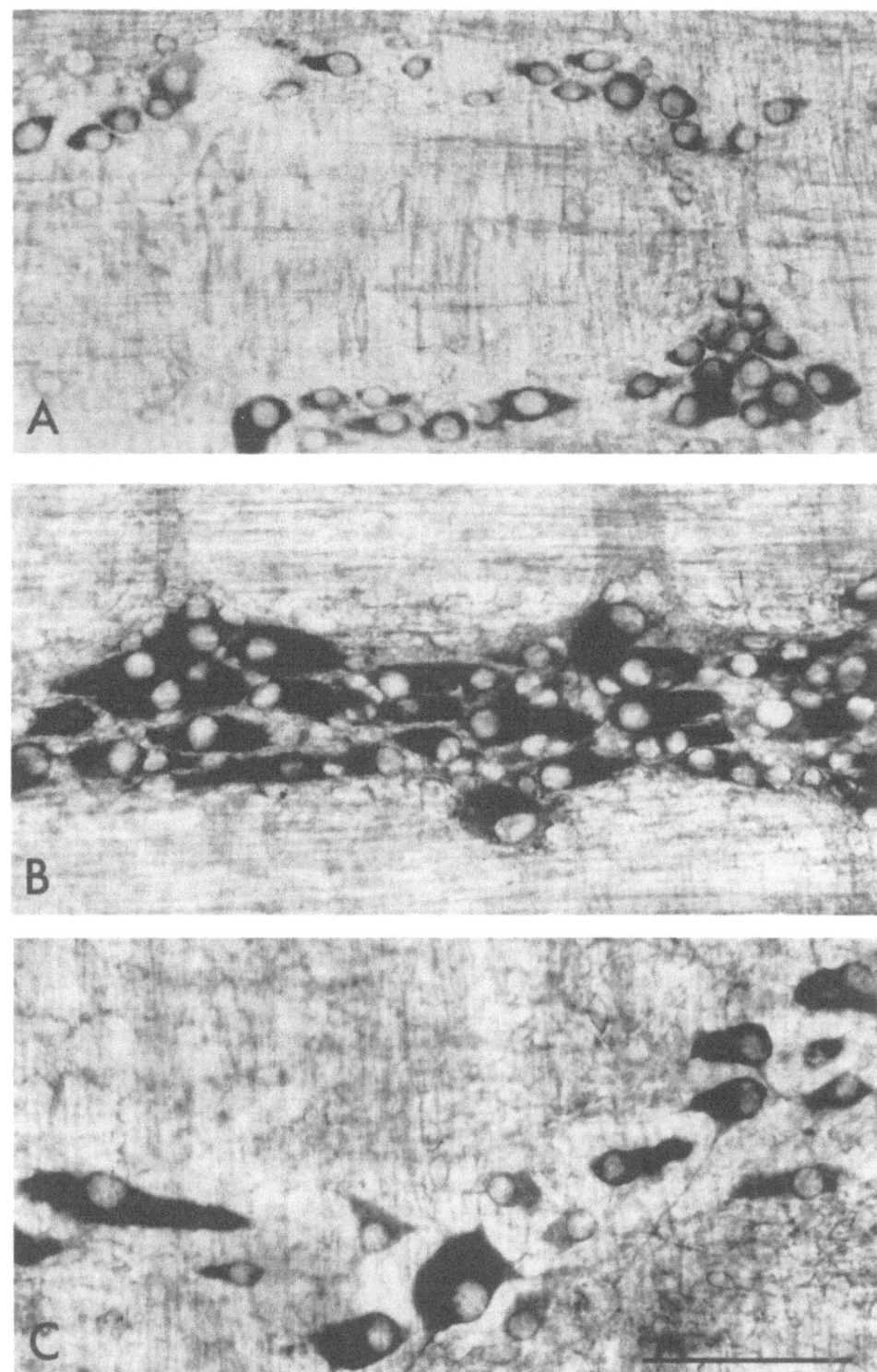

Table 2. Summary of the neuronal cell size measurements in the small intestine of mice, guinea-pigs, rabbits and sheep

Species	Average neuronal cell profile area, $\bar{x}_B$ (μm^2)	Average neuronal nucleus profile area, $\bar{x}_C$ (μm^2)	Average glial nucleus profile area, $\bar{x}_D$ (μm^2)	Percentage of neuropil, Np (%)	Apparent ratio, neurons: glial cells, N:G	Corrected ratio, neurons: glial cells, n:g
Myenteric ganglia						
Mouse	179	56	23	51	100:68	100:109
Guinea-pig	228	74	35	65	100:183	100:266
Rabbit	222	71	26	70	100:157	100:259
Sheep	441	97	29	74	100:244	100:446

small intestine was about 21 metres with an average circumference of 60 mm. The total serosal surface was about 12,600 cm² and the total number of neurones was calculated as 31,500,000 (Table 1).

Neuron Size. In each species studied the myenteric neuronal perikarya are very variable in size. Cell sizes (expressed as the area of the maximal cell profile) vary from 75 um² to about 600 um² in the myenteric ganglia of the mouse. The range of sizes is wider in the guinea-pig (75 to about 800 um²) than in the mouse, and it is wider still in the sheep, where neuron sizes range from 75 to over 1400 um² (Gabella and Trigg 1984). The average neuronal cell profile area is 179 um² in the mouse, 228 um² in the guinea-pig, and 441 um² in the sheep. There are also important differences in the percentage of the ganglion volume occupied by neuropil as opposed to nerve cell bodies, and in the numerical ratio between neurones and glial cells (Table 2). While in the mouse there is approximately the same number of neurones and glial cells, in the sheep glial cells outnumber neurones by more than 4 to 1. Table 1 and Table 2 show that there is an inverse relation between neuronal packing density and average (and maximal) neuronal size. Moreover, where the average neuronal size is larger, the neurones also aggregate in larger ganglia.

Summary of Differences Between Species. The pattern of the myenteric plexus is very different in the three species examined; in all three, however, it shows a distinct polarization in the direction of the circular musculature. The smallest of the three animal species has the smallest number of neurones, the largest species the largest number (Table 2). In spite of this correlation, there is no direct proportionality between body size, or length of the intestine, and number of neurones. The lack of direct proportionality is related to the fact that in a large animal there are fewer neurones per unit surface of the intestine – and, if one takes into account that the musculature is also thicker, there are even fewer neurones per given volume of muscle tissue – than in a small animal. A lower neuronal density is accompanied by the occurrence of larger ganglia, which are located further apart from one another and are made of neurones of markedly larger average size. With high neuronal density, the ganglia are smaller, are closer to one another and are made of markedly smaller neurones.

Discussion

Changes During Post-Natal Growth. Although it is not known at what age the intramural nerve cells cease to divide, it is known that division of nerve cells is still taking place in the enteric ganglia at developmental stages when it has ceased in the central nervous system, with the effects of a substantial increase in the number of ganglion neurones during postnatal life. In the rat small intestine there are about 1,850,000 myenteric neurones in adult animals against about 420,000 at birth (Gabella 1971). Another example of an increase in the number of enteric neurones during postnatal life is found in the chick (Ali and McLelland 1979); 2,058,000 and 276,000 myenteric neurones were counted in the small and large intestine of 18-week-old fowls, but only 1,496,000 and 151,000 in 3-week-old chicks. In the myenteric plexus of the mouse duodenum there is uptake of tritiated thymidine, indicative of cell division, in ganglion neurones at least up to two weeks after birth (Benjamin et al. 1985).

Changes in the Hypertrophic Intestine. While there is no increase in neurone numbers in the myenteric plexus of adult animals, there remains even in adult animals further potential for neuronal growth in the form of neuronal hypertrophy, as can be observed in pathological and experimental conditions. In the present experiments hypertrophy of the gut of guinea-pigs and rats was obtained surgically by placing a narrow strip of cellophane around a loop of intestine. Reduction of the lumen at the site of the operation causes accumulation of ingesta on the oral side; there is a slow increase in the diameter of the intestine (up to 2.5 times the control value) and the distension of the wall is accompanied by thickening of the musculature (up to 5 times the control value). In addition to the growth of the musculature there is also growth of the myenteric plexus, although to a smaller degree. The plexus amounts to about 3% the volume of the muscle in the control gut; after a ten-fold increase in muscle volume the plexus is 0.6% the muscle volume: this represents a two-fold growth in the total volume of the plexus. The number of ganglion neurones in a segment of intestine is unchanged, whereas their density is lower in hypertrophy than in controls. There is a massive increase of neuronal sizes, the average area of neurone outlines being twice as large as in controls. The entire population of neurones grows, and the histogram of sizes is shifted to the right. Nearly half the neuronal profiles measure over 500 um^2 (less than 3% in controls), and less than 1% measure less than 150 um^2 (21% in controls). The neuropil grows less than nerve cell bodies and it falls in percentage from 65% to about 55%. There are only small changes in glial cell number, while there is increase in glial cell size. Ultrastructurally, the general texture of the ganglia is preserved in hypertrophy. The rough endoplasmic reticulum becomes more prominent in the enlarged neurones, and in glial cells gliofilaments become more abundant. The occurrence of extensive neuronal hypertrophy in these experimental conditions indicates that a vast potential for growth remains in the enteric ganglia of adult animals.

Acknowledgement. This work is supported by the Medical Research Council (U.K.).

References

Ali HA, McLelland J (1979) Neuron number in the intestinal myenteric plexus of the domestic fowl (Gallus gallus). Zbl Vet Med 8: 277–283

Benjamin S, Rothman TP, Gershon MD (1985) Postnatal generation of neurons in the developing enteric nervous system. Neurosci Abst 11: 1152

Gabella G (1971) Neuron size and number in the myenteric plexus of the newborn and adult rat. J Anat (Lond) 109: 81–95

Gabella G (1979) Innervation of the gastrointestinal tract. Int Rev Cytol 59: 129–193

Gabella G, Trigg P (1984) Size of neurons and glial cells in the enteric ganglia of mice, guinea pigs, rabbits and sheep. J Neurocytol 13: 449–471

Transmission, Excitation, and Second Messengers

Properties of Intramural Neurones Cultured from the Heart and Bladder

G. Burnstock[1], T. G. J. Allen[1], C. J. S. Hassall[1], and B. S. Pittam[2]

[1] Department of Anatomy and Embryology, University College London, Gower Street, London WC1E 6BT, Great Britain
[2] Gastrointestinal Research Unit, The London Hospital Medical College, Ashfield Street, London E1, Great Britain

Introduction

The use of culture preparations to study the properties and interactions of intramural neurones was first described for enteric neurones (see Jessen and Burnstock 1982). Comparison of the cultures with in situ preparations showed that most of the structural, functional and biochemical properties of enteric neurones were maintained in the culture situation (Baluk et al. 1983; Jessen et al. 1983a,b). This approach has now been extended to intramural neurones in heart and bladder, where direct study of these neurones in situ is limited by their inaccessibility and the difficulties in distinguishing the properties of extrinsic nerves from those of intrinsic nerves. To overcome these problems, we have developed dissociated cell culture preparations from the heart and bladder containing intramural neurones with associated non-neuronal and muscle cells (Hassall and Burnstock 1986; Pittam et al. 1986). In this paper we review and compare the results of work carried out on intramural heart and bladder neurones in culture.

Histochemistry and Immunocytochemistry

The methods used for preparing the cultures from the newborn guinea-pig heart and bladder result in the degeneration of extrinsic nerves and only neurones present within the heart and bladder survive (Fig. 1A,B). Hence, these culture preparations can be used to determine the extrinsic or intrinsic origin of the neuropeptides localised in the heart and bladder in situ. Of the neuropeptide Y (NPY)-, vasoactive intestinal polypeptide (VIP)-, substance P (SP)-, enkephalin (ENK)- and neurotensin-immunoreactive nerve fibres found in the heart, only VIP and NPY have been demonstrated in neurone cell bodies in situ (see Weihe et al. 1983, 1984; Dalsgaard et al. 1986) and in culture (Hassall and Burnstock 1984, 1986) indicating that at least some of the NPY and VIP fibres in the heart are of intrinsic origin. However, SP- and somatostatin (SOM)-immunoreactive neurone cell bodies, as well as VIP- and NPY-containing neurones (Crowe et al. 1986; S James, unpublished observations) are present in bladder cultures, so that a proportion of all of these peptides are intrinsic to the bladder. The presence of high numbers of NPY-containing intramural heart and bladder neurones in culture supports the view that this peptide is not confined to sympathetic neurones in the periphery (Hassall and Burnstock 1984, 1986; Mattiasson et al. 1985; Dalsgaard et al. 1986), since these ganglia are not part of the sympathetic nervous system.

Experimental Brain Research Series 16
© Springer-Verlag Berlin · Heidelberg 1987

Other studies of the heart and bladder cultures have demonstrated neuronal properties not previously detected, such as the presence of 5-hydroxytryptamine (serotonin, 5-HT) and catecholamines in some of the bladder neurones in culture, but not in situ (Crowe et al. 1986). A large population of intracardiac neurones has also been shown to be 5-HT-immunoreactive in culture and we have demonstrated that these neurones are capable of 5-HT uptake and its synthesis from 5-hydroxytryptophan (Hassall and Burnstock, in preparation). These results suggest that the 5-HT present in neurones in the heart in culture, and possibly also the bladder, may be the result of uptake of this amine from the foetal calf serum (FCS)-supplemented culture medium. Since 5-HT uptake is a characteristic of serotonergic neurones, the hypothesis that 5-HT may be utilised as a neurotransmitter by intramural neurones of the heart and bladder in situ is being further investigated.

Ultrastructure

In general, the ultrastructure of the intramural bladder and heart neurones, and the types of cell interrelationships observed in these cultures are similar (Kobayashi et al. 1986a,b; J Cavanagh, unpublished observations). Further, the organisation of the neurones and their non-neuronal supporting cells, resembles that described in cultures of sensory and sympathetic ganglia (see Bunge et al. 1967; Lever and Presley 1971), but is markedly different from the cellular relationships in explant cultures of myenteric plexus (Baluk et al. 1983).

Autoradiographic Localisation of Receptors

Muscarinic acetylcholine receptors have been localised on both intramural heart and bladder neurones in culture by autoradiography using the irreversible ligand (^{3}H)propylbenzilylcholine mustard (CJS Hassall, S James, NJ Buckley, unpublished observations). These findings have important implications for interpretation of data from receptor studies on heart and bladder homogenates, since a proportion of the receptors characterised in this way will be associated with the intramural neurones rather than the heart or bladder muscle.

Fig. 1. Phase contrast micrographs of a group of intramural bladder ▶ neurones (A) and intracardiac neurones (B); examples of neuronal somata are indicated by arrow heads. (C) multiple firing to prolonged (300 ms) depolarising current injection in an intramural bladder neurone. (D) the pronounced slow afterhyperpolarisation following single action potentials in a cultured heart neurone. (E) and (F), depolarisation and firing in response to 20 ms micro-application of 1 mM ATP in intramural bladder and heart neurones respectively; downward deflections in (F) result from passing brief hyperpolarising constant current pulses across the cell membrane and indicate an increase in membrane conductance during the response. (G) long latency depolarisation and multiple firing in response to exogenous application of VIP (16 uM) in an intramural bladder neurone. (H) depolarisation with a decrease in membrane resistance following micro-application of methylcholine (10 uM) for 100 ms in a cultured heart neurone; the increase in the excitability of the neurone during the response leads the generation of 'anodal break' firing

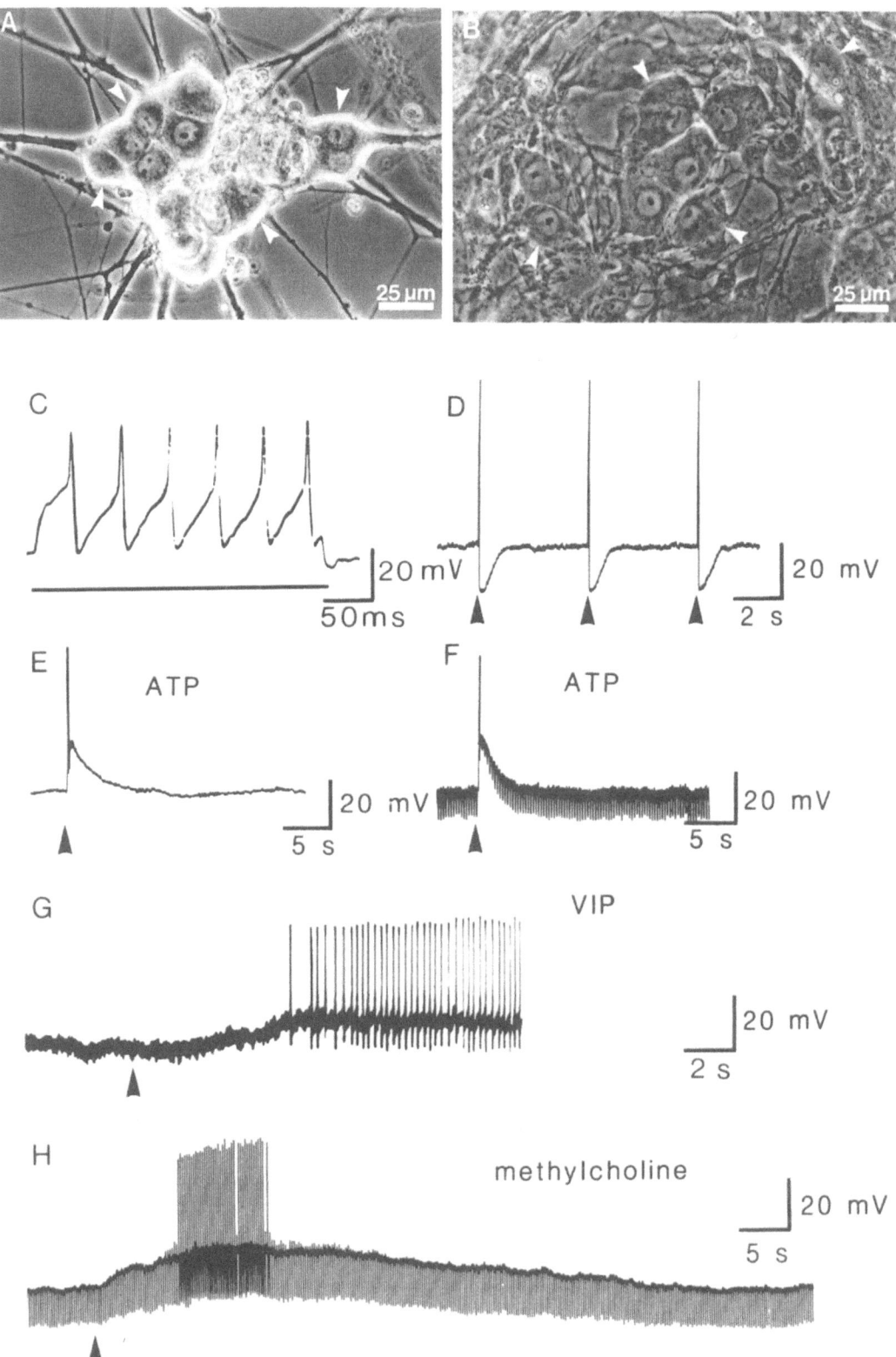

Electrophysiology

Intramural heart and bladder neurones maintained in dissociated cell culture for between 4 and 21 days have been studied using conventional intracellular microelectrodes. Membrane potentials of heart neurones recorded in oxygenated Krebs' solution ranged from −45 to −76 mV (mean approx. −55 mV); those of bladder neurones recorded in medium 199 ranged from −40 to −60 mV. All measurements were made at 37° C.

By using direct intrasomal current injection, different types of neurones could be distinguished. In the majority of intracardiac neurones only one action potential could be elicited, even to prolonged current injection. The action potentials in these cells were frequently followed by pronounced slow afterhyperpolarisations (sAH) (Fig. 1D), the magnitude and duration of which increased with the number of preceding spikes. The sAH was unaffected by tetrodotoxin (0.3 uM)− but abolished by cadmium chloride (20 uM)−containing solutions. In a smaller number of intracardiac neurones, primarily those not exhibiting an sAH, depolarising current injection elicited trains of action potentials at high frequencies, which could be sustained for prolonged periods. Unlike the intracardiac neurones, the responses of intramural bladder neurones to direct stimulation were generally of the latter non−adapting type (Fig. 1C), although in a few cells the train was damped out whilst in others just one action potential was fired at the start of the depolarising current pulse. However, so far there is no evidence that any bladder neurones display a sAH.

Intracardiac neurones were rarely spontaneously active, this is in contrast to the bladder, where many of the neurones exhibited non−synaptic spontaneous activity. Such activity took the form of slow wave changes in excitability, small changes in membrane potential or spontaneously fired action potentials, the frequency of which could be as high as 60 Hz.

Exogenous application of putative neurotransmitters and neuropeptides produced a number of different neuronal responses. Acetylcholine elicited fast nicotinic responses (blocked by 100 uM hexamethonium) in all heart and bladder neurones tested, as well as slower muscarinic responses (selectively blocked by 0.1 uM atropine) in a subpopulation of neurones from both preparations (see Fig. 1H). ATP (adenosine 5'−triphosphate) had a potent excitatory effect upon a large number of both bladder and heart neurones (Fig. 1E,F). In each case it produced a rapid depolarisation of similar latency and time course to that of the cholinergic nicotinic response. This was often large enough (up to 25 mV) to cause firing in many of the cells. In a brief survey of the effects of neuropeptides on the bladder intramural neurones, VIP (Fig. 1G) and SP were predominantly excitatory, depolarising the membrane beyond the threshold for initiation of action potentials so that the slow (50 s to several minutes) depolarisation was accompanied by a train of spikes. Although rare, inhibitory hyperpolarising responses to both these peptides were also seen. Met−ENK produced only small (2−3 mV) hyperpolarisations, however, the strong inhibitory effect of this peptide was demonstrated by its ability to reduce or abolish spontaneous activity. Preliminary results of experiments on peptidergic effects on intracardiac neurones show that SP acts on a subpopulation of these neurones, eliciting a similar depolarising response to that seen in bladder neurones.

From the data reviewed here, it is apparent that there are significant differences in the electrical properties of intramural bladder
and heart neurones. Bladder neurones are in general highly excitable
and frequently exhibit spontaneous activity. In contrast, only a
subpopulation of intracardiac neurones possess these characteristics. In the remainder there are mechanisms which limit the rate and
duration of firing, most notably the existence of a prominent slow
afterhyperpolarisation, during which the cells are refractory. It
has been noted that bladder neurones, unlike those of the heart,
require only small changes in membrane potential or conductance to
affect significant alterations in firing rate. This may in part be
due to the different recording conditions: it is possible that the
medium 199 plus FCS in which recordings from bladder neurones were
made, contained elevated levels of potassium which would raise the
excitability of the cells. However, results from a small number of
experiments carried out in Krebs' solution indicate that the properties under these conditions are not significantly different from
those observed in medium 199. Hence, it is more likely that the
differences in excitability of the neurones in these two preparations reflects the different functional roles played by intramural
bladder and heart ganglia, with bladder neurones being required to
modulate tone whilst heart neurones may act to gate information
transfer from one area of the heart to another, possibly by using a
similar method to that proposed for enteric AH cells (Wood 1984).
The potent and similar action of ATP on both ganglia preparations is
particularly interesting, for it suggests that ATP may act as one of
the major neurotransmitters utilised by intramural neurones in the
local neuronal regulation of heart and bladder.

The intramural ganglia in situ function as sites of relay and possibly integration of the extrinsic innervation to the heart and
bladder. In addition, by analogy with the gastro-intestinal tract,
where the enteric nervous system is thought to mediate local reflexes and where extrinsic nerves act by modulating this activity, the
intramural ganglia of the heart and bladder may also constitute part
of an intrinsic reflex network. Further study of the cultured bladder and heart neurones described here will help determine the nature
of their role in situ.

References

Baluk P, Jessen KR, Saffrey MJ, Burnstock G (1983) The enteric
nervous system in tissue culture. II. Ultrastructural studies of
cell types and their relationships. Brain Res 262: 37-47

Bunge MB, Bunge RP, Peterson ER, Murray MR (1967) A light and electron microscope study of long-term organized cultures of rat dorsal
root ganglia. J Cell Biol 32: 439-466

Crowe RC, Haven AJ, Burnstock G (1986) Intramural neurones of the
guinea-pig urinary bladder: histochemical localization of putative
neurotransmitters in cultures and newborn animals. J Auton Nerv
Syst 15: 319-339

Dalsgaard C-J, Franco-Cereceda A, Saria A, Lundberg JM, Theodorsson-
Norheim E, Hökfelt T (1986) Distribution and origin of substance P-
and neuropeptide Y-immunoreactive nerves in the guinea-pig heart.
Cell Tissue Res 243: 477-485

Hassall CJS, Burnstock G (1984) Neuropeptide Y-like immunoreactivity in cultured intrinsic neurones of the heart. Neurosci Lett 52: 111-115

Hassall CJS, Burnstock G (1986) Intrinsic neurones and associated cells of the guinea-pig heart in culture. Brain Res 364: 102-113

Jessen KR, Burnstock G (1982) The enteric nervous system in tissue culture; a new mammalian model for the study of complex nervous networks. In: Kalsner S (ed) Trends in autonomic pharmacology, vol II. Urban & Schwarzenberg, München, pp 95-115

Jessen KR, Saffrey MJ, Baluk P, Hanani M, Burnstock G (1983a) The enteric nervous system in tissue culture. III. Studies on neuronal survival and the retention of biochemical and morphological differentiation. Brain Res 262: 49-62

Jessen KR, Saffrey MJ, Burnstock G (1983b) The enteric nervous system in tissue culture. I. Cell types and their interactions in explants of the myenteric and submucous plexuses from guinea pig, rabbit and rat. Brain Res 262: 17-35

Kobayashi Y, Hassall CJS, Burnstock G (1986a) Culture of intramural cardiac ganglia of the newborn guinea-pig. I. Neuronal elements. Cell Tissue Res 244: 595-604

Kobayashi Y, Hassall CJS, Burnstock G (1986b) Culture of intramural cardiac ganglia of the newborn guinea-pig. II. Non-neuronal elements. Cell Tissue Res 244: 605-612

Lever JD, Presley R (1971) Studies on the sympathetic neurone in vitro. In: Eränkö O (ed) Histochemistry of nervous transmission. Progr Brain Res Vol 34. Elsevier, Amsterdam, pp 499-512

Mattiasson A, Ekblad E, Sundler F, Uvelius B (1985) Origin and distribution of neuropeptide Y-, vasoactive intestinal polypeptide- and substance P-containing nerve fibers in the urinary bladder of the rat. Cell Tissue Res 239: 141-146

Pittam BS, Burnstock G, Purves RD (1986) Urinary bladder intramural neurones: an electrophysiological study utilizing a tissue culture preparation. Brain Res (in press)

Weihe E, McKnight AT, Corbett AD, Hartschuh W, Reinecke M, Kosterlitz HW (1983) Characterization of opioid peptides in guinea-pig heart and skin. Life Sci Suppl 1, 33: 711-714

Weihe E, Reinecke M, Forssmann WG (1984) Distribution of vasoactive intestinal polypeptide-like immunoreactivity in the mammalian heart; interrelation with neurotensin- and substance P-like immunoreactive nerves. Cell Tissue Res 236: 527-540

Wood JD (1984) Enteric neurophysiology. Am J Physiol 247: G585-G598

Peptides and Transmission in Mammalian Sympathetic Ganglia and Spinal Cord

N. J. Dun[1], N. Mo[1], Z. G. Jiang[2], A. Saria[3], M. Kiraly[4], and R. C. Ma[5]

[1] Department of Pharmacology, Loyola University, Stritch School of Medicine, 2160 S. First Avenue, Maywood, IL 60153, USA
[2] Department of Physiology, Wannan Medical College, Wannan, People's Republic of China
[3] Abteilung für Pharmakologie, Universität Graz, Graz, Austria
[4] Institut für Physiologie, Universität Lausanne, Lausanne, Switzerland
[5] Department of Physiology, Anhui Medical College, Anhui, People's Republic of China

Introduction

The report of immunoreactivity to substance P (SP) in the sympathetic ganglia by Hökfelt et al. (1977) has provided the impetus for
the discovery of immunoreactivities to more than 10 peptides in
various sympathetic ganglia. In particular, the guinea-pig prevertebral ganglia (coeliac-superior and inferior mesenteric) ganglia are
richly endowed with an array of peptides (Schultzberg et al. 1983).
Immunoreactivities to the majority of peptides in sympathetic ganglia are observed in varicose fibers in close proximity to ganglionic
neurons. Further, electron microscopic studies revealed that immunoreactivities to a number of peptides are localized to dense-cored
vesicles contained in nerve endings forming axo-dendritic, axo-
somatic and infrequently axo-axonic synapses (Kondo and Yui 1981,
1982). These are the morphological substrates necessary for a transmitter and/or modulator role of peptides in the ganglia.

Unlike the nicotinic effect of acetylcholine (ACh) on sympathetic
neurons or glutamate on spinal neurons where actions are brisk, the
response of the ganglion cells to peptides is generally slow. On
this basis, peptides would seem to serve primarily as a "gain control" rather than a "switch", thereby providing a modulatory, whether facilitatory or inhibitory, influence on the primary, nicotinic
transmission of the ganglion cells. We shall attempt to give an
overview of several mechanisms, based largely on results obtained
from the guinea-pig inferior mesenteric ganglia (IMG) whereby peptides may modify the electrical activity of sympathetic neurons
and/or ganglionic transmission. A brief comment with respect to the
actions of several peptides on spinal neurons is intended to under-
score the similarity between their central and peripheral effects.

Facilitatory Action of Peptides

Several peptides known to be present in the sympathetic ganglia
including SP, vasoactive intestinal polypeptide (VIP), cholecystokinin (CCK), neurokinin A (NKA), calcitonin gene-related peptide
(CGRP) and vasopressin when administered to IMG neurons caused
characteristically a slow depolarization and enhanced membrane excitability leading to, in many instances, sub- and/or supra-threshold
discharges in otherwise quiescent neurons (Dun and Minota 1981; Mo
and Dun 1984, 1986a; Saria et al. 1985).

The question that is of obvious interest is whether there is a
synaptic pathway(s) subserved by peptides in the IMG. In this
respect, repetitive stimulations of the nerve trunks leading to the

Experimental Brain Research Series 16
© Springer-Verlag Berlin · Heidelberg 1987

IMG evoke, in addition to the fast nicotinic excitatory postsynaptic potentials (EPSPs), a slow depolarizing potential that is resistant to cholinergic antagonists and is termed non-cholinergic EPSP (Dun and Jiang 1982). SP or a related tachykinin was implicated as the mediator underlying the non-cholinergic EPSP (Dun and Jiang 1982; Saria et al. 1985). Interestingly, Kreulen and Peters (1986) showed recently that distension of a piece of colon with its nerves atta-ched to the IMG caused in these neurons a slow depolarization with electrophysiological and pharmacological properties similar to that induced by exogenously applied SP. The significance of their experi-ments is the direct demonstration of a physiological pathway utili-zing peptides, possibly SP, in conveying afferent information from the gastrointestinal tract to the IMG postulated by us several years ago (Dun and Jiang 1982).

It should, however, be stressed that in addition to tachykinins, other peptides and/or indoleamines may serve as the mediators in generating a non-cholinergic EPSP in prevertebral neurons (Dun and Kiraly 1983; Dun et al. 1984). The development of selective antago-nists to various peptides will be necessary to decipher more preci-sely the synaptic pathways subserved by different peptides in the peripheral as well as central nervous systems.

What might be the cellular function of peptides and of non-choliner-gic EPSP in IMG neurons? As most of the central cholinergic inputs to the prevertebral neurons are subthreshold, and as summation of several inputs is necessary to discharge the neuron (Szurszewski 1981), the non-cholinergic EPSP which as a time course of seconds to minutes, would be effective in synchronizing the likelihood of temporal summation of cholinergic EPSPs impinging upon a ganglion cell. Further, an increase of membrane input resistance during the course of non-cholinergic EPSP should increase space constant of the soma-dendritic membrane, thus, maximizing the likelihood of spatial summation of cholinergic EPSPs. These two mechanisms in concert provide an effective means of ensuring the firing of a ganglion cell over a period of time, despite the converging inputs being subthres-hold as depicted in Fig. 1.

Interestingly, this type of facilitation can be demonstrated in central neurons as well, as SP, NKA, VIP and vasopressin caused in rat lateral horn cells a slow depolarization characteristically similar to that evoked by these peptides in IMG neurons. More impor-tantly, the subthreshold EPSPs evoked by dorsal rootlet stimulations were augmented during the course of slow depolarization, resulting in a discharge of the neurons in question (Ma and Dun 1985; Mo and Dun 1986b). Thus, one function common to several peptides in the nervous system may be related to their ability in modulating neuro-nal excitability over a period of time.

A second type of facilitatory action of peptides identified in sympathetic ganglia involves the enhancement of the response caused by another transmitter. An example is VIP. In a portion of IMG neurons, while causing little or no membrane depolarization, VIP enhanced the depolarization or hyperpolarization induced by muscari-nic agonists such as methacholine (Mo and Dun 1984). A similar facilitatory effect of VIP on muscarinic discharges was reported in cat superior cervical ganglia (Kawatani et al. 1985). It should be pointed out that the facilitatory effect of VIP on muscarinic res-ponse appears to be independent of its ability in causing a slow depolarization. The mechanism whereby VIP enhances the muscarinic response in sympathetic neurons is not known. In this respect, VIP was found to increase the affinity of muscarinic agonists to the

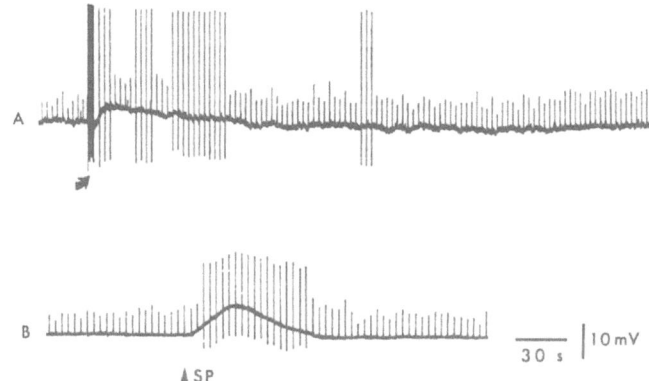

Fig. 1. Potentiation of cholinergic EPSPs during the course of non-cholinergic EPSP and substance P (SP)-induced depolarization in two guinea-pig IMG neurons. Subthreshold cholinergic EPSPs (small upward deflections) were evoked by stimulation of the right hypogastric nerve, and non-cholinergic EPSP was elicited by repetitive stimulation of the left hypogastric nerve (curved arrow, 30 Hz, 2 s). SP-depolarization was elicited by pressure ejection of SP (arrowhead, 100 ms pulse duration). Subthreshold cholinergic EPSPs were clearly potentiated and converted to action potentials at the peak of non-cholinergic EPSP (tracing A) and SP-induced depolarization (tracing B)

receptors in submandibular glands (Lundberg et al. 1982). Whether VIP enhances muscarinic response by a similar mechanism in the sympathetic ganglion remains to be established.

Inhibitory Modulation by Peptides

Contrary to the facilitatory actions of the majority of peptides, enkephalins (ENK) depresses the amplitude of cholinergic EPSPs as well as the non-cholinergic EPSP (Konishi 1985). The findings that the synaptic depression occurred without appreciable change of the resting membrane properties of the ganglion cells or the sensitivity of the postsynaptic membrane to iontophoretically applied ACh or SP suggest that ENK suppresses synaptic transmission principally by a presynaptic mechanism whereby the amount of transmitter release is impaired (Jiang et al. 1982; Konishi 1985).

A physiological role of ENK in dampening the on-going synaptic activity was suggested by the observations that cholinergic EPSPs and non-cholinergic EPSP were reduced by conditioning stimulations and enhanced by naloxone, an opiate antagonist (Jiang et al. 1982; Konishi 1985). It was inferred from these observations that the synaptically released ENK exerts an inhibitory effect on the release of transmitters from presynaptic nerve endings and that naloxone causes disinhibition.

Although presynaptic inhibition appears to be the predominant effect in the sympathetic ganglia, opiate peptides hyperpolarize myenteric neurons and prolong spike afterhyperpolarization, probably by activating a Ca-dependent K conductance (Morita and North 1981). In this case, opiate peptides may alter the input and output ratio of the ganglion cells by prolonging the spike afterhyperpolarization.

The mechanism by which opiate peptides inhibit transmitter release is not entirely clear. Two possible mechanisms should be considered in this context. First, Ca influx associated with presynaptic nerve terminal action potential might be suppressed. There is evidence that opiates shorten the duration of action potentials elicited from cultured sensory neurons by suppressing the Ca current (Mudge et al. 1979). An alternative explanation is that opiates block the invasion of action potential into presynaptic terminals. This is deduced from experiments in which opiates, when applied to the cell processes leading to the myenteric neurons, were found to block action potential invasion into the soma (Morita and North 1981).

Acknowledgements. Supported by NS15848 & NS18710 from the Department of Health and Human Services.

References

Dun NJ, Jiang ZG (1982) Non-cholinergic excitatory transmission in inferior mesenteric ganglia of the guinea-pig: possible mediation by substance P. J Physiol (Lond) 325: 145–159

Dun NJ, Kiraly M (1983) Capsaicin releases a substance P-like peptide in guinea-pig inferior mesenteric ganglia. J Physiol (Lond) 340: 107–120

Dun NJ, Minota S (1981) Effects of substance P on neurones of the inferior mesenteric ganglia of the guinea-pig. J Physiol (Lond) 321: 259–271

Dun NJ, Kiraly M, Ma RC (1984) Evidence for a serotonin-mediated slow excitatory potential in the guinea-pig coeliac ganglia. J Physiol (Lond) 351: 61–76

Hökfelt T, Elfvin LG, Schultzberg M, Goldstein M, Nilsson G (1977) On the occurrence of substance P-containing fibers in sympathetic ganglia: immunohistochemical evidence. Brain Res 132: 29–41

Jiang ZG, Simmons MA, Dun NJ (1982) Enkephalinergic modulation of non-cholinergic transmission in mammalian prevertebral ganglia. Brain Res 235: 185–191

Kawatani M, Rutigliano M, deGroat WC (1985) Depolarization and muscarinic excitation induced in a sympathetic ganglion by vasoactive intestinal polypeptide. Science 229: 879–881

Kondo H, Yui R (1981) An electron microscopic study on substance P-like immunoreactive nerve fibers in the celiac ganglion of guinea pigs. Brain Res 222: 134–137

Kondo H, Yui R (1982) An electron microscopic study on VIP-like immunoreactive nerve fibers in the celiac ganglion of guinea pigs. Brain Res 237: 227–231

Konishi S (1985) Opioid peptides: peripheral nervous system. In: Rogawski MA, Barker JL (eds) Neurotransmitter actions in the vertebrate nervous system. Plenum, New York, pp 365

Kreulen DL, Peters S (1986) Non-cholinergic transmission in a sympathetic ganglion of the guinea-pig by colon distension. J Physiol (Lond) 374: 315-334

Lundberg JM, Hedlund B, Bartfai T (1982) Vasoactive intestinal polypeptide enhances muscarinic ligand binding in cat submandibular salivary gland. Nature 295: 147-149

Ma RC, Dun NJ (1985) Vasopressin depolarizes lateral horn cells of the neonatal rat spinal cord in vitro. Brain Res 348: 36-43

Mo N, Dun NJ (1984) Vasoactive intestinal polypeptide facilitates muscarinic transmission in the mammalian sympathetic neurons. Neurosci Lett 52: 19-24

Mo N, Dun NJ (1986a) Cholecystokinin octapeptide depolarizes guinea-pig inferior mesenteric ganglion cells and facilitates nicotinic transmission. Neurosci Lett 64: 262-268

Mo N, Dun NJ (1986b) Depolarization and facilitation of synaptic transmission in lateral horn neurons of neonatal rat spinal cord slices by substance P. Soc Neurosci Abstr 12: 149

Morita K, North RA (1981) Opiates and enkephalin reduce the excitability of neuronal processes. Neuroscience 6: 1943-1951

Mudge AW, Leeman SE, Fischbach GD (1979) Enkephalin inhibits release of substance P from sensory neurons in culture and decreases action potential duration. Proc Natl Acad Sci USA 76: 526-530

Saria A, Ma RC, Dun NJ (1985) Excitatory effects of neurokinin A on neurones of the guinea-pig inferior mesenteric ganglia. Neurosci Lett 60: 145-150

Schultzberg M, Hökfelt T, Lundberg JM, Dalsgaard CJ, Elfvin LG (1983) Transmitter histochemistry of autonomic ganglia. In: Elfvin LG (ed) Autonomic ganglia. Wiley, Chichester, pp 205

Szurszewski JH (1981) Physiology of mammalian prevertebral ganglia. Ann Rev Physiol 43: 53-68

A Presynaptic Action of Vasopressin in the Superior Cervical Ganglion of the Rat

M. Kiraly[1], E. Tribollet[2], M. Dolivo[1], and J. J. Dreifuss[2]

[1] Institut für Physiologie, Medizinische Fakultät, 1005 Lausanne, Switzerland
[2] Abteilung für Physiologie, Medizinisches Zentrum der Universität, 1211 Genf, Switzerland

Introduction

Vasopressin and oxytocin are neuropeptides which have been known for many years to act as hormones on kidney, liver and smooth muscle cells. It has also been established that both neuropeptides can affect neuronal activity in the central nervous system (Mühlethaler et al. 1985), for example in areas related to the autonomic nervous system such as the nucleus of the solitary tract (Morris et al. 1984), the dorsal motor nucleus of the vagus nerve (Charpak et al. 1984) and the intermediolateral column of the spinal cord (Gilbey et al. 1982; Ma and Dun 1985). More recently, Hanley et al. (1984) detected vasopressin-like immunoreactivity in both nerve fibres and in adrenergic neurones in the rat superior cervical ganglion and Bone et al. (1984) reported that vasopressin causes a receptor-mediated stimulation of the turnover of membrane inositol lipids in this ganglion. It was thus conceivable that vasopressin, in addition to its other nervous actions, may be directly involved in the modulation of peripheral autonomic function. This prompted us to investigate the effects of vasopressin on rat sympathetic neurones and on impulse transmission across the superior cervical ganglion (see also Kiraly et al. 1985, 1986).

Effects of Vasopressin on Ganglionic Transmission

For electrophysiological investigations intracellular recordings were carried out in vitro and synaptic potentials were evoked by stimulation of the preganglionic nerve. Vasopressin superfused for 1 or 2 min at 0.01–10 uM reversibly depressed synaptic transmission by reducing the amplitude of f–EPSPs in 94 of 195 cells tested. A representative experiment is shown in Fig. 1, where vasopressin first blocked the generation of the action potential and then reduced even further the amplitude of the f–EPSPs. These effects were fully reversible.

Quantitative evaluation of the synaptic depression was carried out by measuring the peak amplitude of f–EPSPs recorded before and during vasopressin application from neurons either superfused with modified Krebs solution containing 0.75 mM Ca^{++}/4 mM Mg^{++} or from neurones slightly hyperpolarized by applying continuous anodal current. Synaptic reduction, expressed as percentage of the control value, amounted to 46.9 ± 5.7 (mean ± S.E.M., n = 9), 49.7 ± 5.7 (n = 14), 22.4 ± 4.7 (n = 14) and 5.4 ± 1.9 (n = 5) at 10, 1, 0.1 and 0.05 uM, respectively. This inhibitory effect of vasopressin was reversibly abolished by the structural analog d(CH_2)$_5$Tyr(Et)VAVP, a potent antagonist of the vasopressor effect of vasopressin.

Experimental Brain Research Series 16
© Springer-Verlag Berlin · Heidelberg 1987

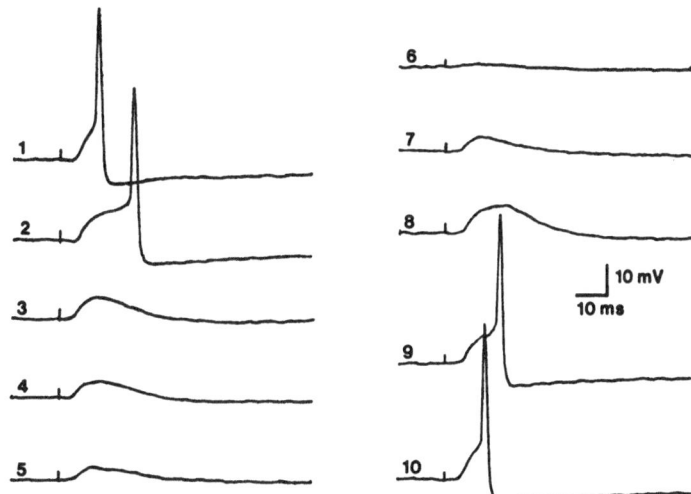

Fig. 1. Effect of vasopressin on the response of a rat ganglion cell to orthodromic stimulation. Records 1–10 are consecutive, but not continuous. Record 1 was taken in Krebs solution; records 2 to 5 during 2 min superfusion with 1 uM vasopressin; records 6 to 10 show progressive recovery after 1, 2, 5, 7 and 15 min washout

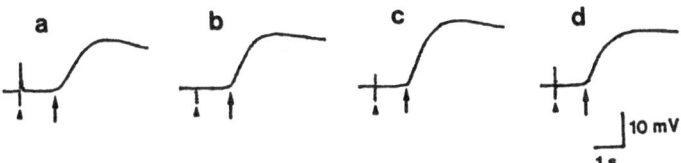

Fig. 2. Effect of vasopressin on f–EPSPs and on ACh–potentials, recorded from a ganglion cell. Arrowheads and arrows indicate artifact of stimulation and ACh pressure ejection (0.2 M, 4 ms, 1.8 bar), respectively. Records a and b were obtained before and 2 min after application of vasopressin 10 uM; c and d were taken 3 and 7 min after vasopressin superfusion was discontinued. Note the selective depression of the f–EPSP

A crucial observation with regard to the site of action of vasopressin was that vasopressin did not reduce the amplitude of the postsynaptic depolarization induced by acetylcholine (ACh) applied either by superfusion or pressure ejection. As shown in Fig. 2, vasopressin depressed the f–EPSP without reducing the ACh potential. This result implies a presynaptic site of action for vasopressin. In line with this contention are our observations that the vasopressin-induced synaptic depression was atropine-resistant and that vasopressin at micromolar concentrations did not produce any marked effect on the postsynaptic membrane potential or conductance

Table 1. Effect of vasopressin, 10 uM, on spontaneous and evoked mEPSPs

| | spontaneous mEPSPs | | evoked mEPSPs | |
	frequency (s^{-1})	amplitude (mV)	Quantal content	Quantal size (mV)
CONTROL	0.21* ± 0.04	0.86 ± 0.04	1.46 ± 0.21	0.97 ± 0.17
VASOPRESSIN	0.13** ± 0.03	0.83 ± 0.05	0.61*** ±.0.25	1.13 ± 0.17

*   All values represent means $\pm$ S.E.M. (n = 4 or 8)
**   p<0.01 vs. control
***   p<0.05 vs. control

Effect of Vasopressin on Presynaptic ACh Release

A study of spontaneous and evoked miniature EPSPs (mEPSP) also corroborated the hypothesis of a presynaptic site of action for vasopressin. The mean quantal content and size of evoked mEPSPs were determined from cells superfused with a modified Krebs solution containing 0.5 mM Ca^{++}/5.5 mM Mg^{++}. Quantal content was obtained by the failure method. Table I shows that vasopressin significantly decreased the mean quantal content without affecting the quantal size. It also shows the mean value for the frequency and amplitude of spontaneous mEPSPs obtained from cells superfused with a K^+ enriched (20 mM) Krebs solution. Vasopressin significantly reduced the frequency of spontaneous mEPSPs without affecting their amplitude.

A presynaptic site of action of vasopressin, as is apparent from these data, is reminiscent of what is known of other neuromodulators affecting ganglionic transmission. Thus, both enkephalins and amines have been previously shown to reduce ACh output in mammalian sympathetic ganglia (Dun and Nishi 1974; Dun and Karczmar 1977; Konishi et al. 1979). In this respect it is noteworthy that vasopressin at equimolar concentrations appears at least as potent as enkephalins and even more potent than catecholamines.

Receptor Characterization

Two types of vasopressin receptors have been differentiated on the basis of their affinities for synthetic structural analogs and of the nature of their second messenger. Vasopressor or V_1 receptors occur in hepatocytes and vascular smooth muscle cells where they stimulate inositol lipid breakdown. Antidiuretic or V_2 receptors are found in renal tubular cells where their action is linked to adenylate cyclase. We therefore attempted to assess, by using autoradiography, if one of these two receptor-subtypes was present in rat superior cervical ganglion. Unfixed, frozen sections were incubated with 1.5 nM tritiated vasopressin of high specific activity as

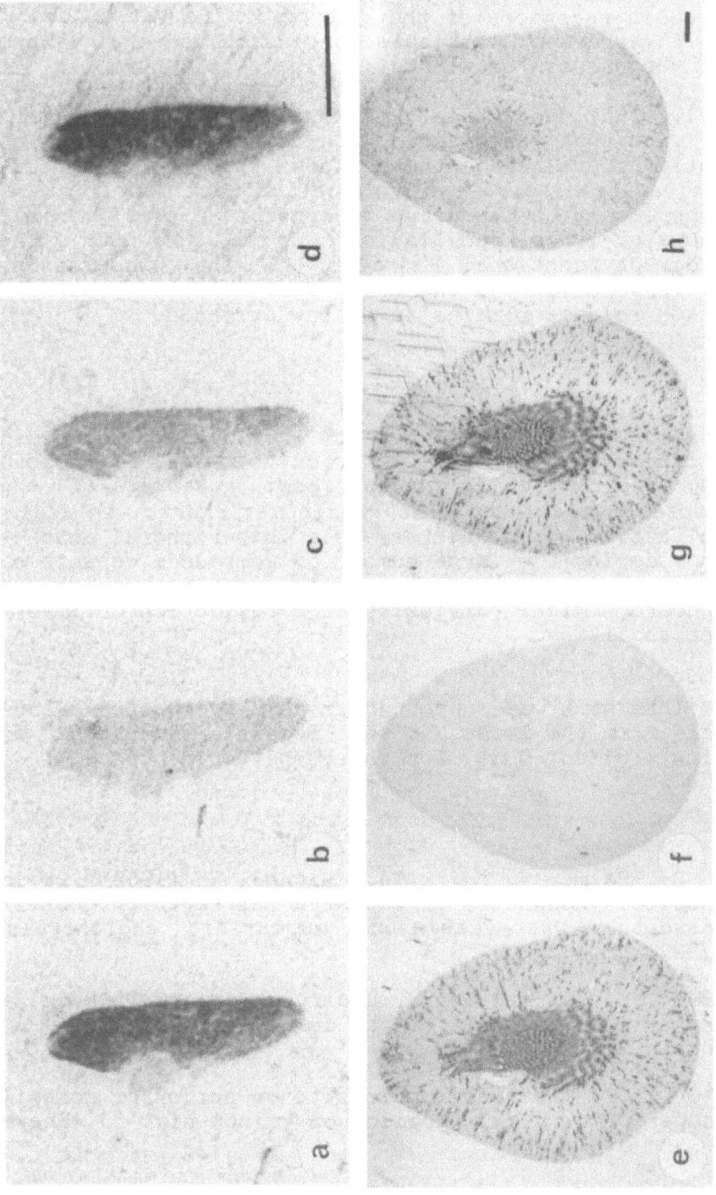

Fig. 3. Vasopressin binding in sections of superior cervical ganglion and of kidney. Photographs a and e show the labelling in sections incubated with 1.5 nM (³H)vasopressin. Most of this labelling was displaced by 10 uM unlabelled vasopressin (b and f). The V₁ against Phe₂Orn₈VT, 100 nM, displaced radioactivity in the superior cervical ganglion (c) but not in the kidney (g). In contrast, the V₂ agonist dTyr(Me)VDAVP displaced vasopressin bound to renal (h) but not to ganglionic tissue (d). Calibration bar = 2 mm

described elsehwere (Kiraly et al. 1986). Displacement of the labelled ligand was explored using a selective vasopressin agonist and a potent antidiuretic agonist. After 4 months of exposure binding of (³H)vasopressin was found throughout the ganglion and in the central parts of the kidney (Fig. 3). Incubation with unlabelled AVP resulted in an almost complete suppression of binding in both tissues. The vasopressor agonist displaced the binding in the ganglion, but not in the kidney. The opposite was observed with the antidiuretic agonist: binding was nearly suppressed in the kidney, but hardly modified in the ganglion.

These results provide evidence for the presence of specific and functional vasopressin receptors of the V_1-type in the rat superior cervical ganglion. In agreement with this conclusion, several groups had reported that vasopressin increases inositol lipid turnover in the ganglion (Bone et al. 1984; Hanley et al. 1984; Horwitz et al. 1986; Kiraly et al. 1986). Recent data of Horwitz et al. (1986) are in accord with the view that vasopressin, in contrast to muscarine, acts at a presynaptic site. These authors showed that the incorporation of (^{3}H)inositol after vasopressin application took mostly place over neuropil, whereas with muscarine incorporation was maximal around the perikarya of large ganglionic neurones. However, the existence of a direct functional link between the vasopressin-induced inositol incorporation and the vasopressin-induced presynaptic inhibition of ACh release remains to be investigated.

Conclusion

In this communication, we have shown that exogenous vasopressin reduces impulse transmission in the superior cervical ganglion, probably by acting presynaptically on vasopressin receptors of the V_1-type. We therefore suggest that vasopressin may exert, in addition to its central action, a modulatory role in peripheral autonomic regulation. Considering the large number of compounds capable of acting as neuromodulators in autonomic ganglia, it is tempting to imagine that this neurochemical complexity is a reflection of important integrative capabilities.

Acknowledgements. This work was supported in part by grants 3.258-0.85 and 3.358-0.86 from the Swiss National Science Foundation. We thank Dr. M. Manning (Toledo, Ohio) for the supply of peptides.

References

Bone EA, Fretten P, Palmer S, Kirk CJ, Michell RH (1984) Rapid accumulation of inositol phosphates in isolated rat superior cervical ganglia exposed to V_1-vasopressin and muscarinic cholinergic stimuli. Biochem J 221: 803-811

Charpak S, Armstrong WE, Mühlethaler M, Dreifuss JJ (1984) Stimulatory action of oxytocin on neurones of the dorsal motor nucleus of the vagus nerve. Brain Res 300: 83-89

Dun N, Karczmar AG (1977) The presynaptic site of action of norepinephrine in the superior cervical ganglion of guinea-pig. J Pharm Exp Ther 200: 328-335

Dun N, Nishi S (1974) Effects of dopamine on the superior cervical ganglion of the rabbit. J Physiol (Lond) 239: 155-164

Gilbey MP, Coote JH, Fleetwood-Walker S, Peterson DF (1982) The influence of the paraventriculo-spinal pathway, and oxytocin and vasopressin on sympathetic preganglionic neurones. Brain Res 251: 283-290

Hanley RM, Benton HP, Lightman SL, Todd K, Bone EA, Fretten P, Palmer S, Kirk CJ, Michell RH (1984) A vasopressin-like peptide in the mammalian sympathetic nervous system. Nature 309: 258-261

Horwitz J, Anderson CH, Perlman RL (1986) Comparison of the effects of muscarine and vasopressin on inositol phospholipid metabolism in the superior cervical ganglion of the rat. J Pharm Exp Ther 237: 312-317

Kiraly M, Audigier S, Tribollet E, Barberis C, Dolivo M, Dreifuss JJ (1986) Biochemical and electrophysiological evidence of functional vasopressin receptors in the rat superior cervical ganglion. Proc Natl Acad Sci USA 83: 5335-5339

Kiraly M, Maillard M, Dreifuss JJ, Dolivo M (1985) Neurohypophysial peptides depress cholinergic transmission in a mammalian sympathetic ganglion. Neurosci Lett 62: 89-95

Konishi S, Tsunoo A, Otsuka M (1979) Enkephalins presynaptically inhibit cholinergic transmission in sympathetic ganglia. Nature 282: 515-516

Ma RC, Dun NJ (1985) Vasopressin depolarizes lateral horn cells of the neonatal rat spinal cord in vitro. Brain Res 348: 36-43

Morris R, Farmery SM, Roberts CJ, Hill RG (1984) The effects of oxytocin and vasotocin analogues on the responses of rat brainstem neurones to oxytocin. Neurosci Lett 48: 161-166

Mühlethaler M, Raggenbass M, Dreifuss JJ (1985) Oxytocin and vaso-pressin. In: Rogawski MA, Barker JL (eds) Neurotransmitter actions in the vertebrate nervous system. Plenum, New York, pp 439

Autonomic Control of Penile Erectile Tissue

W. G. Dail

Department of Anatomy, School of Medicine, University of New Mexico, Albuquerque, NM 87131, USA

Introduction

Regulation of visceral tissues is regarded as the function of the
autonomic nervous system (ANS), yet there is a remarkable hierarchy
in the degree of dependence of tissues on extrinsic autonomic innervation. Some organs rely on the central control of the ANS simply to
modify activity generated and largely regulated within the organ,
such is the case of the gastrointestinal tract. Other tissues, for
example the urinary bladder, are totally reliant on preganglionic
autonomic nerves to subserve any normal function. Complete dependence is also true of penile erectile tissue wherein interruption of
autonomic innervation results in erectile failure. Compared to other
visceral tissues penile erectile tissue has been little explored as
a model of autonomic regulation. This discussion will review the
nature of the effector tissues in the penis, the neuroanatomy of
penile erection, the neurotransmitters present in penile nerves and
their effects on smooth muscle in the penis.

Angioarchitecture of Erectile Tissue

Control of penile blood flow is effected by smooth muscle in two
locations: muscular arteries and columns of muscle lining the cavernous spaces (trabecular muscle). On entering the penis, the deep
artery branches into multiple vessels which empty into the intercon-
necting cavernous spaces. In the proximal portion of the mammalian
penis, the corpora cavernosa penis (CCP) are composed almost entirely of trabecular smooth muscle surrounding endothelial-lined spaces
while in the more distal portion of the penis trabecular muscle is
rare. This conformation of the cavernous spaces is consistent with
the sparse innervation of the shaft of the penis and suggests that
the distal portion of the penis serves as a reservoir, the filling
of which is dictated by the events which occur in the crura. Veins
which drain the CCP are located at the periphery of the cavernous
spaces (Fig. 1) just deep to the collagenous tunica albuginea. Since
in mammals the principal, and in some animals the exclusive, drainage of the cavernous spaces is into the peripherally-located veins,
passive occlusion of venous outflow from the cavernous spaces may
contribute to erection.

Nerves in Erectile Tissue

Neurohistochemical studies show four types of nerve fibers in the
CCP: adrenergic fibers, acetylcholinesterase positive (AChE+) fibers
and nerves immunoreactive for vasoactive intestinal polypeptide

Experimental Brain Research Series 16
© Springer-Verlag Berlin · Heidelberg 1987

(VIP) and neuropeptide Y (NPY). With the possible exception of NPY, on which there have been few studies, all nerves are distributed to vascular and trabecular smooth muscle. Therefore, there is no differential distribution of penile nerves which might imply a unique site of action of one type of neurotransmitter.

Autonomic Ganglia of Penile Erectile Tissue

Dye tracing studies in the rat show that penile ganglion cells are located exclusively in the sympathetic chain and in the pelvic plexus (Dail et al. 1985). In the cat, penile neurons also occur in the inferior mesenteric ganglion (Booth et al. 1986). Sympathetic penile neurons in the rat are concentrated in lower lumbar and sacral ganglia (Fig. 2) but also occur in chain ganglia as far rostral as the diaphragm. Surgical sympathectomy indicates that the sympathetic chain is the major if not the exclusive source of adrenergic innervation of the CCP in the rat. Penile neurons in the major pelvic ganglion (MPG) almost all are AChE+ and are immunoreactive for VIP (Dail et al. 1986a). No penile neurons in the MPG stain for tyrosine hydroxylase, indicating further that penile adrenergic neurons reside in chain ganglia. Therefore, an earlier pharmacological study which suggested that the penis of the rat is innervated by short adrenergic neurons (Dail and Evan 1974) is not supported by tracing studies.

Spinal Control of the Penis

Smooth muscle in the penis is controlled by three autonomic centers in the spinal cord, one at the sacral level and two in the thoracolumbar region (Fig. 2). The sacral spinal cord is the principal center which initiates penile erection. Interruption of the sacral cord outflow does not prevent erection in some animals although impotence results if the hypogastric nerve is included in the lesion. Tracing and histochemical studies of the so-called "suprasacral" center for erection have revealed that some 20% of the penile neurons in the pelvic plexus are innervated by preganglionic fibers in the hypogastric nerve (Fig. 2) (Dail et al. 1985). This neuroanatomical study supports the contention that there is a sympathetic component to penile erection and, in view of the histochemical characteristics of penile neurons in the pelvic plexus, further suggests that it is a sympathetic cholinergic component (Dail et al. 1986a). Neurophysiological studies in the rabbit indicate that the penile pathway in the hypogastric nerve operates through cholinergic ganglion cells (Sjöstrand and Klinge 1979). At present it is not known if the pelvic nerve and hypogastric nerve innervate separate populations of penile neurons or converge in part on a single population. If the parasympathetic outflow from the sacral cord and the sympathetic outflow in the hypogastric nerve subserve the same function, it would be interesting to know if and how the spinal cord is organized to coordinate the vasodilator discharge.

Little is known of the organization, orientation and numbers of penile neurons in the spinal cord. In the sacral spinal cord of the dog, penile neurons are said to occupy the lateral border of the intermediate gray matter (Lue et al. 1984), but this has not been determined for other species. Similar information on preganglionic sympathetic penile neurons is not available; however, preganglionic neurons which form the hypogastric nerve and those which contribute to the sympathetic chain have been studied in the rat (Nadelhaft and McKenna 1985) and presumably this would include preganglionic penile

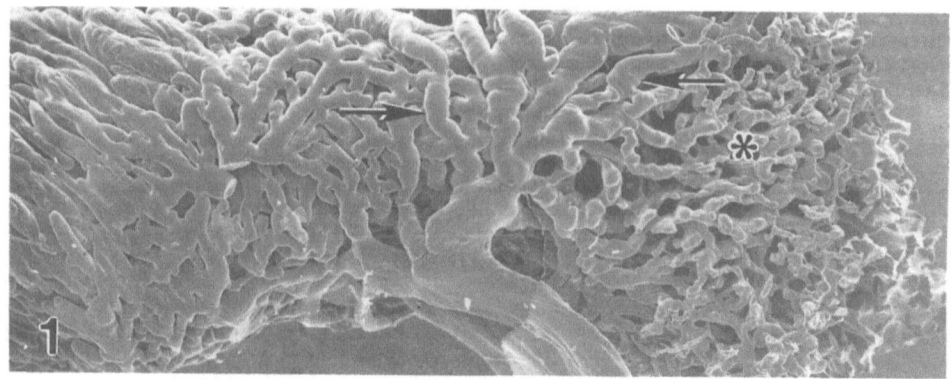

Fig. 1. Scanning electron micrograph of a vascular cast of the crus of the rat penis (Mercox-filled). The cavernous spaces, shown at the fractured surface on the right of the micrograph (asterisk), drain to surface veins (arrows)

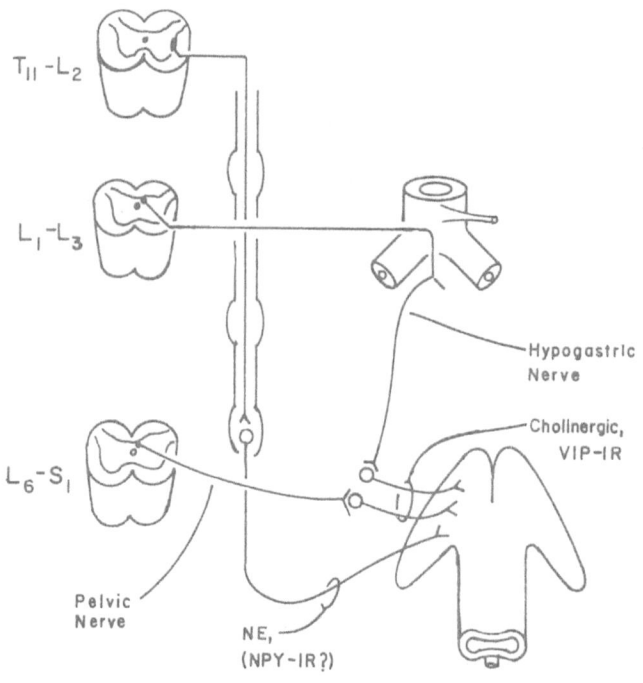

$T_{II}-L_2$

L_1-L_3

Hypogastric
Nerve

Cholinergic,
VIP-IR

L_6-S_1

Pelvic
Nerve

NE,
(NPY-IR?)

2

Fig. 2. Schematic diagram of the autonomic innervation of the penis of the rat. The spinal cord location of preganglionic penile neurons is generalized from studies of the origins of autonomic outflows in these regions. The source of NPY-immunoreactive fibers to the penis is presumed to be the sympathetic chain, but has yet to be determined.

neurons. The hypogastric nerve in the rat originates from spinal levels L_{1-3} while preganglionics which contribute to sacral chain ganglia arise from $T_{11}-L_2$. Moreover, most of the hypogastric neurons are located in the dorsal gray commissure while sympathetic chain preganglionics are found in the intermediolateral edge of the gray matter.

Neurotransmitters of Penile Nerves

In the flaccid state, the cavernous spaces are closed and blood flow through penile vessels is minimal. Adrenergic innervation of alpha receptors may maintain the non-erect state by providing an excitatory tone to penile vessels and to trabecular muscle (Sjöstrand and Klinge 1979). Inhibition of adrenergic tone may be an important component of penile erection, and detumescence may be caused by a reinstatement of tone. Suggestions that adrenergic nerves facilitate penile erection were based in part on the observation that stimulation of the hypogastric nerves in some animals caused erection (for review, see Sjöstrand and Klinge 1979). However, since this pathway is probably cholinergic, the case for an adrenergic contribution to erection seems lessened. Beta adrenergic receptors are present in the CCP and their activation relaxes penile smooth muscle, but at present their physiological significance is unclear.

There is less agreement on the nature of the vasodilator substances which inhibit penile muscle. Challenges to the role of acetylcholine as the inhibitory transmitter of penile nerves are based on the inability of atropine to completely block nerve-induced erection and the failure of exogenous acetylcholine to closely mimic nerve-induced erection (Sjöstrand and Klinge 1979). The presence of VIP in penile nerves, the ability of VIP to relax penile muscle, and evidence that VIP is released from the penis during erection strengthen arguments that this peptide is responsible for the atropine insensitive component of penile erection (Willis et al. 1981; Virag et al. 1983; Andersson et al. 1984). Although the atropine resistance of nerve-induced erection may be due to VIP, it is, however, increasingly clear that penile nerves from the pelvic plexus are cholinergic and more probably, cholinergic-VIPergic. In man, acetylcholine relaxes norepinephrine-contracted muscle strips from the penis (Hedlund and Andersson 1985) and in the rat, acetylcholine inhibits nerve-induced contractions of trabecular muscle (Dail, McGuffee, Minorsky and Little, unpublished observations). The localization of cholinergic muscarinic receptors in the erectile tissue of the rat (Dail et al. 1986b) and the presence of choline acetyltransferase in penile nerves in this species argue for a cholinergic mechanism in erection. Studies indicate that acetylcholine may contribute to erection both by a direct effect on penile smooth muscle and by presynaptic inhibition of the release of norepinephrine (Hedlund and Andersson 1985).

Acknowledgements. Work described in this paper was supported by NIH Grant RO1NS19839 and NIH RRO8139.

References

Andersson P-O, Bloom SR, Mellander S (1984) Haemodynamics of pelvic nerve induced erection in the dog: possible mediation by vasoactive intestinal polypeptide. J Physiol 350: 209-224

Booth JM, Roppolo RD, deGroat WC (1986) Distribution of cells and fibers projecting to the penis of the cat. Soc Neurosci Abstr 12: 1056

Dail WG, Evan AP (1974) Experimental evidence that the penis of the rat is innervated by short adrenergic neurons. Amer J Anat 141: 203–218

Dail WG, Manzanares K, Moll MA, Minorsky N (1985) The hypogastric nerve innervates a population of penile neurons in the pelvic plexus. Neuroscience 16: 1379–1386

Dail WG, Minorsky N, Moll MA, Manzanares K (1986a) The hypogastric nerve pathway to penile erectile tissue: histochemical evidence supporting a vasodilator role. J Auton Nerv Syst 15: 341–349

Dail WG, Savage DD, Minorsky N (1986b) Localization of muscarinic cholinergic receptors in corpora cavernosa penis. Anat Rec 214: 28A

Hedlund H, Andersson K-E (1985) Comparison of the responses to drugs acting on adrenoreceptors and muscarinic receptors in human isolated corpus cavernosum and cavernous artery. J Auton Pharmac 5: 81–88

Lue TF, Zeineh SJ, Schmidt RA, Tanagho EA (1984) Neuroanatomy of penile erection: its relevance to iatrogenic impotence. J Urol 131: 273–280

Nadelhaft I, McKenna KE (1985) Sympathetic afferent and preganglionic neurons labelled by horseradish peroxidase applied to the hypogastric nerve and the sacral sympathetic chain of the rat. Soc Neurosci Abstr 11: 764

Sjöstrand NO, Klinge E (1979) Principal mechanisms controlling penile erection and protrusion in rabbits. Acta Physiol Scand 106: 199–214

Virag R, Ottesen B, Fahrenkrug J, Levy C, Wagner G (1982) Vasoactive intestinal polypeptide release during penile erection in man. Lancet II: 1166

Willis EA, Ottesen B, Wagner G, Sundler F, Fahrenkrug J (1981) Vasoactive intestinal polypeptide (VIP) as a possible neurotransmitter involved in penile erection. Acta Physiol Scand 113: 545–547

Effect of Noradrenaline on the Electrical Activities of Lateral Horn Cells in Cat Spinal Cord Slices

S. Nishi[1], M. Yoshimura[1], and C. Polosa[2]

[1] Department of Physiology, Kurume University School of Medicine, Kurume, Japan
[2] Department of Physiology, McGill University, Montreal, Canada

Introduction

The intermediolateral nucleus of the thoracic spinal cord receives abundant catecholamine-containing nerve terminals, apparently originating from supraspinal structures (Carlsson et al. 1964; Dahlström and Fuxe 1965). The functional role of this catecholaminergic innervation has been controversial. Several workers have reported that microiontophoretically-applied catecholamines inhibit sympathetic preganglionic neurones (SPNs) (Ryall 1967; Coote et al. 1981; Guyenet and Cabot 1981; Kadzielawa 1983). On the other hand, there are also findings which suggest an excitatory role for catecholamines on SPNs (Hare et al. 1972; Simon and Schram 1983; Ross et al. 1984).

With intracellular recording from the SPNs in the slice of the upper thoracic spinal cord of the adult cat, Yoshimura and Nishi (1982) have shown that noradrenaline (NA) causes either depolarization or hyperpolarization or both, depending on individual neurones. Recently, Yoshimura et al. (1986a) have found that regardless of the NA effects on membrane potential, NA commonly causes in the majority of SPNs the appearance of a prominent depolarizing spike-afterpotential. This after-depolarization could result in repetitive firing of the neurone in response to a single intracellular current pulse.

This communication deals with the nature and mechanism of the NA-induced alterations in the action potential and afterpotentials of the SPN.

Alterations of Spike and Afterpotential by NA

When the slice preparation of cat spinal cord (Yoshimura and Nishi 1982) was superfused with Krebs solution containing NA (10–50 uM), the following changes in the action potential of the SPN occurred: the shoulder on the repolarizing phase of the spike was depressed, an afterdepolarization (ADP) which was absent in normal Krebs solution appeared, and the slow component of the afterhyperpolarization (AHP) was depressed or abolished. These changes appeared within one minute of drug application and were quickly reversible upon return to control Krebs solution.

The ADP appeared gradually following initiation of NA superfusion and was well developed by 1 to 2 min (Fig. 1A,B). Subthreshold ADPs had time to peak of approximately 50 ms and decay time of 100 to 600 ms. With increasing time of exposure to an NA-containing medium (20–50 uM) the ADP became larger and reached threshold for spike initia-

Experimental Brain Research Series 16
© Springer-Verlag Berlin · Heidelberg 1987

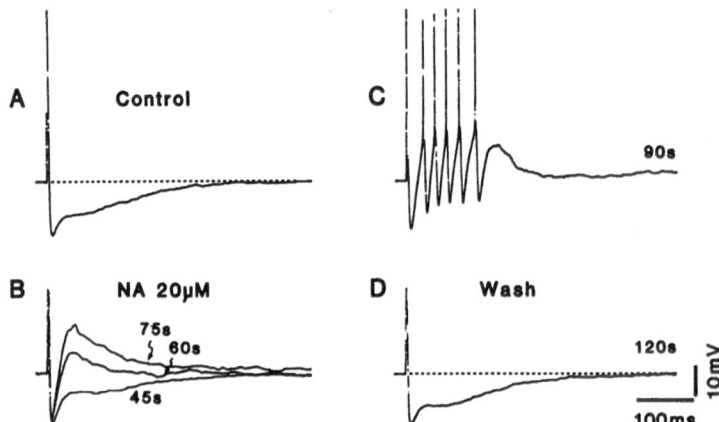

Fig. 1. Sympathetic preganglionic neuron. Depolarizing afterpotential caused by NA superfusion. Records A, B and C: taken before (A), 45, 60 and 75 s after (B) and 90 s after (C) beginning of NA (20 uM) superfusion, respectively. Record D: taken 90 s after discontinuing NA superfusion. All records are from the same neurone. Spike at beginning of sweep evoked by intracellular current injection (duration 2 ms). Spikes are distorted by pen recorder

tion (Fig. 1C). Firing was usually repetitive, and its maximum frequency reached 40 to 50 Hz.

The ADP was associated with a decrease in neurone input resistance (R_N). The decrease was proportional to the size of the ADP. At the ADP peak the R_N was approximately 70% of control.

Ionic Requirements of the NA-Induced ADP

When NA was applied by superfusion with the Ca-channel blocking element Co (2 mM) the ADP was suppressed. As reported elsewhere (Yoshimura et al. 1987), Co blocks the slow component of the AHP, as does NA, and therefore in the presence of Co only the fast component of the AHP is seen. The block was reversed by returning to control Krebs solution. The NA-evoked ADP was markedly depressed or abolished in low Ca (0.25 mM) medium. These results suggest that Ca current is involved in the generation of the ADP. High Ca (10 mM) did not, however, enhance amplitude or duration of the ADP.

The NA-evoked ADP was suppressed by intracellular iontophoresis of EGTA (ethyleneglycol-bis(beta-amino-ethylether)N-N'-tetraacetic acid). Following injection of EGTA from a K EGTA filled electrode with anodal current pulses (0.4 nA, 50 ms pulses at 15 Hz for 30 s) the peak amplitude of ADP decreased to approximately 10% of control. The EGTA block of the ADP could be partially overcome by stimulating the neurones to fire bursts of 20 to 30 spikes at 50 Hz rather than single spikes.

Superfusion of the slice with tetrodotoxin (TTX, 0.6 uM) blocked the action potential elicited by intracellular stimulation of threshold intensity. Increasing stimulus intensity caused the appearance of a spike with threshold at around −30 mV, which slowly arose to a rounded peak of about 70 mV amplitude. This higher threshold, TTX-resistant, spike, which has been shown to be a Ca-spike since it is abolished by Co or low Ca (Yoshimura et al. 1986b), was followed by an ADP. The peak amplitude of the ADP in TTX was insignificantly different from control. Substituting normal Krebs solution with NA-free solution in which NA was replaced with Tris (tris(hydroxyme-thyl)aminomethane) resulted in broad spikes with slow rate of rise and reduced overshoot, and in marked attenuation or suppresion of the NA-evoked ADP. The block of the ADP in NA-free medium occurred even in the presence of tetraethylammonium (TEA, 20 mM), which increases spike duration, presumably by increasing Ca entry into the cell during the spike. These observations suggest that an NA-current is also involved in the generation of the ADP.

A decrease in external K from 3.6 to 2.5 mM or an increase to 10.0 mM influenced the resting potential and AHP amplitude but did not appreciably affect the NA-evoked ADP. When external Cl was partially replaced with isothionate, resulting in a decrease of external Cl to approximately 1/10 of its normal concentration, the NA-evoked ADP did not change appreciably.

Mechanisms Underlying the NA-Evoked Afterdepolarization

The observation that the ADP is suppressed by removal of either NA or Ca from the superfusing medium, by addition of Ca-channel block-ers, and by intracellular injection of EGTA, suggests that both Ca and NA ions are involved in generation of the ionic current under-lying the ADP. Both ions could be the charge carriers of the ADP current. In this case, removal of either ion may be expected to reduce the ADP in proportion to the fraction of the ADP current carried by that ion. Instead, removal of either NA or Ca results in complete or near complete suppression of the ADP. This result suggests that the two ions have different roles, one only being the charge carrier, while the other is involved in some other important step, like channel activation. Therefore, it would seem that the ADP may be generated either by a Ca-activated NA-conductance or by a NA-activated Ca-conductance.

Injection of EGTA into the SPN would reduce intracellular Ca, and bring about an increase in the Ca gradient across the cell membrane. Therefore, if Ca was the charge carrier of the ADP current, the ADP should be increased after EGTA. If, on the other hand, the ADP current was activated by the increased intracellular Ca associated with the spike, the ADP should be suppressed by EGTA. The observed abolition of the ADP following intracellular EGTA injection suggests that the Ca-activated NA-conductance is more likely the ionic mecha-nism underlying the production of the ADP current.

Apparently paradoxical is the fact that NA evokes the ADP which is Ca-dependent, while at the same time causing depression of the Ca-component of the action potential. This apparent contradiction could be explained if the Ca influx required to activate the ADP were small. In this context it is of interest that high Ca (10 mM) did not result in an increased ADP. This finding would be consistent with the hypothesis that Ca acts as an activator of the ADP current, that the amount of Ca influx required is small, and that the spike, in the presence of a normal Ca concentration gradient, provides an influx greater than required for activation of the ADP current.

Conclusion

This study has provided some new findings concerning the action of noradrenaline (NA) on the action potential and afterpotentials of the sympathetic preganglionic neurone (SPN). These new findings are that NA causes: 1) the appearance of an afterdepolarization (ADP); 2) depression of the late component of the afterhyperpolarization (AHP); and 3) suppression of the 'shoulder'of the action potential. The cellular mechanism underlying 2) and 3) appeared to be the reduction by NA of Ca entry during spike action potential, while that underlying 1) seemed to be activation of Ca-activated NA-conductance by NA.

Variations in NA concentration in the SPN environment, as a result of changes in activity level of NA-containing nerve terminals, could modulate the amplitude of the AHP and of the ADP, and result in changes of the neurone responsiveness to synaptic input, and hence in firing frequency. The present data suggest that an increased NA concentration leads to greater responsiveness through the ADP size and the bursting that the ADP generates.

Acknowledgements. This study was supported by a Grant-in-Aid for Scientific Research by the Ministry of Education, Science and Culture of Japan to S.N. and by grants to C.P. by the Medical Council of Canada and the Quebec Heart Foundation.

References

Carlsson A, Falck B, Fuxe K, Hillarp NA (1964) Cellular localization of monoamines in the spinal cord. Acta Physiol Scand 60: 112–119

Coote JH, Macleod VH, Fleetwood-Walker S, Gilbey MP (1981) The response of individual sympathetic preganglionic neurons to micro-electrophoretically applied endogenous monoamines. Brain Res 215: 135–145

Dahlström A, Fuxe K (1965) Evidence for the existence of monoamine neurons in the central nervous system. II. Experimentally induced changes in the intraneuronal amine levels of bulbospinal neuron systems. Acta Physiol Scand 248, Suppl 64: 1–36

Guyenet PG, Cabot JB (1981) Inhibition of sympathetic preganglionic neurons by catecholamines and clonidine: mediation by an alpha-adrenergic receptor. J Neurosci 1: 908–917

Hare BD, Neumayr RJ, Franz DN (1972) Opposite effects of L-DOPA and 5-HTP on spinal sympathetic reflexes. Nature (Lond) 239: 336–337

Kadzielawa K (1983) Inhibition of the activity of sympathetic pre-ganglionic neurons and neurons activated by visceral afferents by alpha-methyl-noradrenaline and endogenous catecholamines. Neurophar-macology 22: 3–17

Ross CA, Ruggiero DA, Park DH, Joh TH, Sved AF, Fernandez-Pardal J, Saavedra JM, Reis DJ (1984) Tonic vasomotor control by the rostral ventro-lateral medulla: effect of electrical or chemical stimulation of the area containing C1 adrenaline neurons on arterial pressure, heart rate and plasma catecholamines and vasopressin. J Neurosci 4: 474–494

Ryall RW (1967) The effect of monoamines on sympathetic preganglionic neurons. Circ Res 21, Suppl 3: 83–87

Simon OR, Schramm LP (1983) Spinal superfusion of dopamine excites renal sympathetic nerve activity. Neuropharmacology 22: 287–293

Yoshimura M, Nishi S (1982) Intracellular recordings from lateral horn cells of the spinal cord in vitro. J Auton Nerv Syst 6: 5–11

Yoshimura M, Polosa C, Nishi S (1986a) Noradrenaline modifies sympathetic preganglionic neuron spike and afterpotential. Brain Res 362: 370–374

Yoshimura M, Polosa C, Nishi S (1986b) Electrophysiological properties of sympathetic preganglionic neurons in the in vitro spinal cord of the cat. Pflügers Arch 406: 91–98

Yoshimura M, Polosa C, Nishi S (1987) Afterhyperpolarization mechanisms in the cat sympathetic preganglionic neuron in vitro. J Neurophysiol (in press)

Muscarinic Transsynaptic Events and Their Functional Significance in Ganglionic Transmission in Mammals

H. Kobayashi and S. Mochida

Department of Physiology, Tokyo Medical College, Shinjuku-ku, Tokyo 160, Japan

Introduction

In mammalian sympathetic ganglia, muscarinic receptors are known to
be present in the principal (postganglionic) neurones as well as in
the interneuronal SIF cells. Activation of these variously located
muscarinic receptors elicits a complex series of slow postsynaptic
potentials either directly or indirectly in the principal neurones
(Eccles and Libet 1961; Libet 1979). Single neurones enzymatically
isolated from superior cervical ganglia of adult rabbits and solita-
rily grown in culture may provide a suitable opportunity to study
the nature of membrane changes induced only directly in those cells
by exogenous muscarinic agonists without further complications arising from any secondary effects through other cells (Mochida and
Kobayashi 1986).

Multiple Conductance Changes

In the neurones prepared as above, 3 different types of muscarinic
changes in membrane conductance (Gm), accompanying the corresponding
changes in membrane potential (Vm), were shown to be induced by
means of a voltage-current (V-I) relationship analysis using conventional intracellular recordings (Fig. 1). These were 1) a Vm-independent increase (A1, B1), 2) a Vm-dependent decrease at Vm levels
positive to about −65 mV (A2, A3, B2), and 3) a Vm-independent
decrease (A4, B3), respectively. As expected from the changes in the

Fig. 1. Changes in the steady-state voltage-current (V-I) relationships (A) and in the membrane conductance (B) induced by muscarine.
V-I curves were obtained by current-clamp method, passing 500 ms
current pulses of various intensities and plotting the amplitudes,
at the end of pulses, of produced electrotonic potentials. In A1-4,
ordinates were membrane potentials (zero point indicating the level
of resting potential in each cell) and abscissae were intensities of
currents passed. In B1-3, membrane conductance was determined from
tangents drawn to the V-I curves in A at each point and its Vm-
dependence was expressed as nS (ordinates) plotted against Vm (abscissae). In A, the curves were obtained from the cell having the
resting Vm (RP) of −64 mV in 1 with 25 uM DL-muscarine, −55 MV in 2
with 25 uM, −55 MV in 3 with 25 uM and −50 MV in 4 with 50 uM,
respectively. In B, the curves in 1 were derived from A1, those in 2
from A2 and those in 3 from A4, respectively. (From Mochida and
Kobayashi 1986, reproduced by permission of the J Auton Nerv Syst.)

Experimental Brain Research Series 16
© Springer-Verlag Berlin · Heidelberg 1987

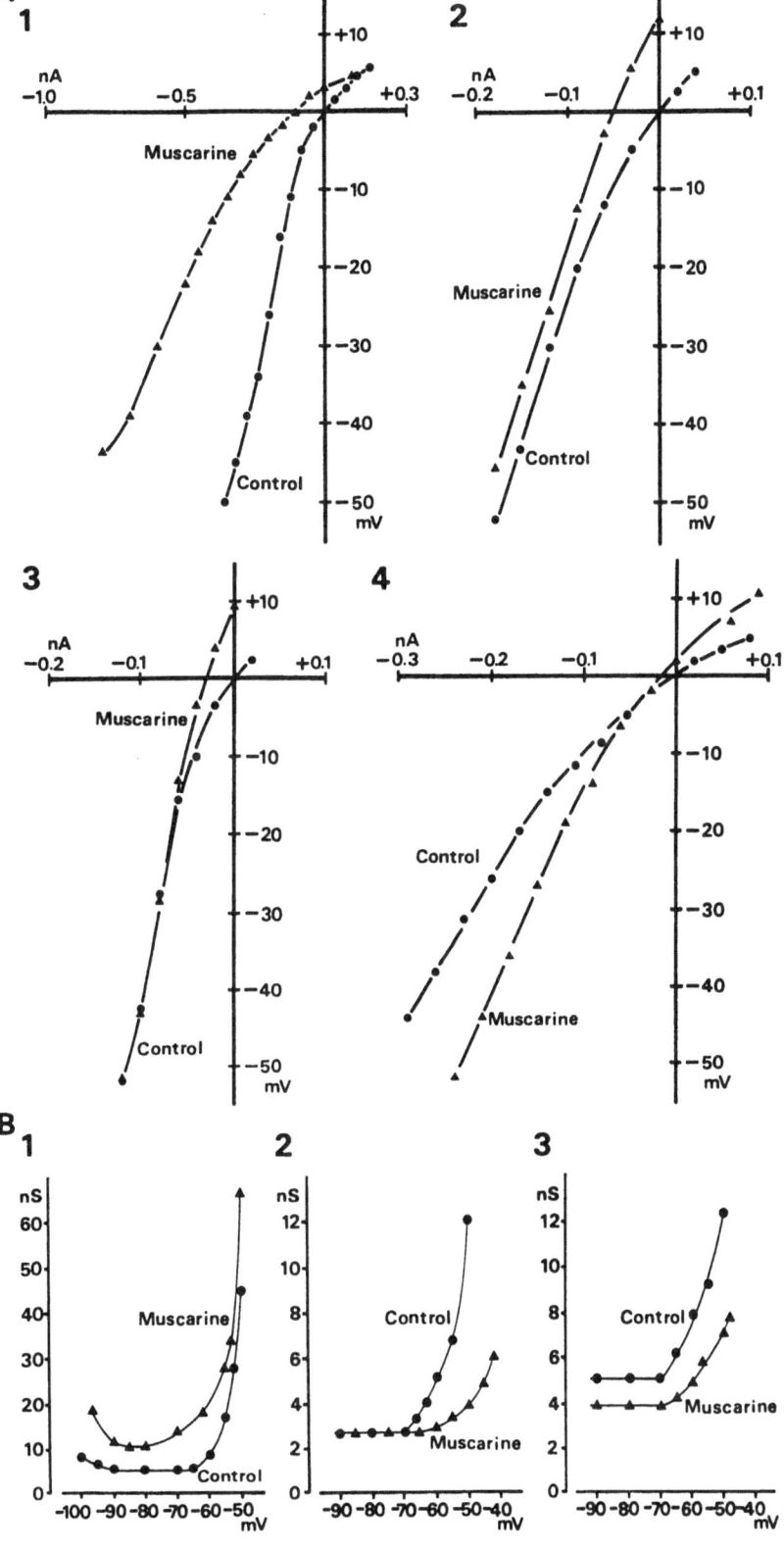

uM, type 1 and type 2 changes were evoked in overlap and with 250 uM
V-I curves, these Gm changes brought about depolarizations except
for the type 3 one (A4, B3) that produced, in addition to a depolarization with a decrease in Gm, a hyperpolarization at a range of Vm
negative to the intersection of the V-I curves obtained before and
after the application of muscarinic agonists. A type of depolarization with neither an increase nor a decrease in Gm (A2; Hashiguchi
et al. 1982) was sometimes evoked at all levels of Vm, in overlap
with the Vm-dependent type 2 change at Vm levels positive to 65 mV,
especially with muscarine (but not with acetylcholine) as the agonist. Changes in Gm and Vm induced by 50 uM DL-muscarine were antagonized in a concentration-dependent fashion by quinuclidinyl benzi-
late (QNB) at a range of 0.1-10 uM. Pirenzepine (10-30 uM) showed
similar antagonism.

The multiple Gm changes were often produced within the same cell but
were found to be discriminately evoked by the different levels of
agonist concentration (Mochida and Kobayashi 1987a). Acetylcholine
(ACh; in the presence of 35 uM d-tubocurarine) induced the type 1
change at and above 10 nM. With much higher concentration of 10-100
or more all 3 types of changes were produced simultaneously in
different time-courses, resulting in the production of a complex
triphasic Vm change.

The ionic mechanism for various types of Gm changes have not yet
been completely clarified. The type 1 change with an increase in Gm
seems to be produced by an increase in Cl conductance, as the reversal potential (-34 mV) for this type of change was close to the
equilibrium potential for Cl ions in this cell species (Woodward et
al. 1969) and was shifted in accordance with the Nernst equation on
replacing the external Cl ions with impermeable anions. The type 2
change is obviously attributable to a reduction of a specific Vm-
dependent K current (M-current; Brown and Adams 1980). The type 3
change may be due to a decrease in a Vm-independent Cl current (Ix;
Brown and Selyanko 1985). Membrane mechanism underlying a type of
depolarization with no apparent change in Gm is to be investigated.

Changes in Action Potentials

Muscarinic agonists not only produce multiple changes in the steady
Vm and Gm (equivalent to the postsynaptic potentials) but also
directly alter the configuration of action potentials in the same
neurone. The characteristic shoulder in the repolarizing phase of
the spike was eliminated while normally long-lasting after-spike
hyperpolarization (especially its late portion) was markedly depressed (Fig. 2). These changes were most readily elicited with low
concentrations of agonists (e.g., ACh at 1-10 nM) before or even
without any small change in the steady Vm, suggesting that muscarinic receptors may differently control the ionic channels involved in
the generation of action potential from those supporting the resting
Vm. An entry of Ca ions during the action potential may be represented by the shoulder in the repolarizing phase (Hashiguchi 1979; Horn
and McAfee 1980) and the Ca-dependent K conductance, which underlies
the after-spike hyperpolarization (Nohmi et al. 1983), may subsequently be activated. Therefore, muscarinic changes in the configuration of action potential may well be due primarily to the reduction of initial Ca entry. In fact, the Ca-action potential (evoked
in the presence of 80 nM tetrodotoxin) was reversibly suppressed by
the muscarinic action of ACh. Changes in action potentials induced
by 10-100 nM ACh were effectively antagonized by 10 nM of a novel

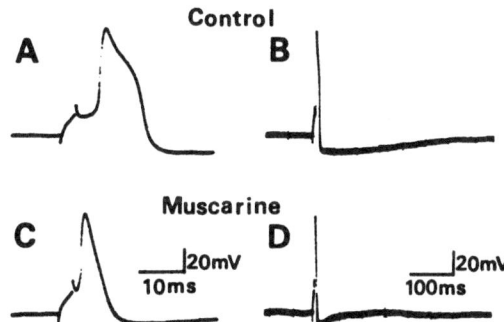

Fig. 2. Effects of muscarine on the action potential. Action potentials were evoked by a depolarizing current pulse (0.6 nA) of 3 ms
duration delivered through the recording electrode. All records were
made in the same cell with the resting Vm (RP) of −55 mV. A and B:
normal action potentials displayed on the cathode-ray oscilloscope
at fast and slow time bases and photographed. C and D: action potentials after the application of DL-muscarine (25 uM in final concentration). (From Mochida and Kobayashi 1986, reproduced by permission
of the J Auton Nerv Syst.)

antagonist AF−DX 116, while they were unaffected by the same concentration of pirenzepine (Mochida and Kobayashi 1987b). (For other
pharmacological characteristics of AF−DX 116, see Hammer et al.
1986).

Functional Significance of Muscarinic Actions

In contrast to the relative simplicity of nicotinic receptor/channel
complex, muscarinic receptors in sympathetic neurones are thus shown
to be coupled with multiple ionic channels even within the single
cell. Muscarinic suppression of the well-known M-current (Brown and
Adams 1980) may <u>partly</u> contribute to the membrane mechanism underlying the slow muscarinic depolarization (equivalent to the slow
excitatory postsynaptic potential) but other ionic mechanisms may
also be involved. Especially in normal (un-impaled) neurones, whose
resting Vm is presumed to be at about −65 mV or even more, the type
1, rather than the type 2, change with an increase in Cl conductance
may be a primary muscarinic response that initiates a depolarization
(Mochida and Kobayashi 1987a). The type 3 change may be to limit an
excessive depolarization and thereby regulate the degree of excitation (Brown and Selyanko 1985). The characteristic changes in the
action potential, in parallel with sustained muscarinic depolarization, may foster generation of repetitive postganglionic firing,
since a shortening especially in the duration of after-spike hyperpolarization enables the cell to rapidly restore the original Vm and
to be quickly ready for next firing. Furthermore, an induction of
muscarinic changes, which was antagonized by AF−DX 116, in the
action potential evoked in the cell body of sympathetic postganglionic neurones indicates an existence of the M2-muscarinic receptor
together with the previously known M1 receptor mediating the slow
muscarinic depolarization (Ashe and Yarosh 1984) in the same cell.
Muscarinic changes thus induced in sympathetic neurones would favour
an economical performance of an important part in the functioning of
the autonomic nervous system, as a sustained postganglionic firing
could occur with lesser dependency on preganglionic inputs.

Acknowledgements. This work was supported by Grants–in–Aid for Special Project Research (nos. 59123001, 60115001 and 61107001) and Cooperative Research (no. 61304031) from the Japanese Ministry of Education, Science and Culture. AF–DX 116 was a gift from Karl Thomae GmbH.

References

Ashe JH, Yarosh CA (1984) Differential and selective antagonism of the slow inhibitory postsynaptic potential and slow excitatory postsynaptic potential by gallamine and pirenzepine in the superior cervical ganglion of the rabbit. Neuropharmacology 23: 1321–1329

Brown DA, Adams PR (1980) Muscarinic suppression of a novel voltage-sensitive K current in a vertebrate neurone. Nature 283: 673–676

Brown DA, Selyanko AA (1985) Two components of muscarine–sensitive membrane current in rat sympathetic neurones. J Physiol (Lond) 358: 335–363

Eccles RM, Libet B (1961) Origin and blockade of the synaptic responses of curarized sympathetic ganglia. J Physiol (Lond) 157: 484–503

Hammer R, Giraldo E, Schiavi GB, Monferini E, Ladinsky H (1986) Binding profile of a novel cardioactive muscarine receptor antagonist, AF–DX 116, to membranes of peripheral tissues and brain in the rat. Life Sci 38: 1653–1662

Hashiguchi T (1979) The Ca–dependent components of action potentials of the rabbit superior cervical ganglion cells. J Tokyo Med Coll 37: 533–544

Hashiguchi T, Kobayashi H, Tosaka T, Libet B (1982) Two muscarinic depolarizing mechanisms in mammalian sympathetic neurons. Brain Res 242: 378–382

Horn JP, McAfee DA (1980) Alpha–adrenergic inhibition of Ca–dependent potentials in rat sympathetic neurones. J Physiol (Lond) 301: 191–204

Libet B (1979) Slow synaptic actions in ganglionic functions. In: Brooks CM, Koizumi K, Sato A (eds) Integrative function of the autonomic nervous system. Univ Tokyo Press, Tokyo, and Elsevier, Amsterdam, pp 203–311

Mochida S, Kobayashi H (1986) Multiple muscarinic responses directly evoked in isolated neurones dissociated from rabbit sympathetic ganglion. J Autonom Nerv Syst 17: 289–301

Mochida S, Kobayashi H (1987a) Three types of muscarinic conductance changes in sympathetic neurons discriminately evoked by the different concentrations of acetylcholine. Brain Res (in press)

Mochida S, Kobayashi H (1987b) Activation of M2–muscarinic receptors causes an alteration of action potentials by modulation of Ca entry in isolated sympathetic neurones of rabbits. Neurosci Lett (in press)

Nohmi M, Kuba K, Morita K (1983) Does intracellular release of Ca participate in the after–hyperpolarization of a sympathetic neuron? Brain Res 268: 158–161

Woodward JK, Bianchi CP, Erulkar SD (1969) Electrolyte distribution in rabbit superior cervical ganglion. J Neurochem 16: 289–299

Immunocytochemistry of Cyclic GMP in the Superior Cervical Ganglion of the Rat: A Combined Quantitative Immunofluorescence and Pharmacological In-Vitro Study

H. W. M. Steinbusch and J. De Vente

Department of Pharmacology, Faculty of Medicine, Free University, Van der Boechorststraat 7, 1081 BT Amsterdam, The Netherlands

Introduction

Cyclic nucleotides have been accepted to be second messengers of neuronal transmission. Recently, we described a new cGMP-antiserum (De Vente et al. 1987a) which was developed on the following assumptions: a) adequate fixation of cGMP to cellular proteins is essential for a consistent demonstration of cGMP, if the soluble pool of cGMP is to be visualized; b) the specificity of the immunocytochemical (ICC) technique may be affected by the fixation procedure (Steinbusch et al. 1986). It was found that formaldehyde adequately cGMP fixed to a protein matrix. Therefore, antisera were raised against a cGMP-protein conjugate which was prepared using formaldehyde as coupling reagent in a way that equals tissue fixation. The specificity and results of a first application in rat brain sections were reported elsewhere (De Vente et al. 1987b).

Quantitative determination of fluorescence intensity in biological as well as in non-biological preparations have been accomplished by use of different types of microfluorimeters. In this paper, we will focus upon 'static-' rather than 'scanning microfluorimetry' for which the reader is referred to Schipper and Tilders (1982). Using static microfluorimeters fluorescence intensity of objects of comparative large dimensions, such as cell bodies, can be determined. Using a gelatin model system, containing known quantities of cGMP fixed to a gelatin matrix with formaldehyde, we found a linear ratio between the concentration of fixed cGMP and the intensity of the cGMP immunofluorescence as measured with a Leitz MPV-II system. This observation indicated the prospect of a quantitation of immunostaining of cGMP in tissue sections, using the intensity of the cGMP immunofluorescence as an indicator for the local concentration of cGMP. For the examination of this probability, using our new antiserum, we chose the superior cervical ganglion (SCG) of the rat since this preparation, as a biological model, has the advantage that: A) cGMP levels in the SCG have been known to increase after neurochemical/pharmacological stimuli (Kebabian et al. 1975b; Wamsley et al. 1979); B) it is a relative consistent structure, containing well distinguishable elements, and C) information about the ICC demonstration of cGMP in this tissue is accessible from the literature (Kebabian et al. 1975a; Steiner et al. 1976; Ariano et al. 1982).

Materials and Methods

The preparation of the cGMP antiserum has been described elsewhere (De Vente et al. 1987a). The antiserum obtained did not cross-react

Experimental Brain Research Series 16
© Springer-Verlag Berlin · Heidelberg 1987

with cAMP, AMP, ATP, adenosine, GMP, GTP, guanosine, either free or
formaldehyde fixed in a gelatin model system in concentrations up to
1 mM.

In-Vitro Incubation of Superior Cervical Ganglia

From female rats (180-200 g), anaesthetized with sodium pentobarbital (Nembutal, 60 mg/kg body weight, i.p.) SGC were removed, deshea-
thed and incubated in a waterbath in Locke solution aerated with 5%
CO_2/95% O_2 at 37°C, pH 7.2. After equilibration for 30 min the SCG
were transferred to fresh media, containing drugs or iso-osmolar
high potassium (details are described under Results) for a period of
6 min. Subsequently, the SCG were immersion-fixed in icecold 4%
(wt/vol) paraformaldehyde in 0.1 M phosphate buffer (pH 7.4), containing 0.2% (wt/vol) picric acid for 90 min, thereafter they were
rinsed in cold 5% sucrose in the same buffer for 1 h. Directly
afterward 10 um thick cryostat sections were cut at -20°C, thawed
onto chromalum-gelatin coated slides and processed for immunofluorescence (Steinbusch and Tilders 1987).

Microfluorimetric Measurements

The intensity of emitted fluorescence light was measured using an
automated microfluorimeter (MPV II, Leitz), as described by Schipper
and Tilders (1982). For orientation and preparation we· used the
transmitted excitation pathway, which intensity was limited to a
tolerable low level to minimize photodecomposition of the fluorophore. Routinely the intensity of the cGMP immunofluorescence was
quantitatively estimated in the cytoplasm of 100 individual large
postganglionic cells using a measuring spot of 113 um² (Fig. 1g).
These cells were chosen at random from areas from one SCG section.
Mean S.E.M. from each experiment were calculated on line and the
mean S.E.M. from the pooled experiments were calculated according to
standard statistical methods. Blank values obtained with pre-immune
serum on sections from the same SCG were subtracted from the values
obtained with the cGMP antiserum.

Results and Discussion

When the SCG were incubated in-vitro for 30 min in normal Locke
solution, cGMP immunoreactivity was found mainly in the large postganglionic cells; a very faint immunostaining was found throughout
the SCG. Large postganglionic cells were recognized using tyrosine-
hydroxylase (TH) antiserum. Small intensely fluorescent (SIF) cells
were identified using TH as well as serotonin antiserum (Verhofstad
et al. 1981; Soinila 1986) (Fig. 1a). In the large postganglionic
neuronal cells as well as in the small fibroblast-like cells cGMP

Fig. 1. Photomicrographs of cryostat sections through the superior ▶
cervical ganglion of the rat incubated with an antibody to serotonin
(a) or cGMP (b-g) and processed for immunohistochemistry or immunofluorescence. a) Shows serotonin-immunoreactive SIF-cells. The other
sections were incubated in-vitro and processed for cGMP. b) Basal
unstimulated ganglion (5.6 mM K⁺), c) 60 mM K⁺; d) 100 mM K⁺;
e) 10^{-8} M and f and g) 10^{-4} M carbachol stimulation. Note in g) the
size of the measuring spot which has been used for the quantitative
immunofluorescence studies. Bars indicate 50 um

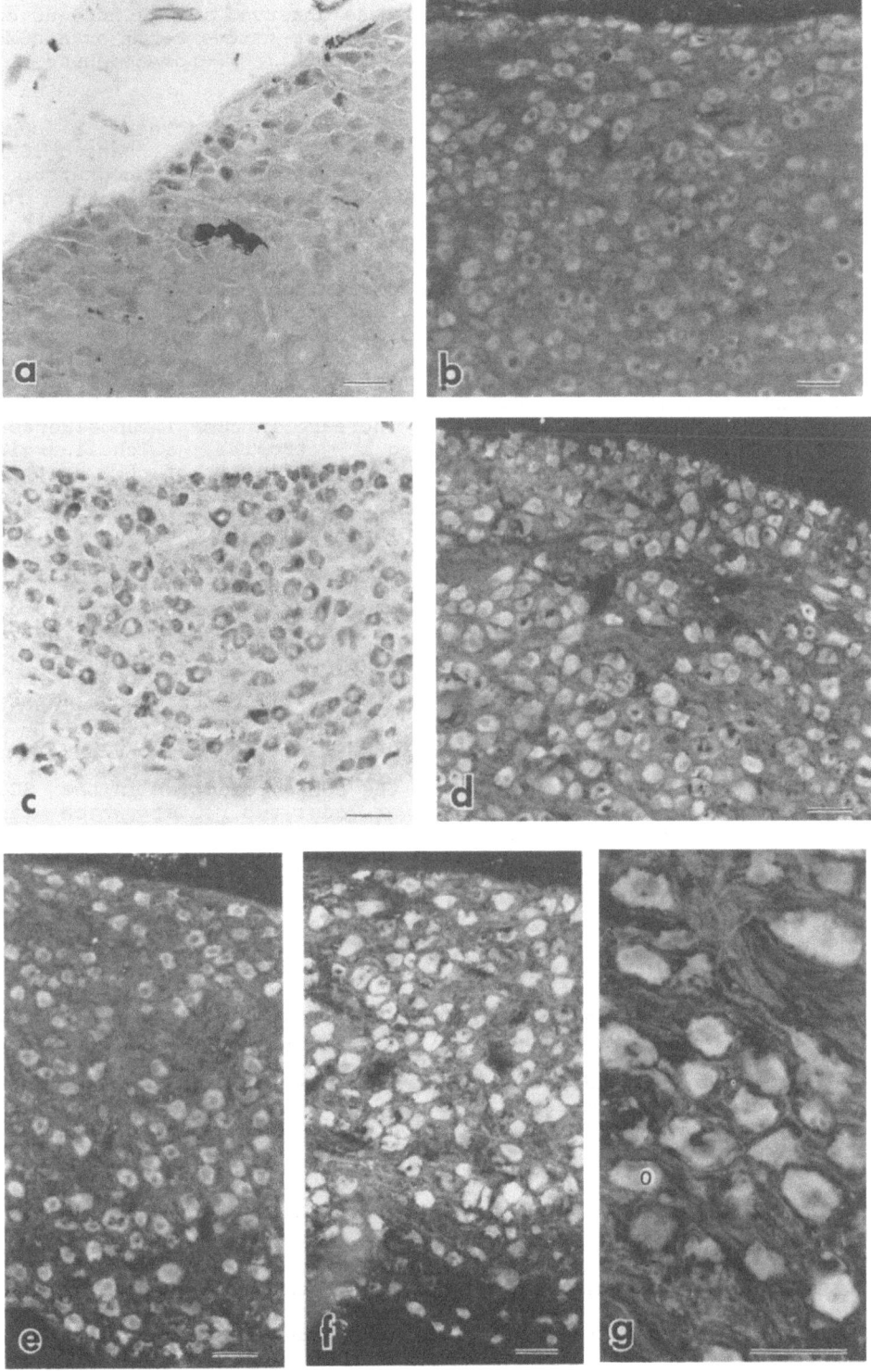

immunoreactivity was observed as equally distributed over the cytoplasm, whereas no cGMP-immunostaining was observed in the nucleus or
the nucleolus. No staining was seen using pre-immune serum or a cGMP
antiserum which had been pre-incubated with cGMP-formaldehyde-THY
conjugate at a concentration of 1.9×10^{-4} M cGMP.

Incubation of SCG with 100 mM KCl led to a strong increase in cGMP
immunostaining. Carbachol, a cholinergic agonist, also enhanced cGMP
immunofluorescence intensity in the large postganglionic neuronal
cells in a concentration range of 10^{-8} M to 10^{-6} M (Fig. 1e-g). The
apperance of cGMP immunofluorescence after K$^+$ or carbachol stimulation of the SCG was qualitative the same: it shows up equally distributed only in the cytoplasm of the large postganglionic neuronal
cells (Fig. 1 b-d). Quantitation of cGMP immunofluorescence intensity in the large postganglionic neuronal cells with this system
provided an objective measure for the results of the pharmacological
manipulation of SCG. Substitution of increasing concentrations of
NaCl for KCl in the incubation medium of the SCG resulted in a dose-
dependent increase of cGMP immunofluorescence intensity in the large
postganglionic neuronal cells. This increase in cGMP immunofluorescence intensity could not be blocked by atropine, a cholinergic
antagonist, up to 10^{-5} M. Biochemical determination of cGMP in the
SCG gave 0.14 pmol/mg protein for basal cGMP levels and 1.15 pmol/mg
protein for cGMP levels stimulated with 100 mM KCl. Incubation of
SCG with increasing concentrations of carbachol (10^{-8} to 10^{-6} M)
resulted in a dose-dependent increase of cGMP immunofluorescence in
the large postganglionic neuronal cells. This carbachol-stimulated
increase in cGMP immunofluorescence could be competitively antagonized by atropine. A high concentration of atropine (10^{-5} M) reduced the carbachol stimulated cGMP immunofluorescence to basal levels. Hexamethonium, a cholinergic antagonist of the nicotinic type,
did not affect the carbachol-induced increase in cGMP-immunostaining
at a concentration of 10^{-6} M: no effect on the maximum immunofluorescence, nor any shift of the dose-response curve for carbachol.

The mechanism of the stimulation of the cGMP production in the SCG
is yet unclear. There appear to be at least two routes which can
elevate cGMP levels. As also illustrated by us, cGMP levels in the
SCG can be enhanced by cholinergic agonists (Brown et al. 1980).
Although the precise mechanism of activation of guanylate cyclase by
cholinergic agonists is yet unknown, it is at least a clear cut
receptor mediated response. The second route, i.e. stimulation of
cGMP levels by high K$^+$, is not mediated by a muscarinic receptor, as
the K$^+$-mediated increase in cGMP cannot be blocked by atropine and
is absent in the denervated SCG (Briggs et al. 1982). This suggests
the involvement of other neurotransmitters or -modulators, released
upon depolarization of the cholinergic nerve terminals (Bone et al.
1985; Volle et al. 1981). Our experiments clearly showed that cGMP
accumulation in the SCG after in-vitro high K$^+$ or carbachol stimulation occurs in the same cell types, i.e. the large postganglionic
neuronal cells and no evidence was found for a presynaptic accumulation of cGMP.

Conclusion

The presented data indicate the presence of cGMP-producing large
postganglionic neuronal cells in the SCG of the rat. The cGMP immunoreactivity intensities could be increased or decreased by stimulation of in-vitro slices with cholingergic agonist or antagonist,
respectively. It seems to be warranted that quantification of immunostaining intensities by antisera directed to second messengers

permits to study postsynaptic parameters in individual cell bodies in heterogeneous tissue.

References

Ariano MA, Briggs CA, McAffee DA (1982) Cellular localization of cyclic nucleotide charges in rat superior cervical ganglion. Cell Mol Neurobiol 2: 143–156

Bone EA, Mitchell RH (1985) Accumulaiton of inositol phosphates in sympathetic ganglia. Effects of depolarization and of amine and peptide neurotransmitters. Biochem J 227: 263–269

Briggs CA, Whiting GJ, Ariano MA, McAffee DA (1982) Cyclic nucleotide metabolism in the sympathetic ganglion. Cell Mol Neurobiol 2: 129–141

Brown DA, Fatherazi S, Garthwaite J, White RD (1980) Muscarinic receptors in rat sympathetic ganglia. Br J Pharmac 70: 577–592

De Vente J, Steinbusch HWM, Schipper J (1987a) A new approach to immunocytochemistry of cGMP: preparation, specificity, and initial application of a new antiserum against formaldehyde-fixed cGMP. Neuroscience (in press)

De Vente J, Garssen J, Steinbusch HWM, Tilders FJH, Schipper J (1987b) Single cell quantitative immunohistochemistry of cyclic GMP in the superior cervical ganglion of the rat. Brain Res (in press)

Kebabian JW, Bloom FE, Steiner AL (1975a) Neurotransmitters increase cyclic nucleotides in postganglionic neurons. Immunocytochemical demonstration. Science 190: 157–159

Kebabian JW, Steiner AL, Greengard P (1975b) Muscarinic cholinergic regulation of cyclic guanosine 3':5'-monophosphate in autonomic ganglia: possible role in synaptic transmission. J Pharmacol Exp Ther 193: 474–488

Schipper J, Tilders FJH (1982) Quantification of formaldehyde induced fluorescence and its application in neurobiology. Brain Res Bull 9: 69–80

Soinila S (1986) Differentiation of the small intensely fluorescent sympathetic cells. In: Panula P, Päivärinta H, Soinila S (eds) Neurohistochemistry: modern methods and applications. Neurology and neurobiology, Vol 16. Alan R Liss, New York, pp 207–226

Steinbusch HWM, De Vente J, Schipper J (1986) Immunohistochemistry of monoamines in the central nervous system. In: Panula P, Päivärinta H, Soinila S (eds) Neurohistochemistry: modern methods and applications. Neurology and neurobiology, Vol 16. Alan R Liss, New York, pp 75–106

Steinbusch HWM, Tilders FJH (1987) Immunohistochemical techniques for light-microscopical localization of dopamine, noradrenaline, adrenaline, serotonin and histamine in the central nervous system. In: Steinbusch HWM (ed) Monoaminergic neurons: lightmicroscopy and ultrastructure. IBRO handbook series: Methods in the neurosciences, Vol 10. Wiley, Chichester, pp 125–166

Steiner AL, Ong S, Wedner HJ (1976) Cyclic nucleotide immunocytochemistry. Adv Cycl Nucl Res 7: 115–155

Verhofstad AAJ, Steinbusch HWM, Penke B, Varga J, Joosten HWJ (1981) Serotonin–immunoreactive cells in the superior cervical ganglion of the rat. Evidence for the existencee of separate ganglionic cells. Brain Res 212: 39–49

Volle RL, Quenzer LF, Patterson BA, Alkadhi KA, Henderson EG (1981) Cyclic guanosine 3':5'–monophosphate accumulation and ^{45}Ca–uptake by rat superior cervical ganglia during preganglionic stimulation. J Pharmacol Exp Ther 219: 338–343

Wamsley JK, West JR, Black AC, Williams RH (1979) Muscarinic cholinergic and preganglionic physiological stimulation increases cGMP levels in guinea pig superior cervical ganglia. J Neurochem 32: 1034–1035